EDEXCEL FURTHER MATHS

CORE PURE
YEAR 2

Series Editor
David Baker

Authors
David Bowles, Eddie Mullan, Garry Wiseman, Katie Wood
Brian Jefferson, John Rayneau, Mike Heylings, Rob Wagner

OXFORD
UNIVERSITY PRESS

OXFORD
UNIVERSITY PRESS

Great Clarendon Street, Oxford, OX2 6DP, United Kingdom

Oxford University Press is a department of the University of Oxford.

It furthers the University's objective of excellence in research, scholarship, and education by publishing worldwide. Oxford is a registered trade mark of Oxford University Press in the UK and in certain other countries.

British Library Cataloguing in Publication Data
Data available

978-019-841524-4

10 9 8 7 6 5 4 3 2

Paper used in the production of this book is a natural, recyclable product made from wood grown in sustainable forests.
The manufacturing process conforms to the environmental regulations of the country of origin.

Printed and bound by CPI Group (UK) Ltd, Croydon, CR0 4YY

Acknowledgements

Authors
David Bowles, Eddie Mullan, Garry Wiseman, Katie Wood
Brian Jefferson, John Rayneau, Mike Heylings, Rob Wagner

Editorial team
Dom Holdsworth, Ian Knowles, Matteo Orsini Jones, Felicity Ounsted

With thanks also to Geoff Wake, Matt Woodford, Susan Lyons, Deborah Dobson, Keith Gallick and Amy Ekins-Coward for their contribution.

Index compiled by Marian Preston, Preston Indexing.

Although we have made every effort to trace and contact all copyright holders before publication, this has not been possible in all cases. If notified, the publisher will rectify any errors or omissions at the earliest opportunity.

p1: guvendemir/iStock; **p29:** Atosan/Shutterstock; **p63:** MicroStockHub/iStock; **p117:** Mikephotos/Dreamstime; **p25:** New York Public Library/SPL; **p59:** Manzotte Photography/Shutterstock; **p59:** Soleil Nordic; **p113:** AlanGH/iStock; **p157:** VisualCommunications/iStock; **p157:** 3Dsculptor/Shutterstock;

Contents

About this book

This book has been specifically created for those studying the Edexcel 2017 Further Mathematics AS and A Level. It's been written by a team of experienced authors and teachers, and it's packed with questions, explanation and extra features to help you get the most out of your course.

Every section starts by covering the basic **Fluency and skills** (A01).

Key points highlight important concepts, and make the information easier to digest.

Worked examples provide a model answer and commentary to realistic practice questions.

There is a Fluency and skills exercise for each section, to practise the skills before moving on to the Reasoning and problem-solving section.

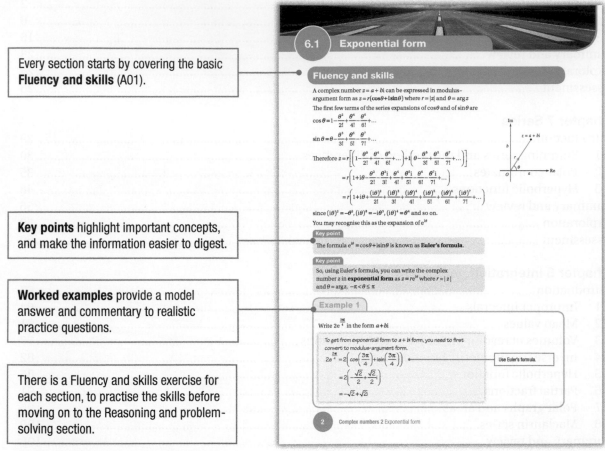

On the chapter **Introduction page**, the **Orientation box** explains what you should already know, what you will learn, and what this leads to.

At the end of every chapter, an **Exploration page** gives you an opportunity to explore the subject beyond the specification.

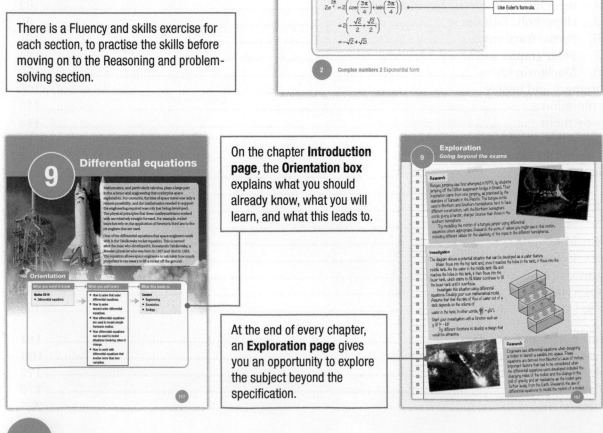

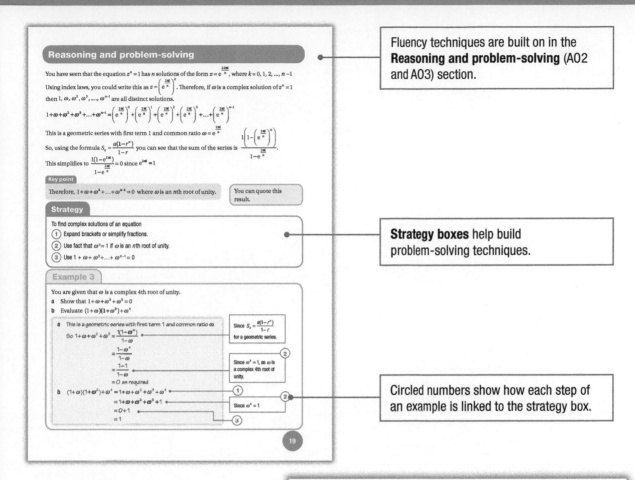

Reasoning and problem-solving

You have seen that the equation $z^n = 1$ has n solutions of the form $z = e^{\frac{2k\pi i}{n}}$, where $k = 0, 1, 2, ..., n-1$

Using index laws, you could write this as $z = \left(e^{\frac{2\pi i}{n}}\right)^k$. Therefore, if ω is a complex solution of $z^n = 1$ then $1, \omega, \omega^2, \omega^3, ..., \omega^{n-1}$ are all distinct solutions.

$1 + \omega + \omega^2 + \omega^3 + ... + \omega^{n-1} = \left(e^{\frac{2\pi i}{n}}\right)^0 + \left(e^{\frac{2\pi i}{n}}\right)^1 + \left(e^{\frac{2\pi i}{n}}\right)^2 + \left(e^{\frac{2\pi i}{n}}\right)^3 + ... + \left(e^{\frac{2\pi i}{n}}\right)^{n-1}$

This is a geometric series with first term 1 and common ratio $\omega = e^{\frac{2\pi i}{n}}$

So, using the formula $S_n = \frac{a(1-r^n)}{1-r}$ you can see that the sum of the series is $\frac{1\left(1-\left(e^{\frac{2\pi i}{n}}\right)^n\right)}{1-e^{\frac{2\pi i}{n}}}$.

This simplifies to $\frac{1(1-e^{2\pi i})}{1-e^{\frac{2\pi i}{n}}} = 0$ since $e^{2\pi i} = 1$

Key point

Therefore, $1 + \omega + \omega^2 + ... + \omega^{n-1} = 0$ where ω is an nth root of unity.

> You can quote this result.

Strategy

To find complex solutions of an equation
① Expand brackets or simplify fractions.
② Use fact that $\omega^n = 1$ if ω is an nth root of unity.
③ Use $1 + \omega + \omega^2 + ... + \omega^{n-1} = 0$

Example 3

You are given that ω is a complex 4th root of unity.
a Show that $1 + \omega + \omega^2 + \omega^3 = 0$
b Evaluate $(1+\omega)(1+\omega^2) + \omega^4$

a This is a geometric series with first term 1 and common ratio ω.

So $1 + \omega + \omega^2 + \omega^3 = \frac{1(1-\omega^4)}{1-\omega}$

> Since $S_n = \frac{a(1-r^n)}{1-r}$ for a geometric series.

$= \frac{1-\omega^4}{1-\omega}$
$= \frac{1-1}{1-\omega}$

> ② Since $\omega^4 = 1$, as ω is a complex 4th root of unity.

$= 0$ as required

b $(1+\omega)(1+\omega^2) + \omega^4 = 1 + \omega + \omega^2 + \omega^3 + \omega^4$ ①
$= 1 + \omega + \omega^2 + \omega^3 + 1$ ②

> ② Since $\omega^4 = 1$

$= 0 + 1$ ③
$= 1$

19

Exercise 9.4B Reasoning and problem-solving

1 The rate, in $\text{cm}^3\,\text{s}^{-1}$, at which air is escaping from a balloon at time t seconds is proportional to the volume, $V\,\text{cm}^3$, of air in the balloon at that time. Initially $V = 2000$

a Show that $V = 2000e^{-kt}$, where k is a positive constant.

Given that $V = 1000$ when $t = 4$

b Show that $k = \frac{1}{4}\ln 2$
c Calculate the value of V when $t = 8$

2 A glass of boiling water is placed in a room. The temperature of the room is 20 °C. At time t minutes the rate of change of temperature of a glass of water is proportional to the difference between the temperature, T °C,

4

The diagram shows a water tank in the shape of a cuboid of base 4 m by 4 m, and height 6 m. Water flows into the tank at a constant rate of $2\,\text{m}^3\,\text{min}^{-1}$. At time t minutes the depth of the water in the tank is x metres. There is a tap at the point T at the base of the tank. When the tap is open water leaves the tank at a rate of $0.4x\,\text{m}^3\,\text{min}^{-1}$. Initially the tank is empty.

a Show that $\dfrac{dx}{dt} = \dfrac{5-x}{}$

---

8 **Assessment**

1 You are given that $f(x) = \dfrac{x^3 - 3\sqrt{x}}{2x^2}$
a Write $f(x)$ in the form $Ax^m + Bx^n$, where A, B, m and n are constants to be found. **[2]**
b Calculate the mean value of $f(x)$ in the interval [1, 4]. **[5]**

2 Find the mean value of the function $g(x) = x\sqrt{x} + \dfrac{3}{x^2}$ for $1 \leq x \leq 3$
Give your answer in the form $a + b\sqrt{3}$, where a and b are constants to be found. **[4]**

3 The mean value of the function $1 + 2x^3$ in the interval $[0, a]$ is 109
Evaluate a **[5]**

9 Find $\displaystyle\int \frac{3}{5\sinh x - 4\cosh x}\,dx$ **[10]**

10 The diagram shows the polar curve with equations $r = 1 + \cos\theta$ for $0 \leq \theta \leq \pi$, along with the line $\theta = \dfrac{\pi}{3}$. The curve and the line intersect at the origin, O, and at the point P
a Find the coordinates of the point P
b Find the area of the shaded region that is bounded by the line and the curve. **[8]**

11 a Find $\displaystyle\int_0^{\frac{1}{2}} \frac{1}{\sqrt{1-x^2}}\,dx$ **b** Find $\displaystyle\int_1^{\frac{3}{2}} \frac{1}{\sqrt{x^2-1}}\,dx$ **[12]**

12 a Sketch the curve with polar equation $r = 4\sin 3\theta$, $0 \leq \theta \leq \pi$
b Find the area enclosed by one loop of this curve. **[9]**

6 Complex numbers 2

Liquids such as water, and gases such as air, are known as fluids. In many ways, the flow of water can be treated the same as the flow of air. The study of such flow is known as fluid dynamics, where complex functions are used to model flow. Aerodynamics is the application of fluid dynamics to the flow of air. Hydrodynamics is the application of fluid dynamics to the flow of liquids. The scientific principles and the underpinning mathematics of fluid dynamics are important in many areas. For example, in the design of vehicles that move in gases and liquids. However, understanding gas and liquid flow is also important when planning how water and gas will reach homes and businesses.

An aircraft that is full of passengers, their luggage, and maybe some other cargo, can lift off from the runway at an airport primarily as a result of the motion of the wing through the air. The mathematics of fluid dynamics allows this motion of the aircraft relative to the air, a fluid, to be analysed in detail. The mathematics relies on complex numbers to provide insight into the flow of the air. It is used by aeronautical engineers when they are designing the wings of an aircraft.

Orientation

What you need to know	What you will learn	What this leads to
Ch1 Complex numbers 1	• How to use exponential form. • How to use de Moivre's theorem. • How to use roots of unity.	**Careers** • Electrical engineering. • Aeronautical engineering. • Mechanical engineering.

Exponential form

Fluency and skills

A complex number $z = a + b\mathrm{i}$ can be expressed in modulus–
argument form as $z = r(\cos\theta + \mathrm{i}\sin\theta)$ where $r = |z|$ and $\theta = \arg z$

The first few terms of the series expansions of $\cos\theta$ and of $\sin\theta$ are

$$\cos\theta = 1 - \frac{\theta^2}{2!} + \frac{\theta^4}{4!} - \frac{\theta^6}{6!} + \dots$$

$$\sin\theta = \theta - \frac{\theta^3}{3!} + \frac{\theta^5}{5!} - \frac{\theta^7}{7!} + \dots$$

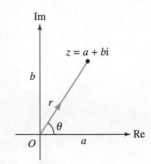

Therefore
$$z = r\left[\left(1 - \frac{\theta^2}{2!} + \frac{\theta^4}{4!} - \frac{\theta^6}{6!} + \dots\right) + \mathrm{i}\left(\theta - \frac{\theta^3}{3!} + \frac{\theta^5}{5!} - \frac{\theta^7}{7!} + \dots\right)\right]$$

$$= r\left(1 + \mathrm{i}\theta - \frac{\theta^2}{2!} - \frac{\theta^3\mathrm{i}}{3!} + \frac{\theta^4}{4!} + \frac{\theta^5\mathrm{i}}{5!} - \frac{\theta^6}{6!} - \frac{\theta^7\mathrm{i}}{7!} + \dots\right)$$

$$= r\left(1 + \mathrm{i}\theta + \frac{(\mathrm{i}\theta)^2}{2!} + \frac{(\mathrm{i}\theta)^3}{3!} + \frac{(\mathrm{i}\theta)^4}{4!} + \frac{(\mathrm{i}\theta)^5}{5!} + \frac{(\mathrm{i}\theta)^6}{6!} + \frac{(\mathrm{i}\theta)^7}{7!} + \dots\right)$$

since $(\mathrm{i}\theta)^2 = -\theta^2$, $(\mathrm{i}\theta)^3 = -\mathrm{i}\theta^3$, $(\mathrm{i}\theta)^4 = \theta^4$ and so on.

You may recognise this as the expansion of $\mathrm{e}^{\mathrm{i}\theta}$

Key point

The formula $\mathrm{e}^{\mathrm{i}\theta} = \cos\theta + \mathrm{i}\sin\theta$ is known as **Euler's formula**.

Key point

So, using Euler's formula, you can write the complex
number z in **exponential form** as $z = r\mathrm{e}^{\mathrm{i}\theta}$ where $r = |z|$
and $\theta = \arg z$, $-\pi < \theta \le \pi$

Example 1

Write $2\mathrm{e}^{\frac{3\pi\mathrm{i}}{4}}$ in the form $a + b\mathrm{i}$

To get from exponential form to $a + b\,i$ form, you need to first
convert to modulus-argument form.

$$2\mathrm{e}^{\frac{3\pi\mathrm{i}}{4}} = 2\left(\cos\left(\frac{3\pi}{4}\right) + \mathrm{i}\sin\left(\frac{3\pi}{4}\right)\right)$$ ● ——— Use Euler's formula.

$$= 2\left(-\frac{\sqrt{2}}{2} + \frac{\sqrt{2}}{2}\mathrm{i}\right)$$

$$= -\sqrt{2} + \sqrt{2}\mathrm{i}$$

Example 2

Write $z = 3 - i$ in exponential form.

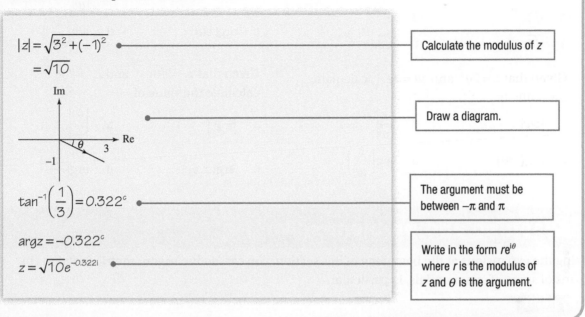

$|z| = \sqrt{3^2 + (-1)^2}$ — Calculate the modulus of z

$= \sqrt{10}$

Draw a diagram.

$\tan^{-1}\left(\dfrac{1}{3}\right) = 0.322^c$ — The argument must be between $-\pi$ and π

$\arg z = -0.322^c$

$z = \sqrt{10}\,e^{-0.322i}$ — Write in the form $re^{i\theta}$ where r is the modulus of z and θ is the argument.

Calculator

Try it on your calculator

Calculators can be used to convert to and from modulus–argument form.

Find out how to convert $\sqrt{2}e^{-\frac{\pi}{4}i}$ to Cartesian form on your calculator.

$\sqrt{2}\angle -\dfrac{\pi}{4} \blacktriangleright a + bi$

$1 - i$

Exercise 6.1A Fluency and skills

1 Write each of these numbers in modulus-argument form.

a $3 + 4i$ b $2 - i$

c 10 d -5

e $2i$ f $-6i$

g $-5 + 12i$ h $-4 - 8i$

i $\sqrt{3} + i$ j $5 - 5i$

2 Write each of these complex numbers in exponential form.

a $2\left(\cos\left(\dfrac{\pi}{12}\right) + i\sin\left(\dfrac{\pi}{12}\right)\right)$

b $4\left(\cos\left(-\dfrac{2\pi}{3}\right) + i\sin\left(-\dfrac{2\pi}{3}\right)\right)$

c $3\left(\cos\left(\dfrac{5\pi}{6}\right) - i\sin\left(\dfrac{5\pi}{6}\right)\right)$

d $6\left(\cos\left(\dfrac{\pi}{7}\right) - i\sin\left(\dfrac{\pi}{7}\right)\right)$

e $\cos\left(\dfrac{7\pi}{5}\right) + i\sin\left(\dfrac{7\pi}{5}\right)$

f $\sqrt{2}\left(\cos\left(-\dfrac{15\pi}{8}\right) + i\sin\left(-\dfrac{15\pi}{8}\right)\right)$

g $\sqrt{3}\left(\cos\left(-\dfrac{5\pi}{6}\right) - i\sin\left(-\dfrac{5\pi}{6}\right)\right)$

h $8\left(\cos\left(-\dfrac{17\pi}{12}\right) - i\sin\left(-\dfrac{17\pi}{12}\right)\right)$

3 Write each of these complex numbers in the form $a + bi$

   a $2e^{\frac{\pi}{2}i}$ **b** $7e^{-\frac{\pi}{3}i}$

   c $\sqrt{2}e^{\frac{\pi}{4}i}$ **d** $e^{-\frac{\pi}{6}i}$

   e $\sqrt{8}e^{-\pi i}$ **f** $\sqrt{3}e^{\frac{5\pi}{6}i}$

4 Given that $z = 2e^{\frac{\pi}{3}i}$ and $w = 3e^{-\frac{\pi}{3}i}$, calculate the value of

   a $|zw|$ **b** $\left|\dfrac{z}{w}\right|$

   c $\arg(zw)$ **d** $\arg\left(\dfrac{z}{w}\right)$

5 Given that $z = 5e^{\frac{2\pi}{7}i}$ and $w = \dfrac{1}{5}e^{-\frac{\pi}{7}i}$ calculate the value of

   a $|zw|$ **b** $\left|\dfrac{z}{w}\right|$

   c $\arg(zw)$ **d** $\arg\left(\dfrac{z}{w}\right)$

6 Given that $z_1 = \sqrt{6}e^{-\frac{\pi}{4}i}$ and $z_2 = \sqrt{3}e^{-\frac{5\pi}{6}i}$, calculate the value of

   a $|z_1 z_2|$ **b** $\left|\dfrac{z_1}{z_2}\right|$

   c $\arg(z_1 z_2)$ **d** $\arg\left(\dfrac{z_1}{z_2}\right)$

Reasoning and problem-solving

Using the expansions at the beginning of this section, you can define trigonometric functions in terms of sums of exponentials, in particular,

Key point

$$\cos\theta = \frac{e^{i\theta} + e^{-i\theta}}{2} \quad \text{and} \quad \sin\theta = \frac{e^{i\theta} - e^{-i\theta}}{2i}$$

Strategy

These results can then be used to prove trigonometric identities.

  (1) Use Euler's formula: $e^{i\theta} \equiv \cos\theta + i\sin\theta$

  (2) Use the facts that $\cos(-\theta) \equiv \cos\theta$ and $\sin(-\theta) \equiv -\sin\theta$

  (3) Use $\cos\theta \equiv \dfrac{e^{i\theta} + e^{-i\theta}}{2}$ or $\sin\theta \equiv \dfrac{e^{i\theta} - e^{-i\theta}}{2i}$

  (4) Use index laws.

Example 3

Prove that $\cos\theta = \dfrac{e^{i\theta} + e^{-i\theta}}{2}$

$e^{i\theta} = \cos\theta + i\sin\theta$ **(1)** Start with Euler's formula.

$e^{-i\theta} = \cos(-\theta) + i\sin(-\theta)$

$\quad\;\; = \cos\theta - i\sin\theta$ **(2)** Since $\cos(-\theta) = \cos(\theta)$ and $\sin(-\theta) = -\sin(\theta)$

$e^{i\theta} + e^{-i\theta} = \cos\theta + i\sin\theta + \cos\theta - i\sin\theta$

$\qquad\qquad = 2\cos\theta$

Therefore $\cos\theta = \dfrac{e^{i\theta} + e^{-i\theta}}{2}$, as required.

Example 4

Prove that $\sin 2\theta \equiv 2\sin\theta\cos\theta$

$$2\sin\theta\cos\theta \equiv 2\left(\frac{e^{i\theta}-e^{-i\theta}}{2i}\right)\left(\frac{e^{i\theta}+e^{-i\theta}}{2}\right)$$

> **3** Write $\sin(\theta)$ and $\cos(\theta)$ in terms of exponentials.

$$\equiv \frac{(e^{i\theta}-e^{-i\theta})(e^{i\theta}+e^{-i\theta})}{2i}$$

$$\equiv \frac{e^{2i\theta}+1-1-e^{-2i\theta}}{2i}$$

> **4** Expand brackets.

$$\equiv \frac{e^{2i\theta}-e^{-2i\theta}}{2i}$$

$$\equiv \sin 2\theta, \text{ as required}$$

> **3** Since $\sin 2\theta = \dfrac{e^{i(2\theta)}-e^{-i(2\theta)}}{2i}$

Exercise 6.1B Reasoning and problem-solving

1 Use Euler's formula to show that $\sin\theta = \dfrac{e^{i\theta}-e^{-i\theta}}{2i}$

2 A complex number z has modulus 1 and argument θ

    a Show that $z^n + \dfrac{1}{z^n} = 2\cos(n\theta)$ **b** Show that $z^n - \dfrac{1}{z^n} = 2i\sin(n\theta)$

3 Given that $z_1 = r_1 e^{\theta_1 i}$ and $z_2 = r_2 e^{\theta_2 i}$, show that

    a $|z_1 z_2| = |z_1||z_2|$ and $\arg(z_1 z_2) = \arg z_1 + \arg z_2$ **b** $\left|\dfrac{z_1}{z_2}\right| = \dfrac{|z_1|}{|z_2|}$ and $\arg\left(\dfrac{z_1}{z_2}\right) = \arg z_1 - \arg z_2$

4 Use $\cos\theta = \dfrac{e^{i\theta}+e^{-i\theta}}{2}$ and $\sin\theta = \dfrac{e^{i\theta}-e^{-i\theta}}{2i}$ to show that

    a $\sin(A+B) \equiv \sin A\cos B + \sin B\cos A$ **b** $\cos(A+B) \equiv \cos A\cos B - \sin A\sin B$

5 Use exponentials to show that

    a $\cos 2x \equiv \cos^2 x - \sin^2 x$ **b** $\cos^2 x + \sin^2 x \equiv 1$

6 Use exponentials to show that

    a $(\cos\theta + i\sin\theta)^2 \equiv \cos 2\theta + i\sin 2\theta$ **b** $(\cos\theta + i\sin\theta)^n \equiv \cos(n\theta) + i\sin(n\theta)$

7 Given that $z = 4\left(\cos\left(\dfrac{\pi}{9}\right) + i\sin\left(\dfrac{\pi}{9}\right)\right)$ and $w = 3\left(\cos\left(\dfrac{2\pi}{9}\right) + i\sin\left(\dfrac{2\pi}{9}\right)\right)$, show that $zw = 6 + 6\sqrt{3}i$

8 Given that $z = 8\left(\cos\left(\dfrac{5\pi}{12}\right) + i\sin\left(\dfrac{5\pi}{12}\right)\right)$ and $w = 6\left(\cos\left(-\dfrac{\pi}{3}\right) + i\sin\left(\dfrac{\pi}{3}\right)\right)$, show that

    a $\dfrac{z}{w} = \dfrac{2\sqrt{2}}{3}(i-1)$ **b** $z^2 = -32\sqrt{3} + 32i$

9 The complex number z is such that $|z| = k$ and $\arg(z) = \theta$ for $k > 0$ and $-\pi < \theta \leq \pi$

    Another complex number is defined as $w = 1 - i$

    Find expressions in terms of k and θ for the modulus and the argument of

    a zw **b** $\dfrac{z}{w}$

De Moivre's theorem

Fluency and skills

If you write a complex number in the form $z = r(\cos\theta + i\sin\theta)$, then you can see that $z^n = [r(\cos\theta + i\sin\theta)]^n$. You can write this as $r^n(\cos\theta + i\sin\theta)^n$

Therefore $z^n = r^n(e^{i\theta})^n$ since $e^{i\theta} = \cos\theta + i\sin\theta$ (using Euler's formula).

You can then use index laws to write $r^n(e^{i\theta})^n = r^n e^{in\theta}$

Using Euler's formula again, this becomes $r^n(\cos(n\theta) + i\sin(n\theta))$

Putting these two results together gives

Key point

$[r(\cos\theta + i\sin\theta)]^n = r^n(\cos(n\theta) + i\sin(n\theta))$, for all integers n, which is known as **de Moivre's theorem**.

You can prove this result using proof by induction.

De Moivre's theorem can be used to simplify powers of complex numbers.

For example, $(\cos\theta + i\sin\theta)^3 = \cos 3\theta + i\sin 3\theta$

$$\frac{1}{\cos\theta + i\sin\theta} = (\cos\theta + i\sin\theta)^{-1} = \cos(-\theta) + i\sin(-\theta)$$

Example 1

Write each of these numbers in the form $a + bi$

a $\left(\cos\dfrac{\pi}{3} + i\sin\dfrac{\pi}{3}\right)^4$ **b** $\left(\cos\dfrac{\pi}{4} - i\sin\dfrac{\pi}{4}\right)^6$

a $\left(\cos\dfrac{\pi}{3} + i\sin\dfrac{\pi}{3}\right)^4 = \cos\dfrac{4\pi}{3} + i\sin\dfrac{4\pi}{3}$ Use de Moivre's theorem.

$= -\dfrac{1}{2} - \dfrac{\sqrt{3}}{2}i$ In the form $a + bi$

b $\left(\cos\dfrac{\pi}{4} - i\sin\dfrac{\pi}{4}\right)^6 = \left(\cos\left(-\dfrac{\pi}{4}\right) + i\sin\left(-\dfrac{\pi}{4}\right)\right)^6$ Using $\cos(-\theta) = \cos\theta$ and $\sin(-\theta) = -\sin\theta$

$= \cos\left(-\dfrac{3\pi}{2}\right) + i\sin\left(-\dfrac{3\pi}{2}\right)$ Needs to be in modulus–argument form and then you can apply de Moivre's theorem.

$= i$

Example 2

Given the complex number $z = -\sqrt{3} + i$, use de Moivre's theorem to find z^{-2} in the form $a + bi$

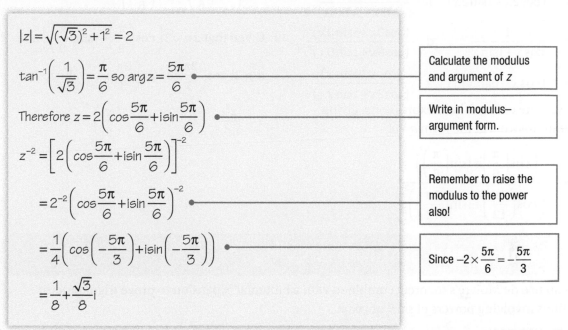

$$|z| = \sqrt{(\sqrt{3})^2 + 1^2} = 2$$

$$\tan^{-1}\left(\frac{1}{\sqrt{3}}\right) = \frac{\pi}{6} \text{ so } \arg z = \frac{5\pi}{6}$$

Calculate the modulus and argument of z

$$\text{Therefore } z = 2\left(\cos\frac{5\pi}{6} + i\sin\frac{5\pi}{6}\right)$$

Write in modulus–argument form.

$$z^{-2} = \left[2\left(\cos\frac{5\pi}{6} + i\sin\frac{5\pi}{6}\right)\right]^{-2}$$

$$= 2^{-2}\left(\cos\frac{5\pi}{6} + i\sin\frac{5\pi}{6}\right)^{-2}$$

Remember to raise the modulus to the power also!

$$= \frac{1}{4}\left(\cos\left(-\frac{5\pi}{3}\right) + i\sin\left(-\frac{5\pi}{3}\right)\right)$$

Since $-2 \times \frac{5\pi}{6} = -\frac{5\pi}{3}$

$$= \frac{1}{8} + \frac{\sqrt{3}}{8}i$$

Exercise 6.2A Fluency and skills

1   Express each of these numbers in the form $a + bi$

   a $\left(\cos\frac{\pi}{3} + i\sin\frac{\pi}{3}\right)^6$ b $\left(\cos\frac{\pi}{6} + i\sin\frac{\pi}{6}\right)^5$

   c $\left(\cos\left(-\frac{\pi}{12}\right) + i\sin\left(-\frac{\pi}{12}\right)\right)^4$

   d $\left(\cos\frac{\pi}{14} - i\sin\frac{\pi}{14}\right)^7$ e $\left(\cos\frac{\pi}{4} + i\sin\frac{\pi}{4}\right)^{-3}$

   f $\left(\cos\frac{\pi}{8} - i\sin\frac{\pi}{8}\right)^{-6}$

2   Given that $z = 3\left(\cos\left(\frac{\pi}{24}\right) + i\sin\left(\frac{\pi}{24}\right)\right)$, express in exact Cartesian form

   a z^4 b z^{-6}

3   Given that $z = 2\left(\cos\left(\frac{\pi}{12}\right) + i\sin\left(\frac{\pi}{12}\right)\right)$, express in exact Cartesian form

   a z^3 b z^{-2}

   c z^6 d z^{-4}

4   Given that $z = 4\left(\cos\left(\frac{3\pi}{2}\right) + i\sin\left(\frac{3\pi}{2}\right)\right)$, express in exact Cartesian form

   a z^2 b z^3

   c $\frac{1}{z}$ d $16z^{-4}$

5   Given that $z = 1 - i$, use de Moivre's theorem to write the following powers of z in the form $a + bi$

   a z^3 b z^7

   c z^{-5} d z^{-6}

6   Given that $z = 3i$, use de Moivre's theorem to write the following in Cartesian form.

   a z^2 b z^{-1}

   c z^{-3} d $\frac{3}{z^3}$

7   Given that $z = -\sqrt{3} + i$, use de Moivre's theorem to write the following in Cartesian form.

   a z^4 b z^{-3}

   c z^{-2} d $\frac{8}{z^6}$

8 Simplify each of these expressions into the form $\cos(ax)+i\sin(ax)$, where a is an integer to be found.

a $(\cos 2x+i\sin 2x)^3$ b $\dfrac{(\cos x+i\sin x)^7}{\cos 5x+i\sin 5x}$

c $\dfrac{1}{\cos x+i\sin x}$ d $\dfrac{(\cos 3x+i\sin 3x)^5}{(\cos 6x+i\sin 6x)^2}$

e $(\cos x-i\sin x)^{-2}$ f $\dfrac{(\cos x+i\sin x)^4}{(\cos 2x-i\sin 2x)^3}$

9 Evaluate each of these expressions, giving your answers in the form $a+bi$

a $\dfrac{\left(\cos\left(\dfrac{\pi}{6}\right)+i\sin\left(\dfrac{\pi}{6}\right)\right)^3}{\left(\cos\left(\dfrac{5\pi}{12}\right)+i\sin\left(\dfrac{5\pi}{12}\right)\right)^2}$

b $\dfrac{\left(\cos\left(\dfrac{\pi}{8}\right)+i\sin\left(\dfrac{\pi}{8}\right)\right)^6}{\left(\cos\left(\dfrac{\pi}{6}\right)+i\sin\left(\dfrac{\pi}{6}\right)\right)^3}$

10 Given that $z=\sqrt{2}\left(\cos\left(\dfrac{\pi}{9}\right)+i\sin\left(\dfrac{\pi}{9}\right)\right)$ and $w=2\left(\cos\left(\dfrac{2\pi}{3}\right)+i\sin\left(\dfrac{2\pi}{3}\right)\right)$, evaluate

a $\dfrac{z^6}{w^2}$ b $\left(\dfrac{z}{w}\right)^{-12}$

Reasoning and problem-solving

You can use de Moivre's theorem combined with a binomial expansion to prove trigonometric identities involving powers of $\sin\theta$ or $\cos\theta$

Strategy 1

To write $\cos(n\theta)$ or $\sin(n\theta)$ in terms of powers of $\cos\theta$ or $\sin\theta$

(1) Use de Moivre's theorem.

(2) Write out the binomial expansion.

(3) Simplify powers of i.

(4) Equate coefficients of real or imaginary parts.

(5) Use $\cos^2\theta+\sin^2\theta\equiv1$ to write the expression as powers of either $\cos\theta$ or $\sin\theta$

Example 3

Express $\sin 5x$ in the form $A\sin x+B\sin^3 x+C\sin^5 x$, where A, B and C are constants to be found.

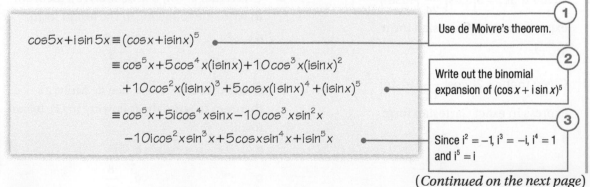

$\cos 5x+i\sin 5x\equiv(\cos x+i\sin x)^5$

① Use de Moivre's theorem.

$\equiv\cos^5 x+5\cos^4 x(i\sin x)+10\cos^3 x(i\sin x)^2$
$+10\cos^2 x(i\sin x)^3+5\cos x(i\sin x)^4+(i\sin x)^5$

② Write out the binomial expansion of $(\cos x+i\sin x)^5$

$\equiv\cos^5 x+5i\cos^4 x\sin x-10\cos^3 x\sin^2 x$
$-10i\cos^2 x\sin^3 x+5\cos x\sin^4 x+i\sin^5 x$

③ Since $i^2=-1$, $i^3=-i$, $i^4=1$ and $i^5=i$

(*Continued on the next page*)

$\text{Im}: \sin 5x = 5\cos^4 x \sin x - 10\cos^2 x \sin^3 x + \sin^5 x$

4

Consider only the imaginary parts since you are interested in $\sin 5x$. You can always equate real and imaginary parts in any equation.

$\sin 5x = 5(1-\sin^2 x)^2 \sin x - 10(1-\sin^2 x)\sin^3 x + \sin^5 x$

$\qquad = 5(1-2\sin^2 x + \sin^4 x)\sin x - 10(1-\sin^2 x)\sin^3 x + \sin^5 x$

$\qquad = 5\sin x - 10\sin^3 x + 5\sin^5 x - 10\sin^3 x + 10\sin^5 x + \sin^5 x$

$\qquad = 5\sin x - 20\sin^3 x + 16\sin^5 x$

So $A = 5, B = -20, C = 16$

5

Replace $\cos^2 x$ by $1 - \sin^2 x$

If $z = \cos\theta + i\sin\theta$ then $z + \dfrac{1}{z} = \cos\theta + i\sin\theta + \dfrac{1}{\cos\theta + i\sin\theta}$

Now $\dfrac{1}{\cos\theta + i\sin\theta} = (\cos\theta + i\sin\theta)^{-1}$

$\qquad\qquad\qquad = \cos(-\theta) + i\sin(-\theta)$ by de Moivre's theorem.

So $z + \dfrac{1}{z} = (\cos\theta + i\sin\theta) + (\cos(-\theta) + i\sin(-\theta))$

$\qquad\quad = (\cos\theta + i\sin\theta) + (\cos(\theta) - i\sin(\theta))$ since $\cos\theta = \cos(-\theta)$ and $\sin\theta = -\sin(-\theta)$

$\qquad\quad = 2\cos\theta$

Similarly, $z - \dfrac{1}{z} = (\cos\theta + i\sin\theta) - (\cos(-\theta) + i\sin(-\theta))$

$\qquad\qquad\quad = (\cos\theta + i\sin\theta) - (\cos(\theta) - i\sin(\theta))$

$\qquad\qquad\quad = 2i\sin\theta$

Key point

If $z = \cos\theta + i\sin\theta$, then

$z + \dfrac{1}{z} = 2\cos\theta \quad$ and $\quad z - \dfrac{1}{z} = 2i\sin\theta$

You can use de Moivre's theorem to generalise this result for powers of z

If $z = \cos\theta + i\sin\theta$, then $z^n + \dfrac{1}{z^n} = (\cos\theta + i\sin\theta)^n + \dfrac{1}{(\cos\theta + i\sin\theta)^n}$

$\qquad\qquad\qquad\qquad = (\cos n\theta + i\sin n\theta) + (\cos(-n\theta) + i\sin(-n\theta))$

$\qquad\qquad\qquad\qquad = (\cos n\theta + i\sin n\theta) + (\cos n\theta - i\sin n\theta)$

$\qquad\qquad\qquad\qquad = 2\cos n\theta$

Similarly, $z^n - \dfrac{1}{z^n} = (\cos\theta + i\sin\theta)^n - \dfrac{1}{(\cos\theta + i\sin\theta)^n}$

$\qquad\qquad\qquad = (\cos n\theta + i\sin n\theta) - (\cos(-n\theta) + i\sin(-n\theta))$

$\qquad\qquad\qquad = (\cos n\theta + i\sin n\theta) - (\cos n\theta - i\sin n\theta)$

$\qquad\qquad\qquad = 2i\sin n\theta$

Key point

If $z = \cos\theta + i\sin\theta$, then

$z^n + \dfrac{1}{z^n} = 2\cos n\theta \quad$ and $\quad z^n - \dfrac{1}{z^n} = 2i\sin n\theta$

Strategy 2

To write a power of $\cos\theta$ or $\sin\theta$ as a series involving $\cos(n\theta)$ or $\sin(n\theta)$

(1) Write in terms of z using $z+\dfrac{1}{z}=2\cos\theta$ or $z-\dfrac{1}{z}=2i\sin\theta$

(2) Write out the binomial expansion.

(3) Use rules of indices to simplify.

(4) Group terms together and use $z^n+\dfrac{1}{z^n}=2\cos n\theta$ or $z^n-\dfrac{1}{z^n}=2i\sin n\theta$

Example 4

Prove that $8\cos^4\theta \equiv \cos4\theta+4\cos2\theta+3$

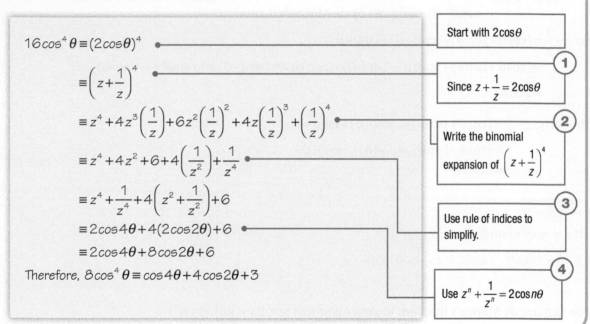

$16\cos^4\theta \equiv (2\cos\theta)^4$

Start with $2\cos\theta$ — (1)

$\equiv \left(z+\dfrac{1}{z}\right)^4$

Since $z+\dfrac{1}{z}=2\cos\theta$

$\equiv z^4+4z^3\left(\dfrac{1}{z}\right)+6z^2\left(\dfrac{1}{z}\right)^2+4z\left(\dfrac{1}{z}\right)^3+\left(\dfrac{1}{z}\right)^4$ — (2)

Write the binomial expansion of $\left(z+\dfrac{1}{z}\right)^4$

$\equiv z^4+4z^2+6+4\left(\dfrac{1}{z^2}\right)+\dfrac{1}{z^4}$

$\equiv z^4+\dfrac{1}{z^4}+4\left(z^2+\dfrac{1}{z^2}\right)+6$ — (3)

Use rule of indices to simplify.

$\equiv 2\cos4\theta+4(2\cos2\theta)+6$

$\equiv 2\cos4\theta+8\cos2\theta+6$ — (4)

Therefore, $8\cos^4\theta \equiv \cos4\theta+4\cos2\theta+3$

Use $z^n+\dfrac{1}{z^n}=2\cos n\theta$

Example 5

Show that $\sin^5 x \equiv a\sin x+b\sin3x+c\sin5x$, where a, b and c are constants to be found.

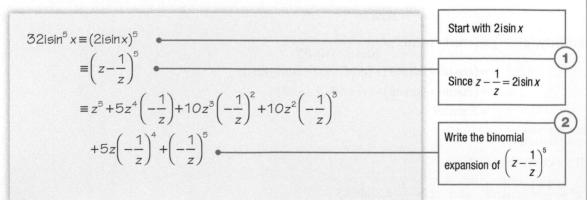

$32i\sin^5 x \equiv (2i\sin x)^5$

Start with $2i\sin x$ — (1)

$\equiv \left(z-\dfrac{1}{z}\right)^5$

Since $z-\dfrac{1}{z}=2i\sin x$

$\equiv z^5+5z^4\left(-\dfrac{1}{z}\right)+10z^3\left(-\dfrac{1}{z}\right)^2+10z^2\left(-\dfrac{1}{z}\right)^3$ — (2)

$+5z\left(-\dfrac{1}{z}\right)^4+\left(-\dfrac{1}{z}\right)^5$

Write the binomial expansion of $\left(z-\dfrac{1}{z}\right)^5$

(Continued on the next page)

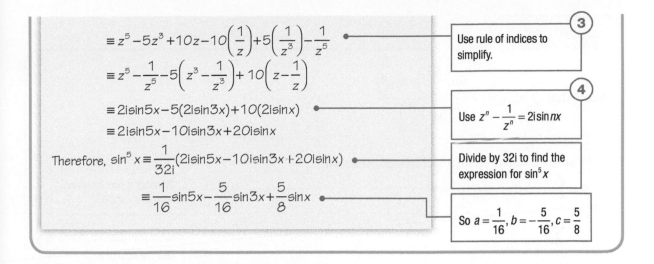

$$\equiv z^5 - 5z^3 + 10z - 10\left(\frac{1}{z}\right) + 5\left(\frac{1}{z^3}\right) - \frac{1}{z^5}$$

3 Use rule of indices to simplify.

$$\equiv z^5 - \frac{1}{z^5} - 5\left(z^3 - \frac{1}{z^3}\right) + 10\left(z - \frac{1}{z}\right)$$

$$\equiv 2i\sin 5x - 5(2i\sin 3x) + 10(2i\sin x)$$

4 Use $z^n - \dfrac{1}{z^n} = 2i\sin nx$

$$\equiv 2i\sin 5x - 10i\sin 3x + 20i\sin x$$

Therefore, $\sin^5 x \equiv \dfrac{1}{32i}(2i\sin 5x - 10i\sin 3x + 20i\sin x)$

Divide by 32i to find the expression for $\sin^5 x$

$$\equiv \frac{1}{16}\sin 5x - \frac{5}{16}\sin 3x + \frac{5}{8}\sin x$$

So $a = \dfrac{1}{16}$, $b = -\dfrac{5}{16}$, $c = \dfrac{5}{8}$

You can sum series involving complex numbers.

If $z = \cos\theta + i\sin\theta$, then the series $1 + z + z^2 + \ldots + z^{n-1}$ is a geometric series with common ratio z

Therefore $\displaystyle\sum_{r=0}^{n-1} z^r = \frac{z^n - 1}{z - 1}$, using the result for the sum of a geometric series from Maths A Level.

When $|z| < 1$, the infinite series $1 + z + z^2 + \ldots + z^r + \ldots$ will converge, so $\displaystyle\sum_{r=0}^{\infty} z^r = \frac{1}{1-z}$

Key point

For the complex number z, $\displaystyle\sum_{r=0}^{n-1} z^r = \frac{z^n - 1}{z - 1}$ and $\displaystyle\sum_{r=0}^{\infty} z^r = \frac{1}{1-z}$

Strategy 3

To find the sum of a series involving $\sin(r\theta)$ or $\cos(r\theta)$

1 Consider a sum involving $z = \cos(r\theta) + i\sin(r\theta)$

2 Use the formula for the sum of a geometric series, S, with first term a and common ratio r

$S_n = \dfrac{a(r^n - 1)}{r - 1}$ or $S_\infty = \dfrac{a}{1-r}$, where $|r| < 1$

3 Use the exponential form to simplify.

4 Use $\cos\theta = \dfrac{e^{i\theta} + e^{-i\theta}}{2}$ or $\sin\theta = \dfrac{e^{i\theta} - e^{-i\theta}}{2i}$

5 Select only the real or the imaginary parts, as required.

Example 6

Show that $\displaystyle\sum_{r=0}^{5}\sin(r\theta)=\dfrac{\sin\left(\dfrac{5\theta}{2}\right)\sin(3\theta)}{\sin\left(\dfrac{\theta}{2}\right)}$

First, let $z=\cos\theta+i\sin\theta$ and consider the sum

$S=\displaystyle\sum_{r=0}^{5}\cos(r\theta)+i\sin(r\theta)$ •———

> ① Consider a sum involving $\cos(r\theta)+i\sin(r\theta)$

$=\displaystyle\sum_{r=0}^{5}z^{r}$

> ② Use the sum of a geometric series with 6 terms ($a=1$, $r=z$)

$=\dfrac{z^{6}-1}{z-1}$ •———

$=\dfrac{e^{6i\theta}-1}{e^{i\theta}-1}$ •———

> ③ Use the exponential form.

$=\dfrac{e^{-\frac{i\theta}{2}}(e^{6i\theta}-1)}{e^{\frac{i\theta}{2}}-e^{-\frac{i\theta}{2}}}$ •———

> Multiply numerator and denominator by $e^{-\frac{i\theta}{2}}$

$=\dfrac{e^{-\frac{i\theta}{2}}(e^{6i\theta}-1)}{2i\sin\left(\dfrac{\theta}{2}\right)}$ •———

> ④ Since $\dfrac{e^{\frac{i\theta}{2}}-e^{-\frac{i\theta}{2}}}{2i}=\sin\left(\dfrac{\theta}{2}\right)$

$=\dfrac{e^{-\frac{i\theta}{2}}e^{3i\theta}(e^{3i\theta}-e^{-3i\theta})}{2i\sin\left(\dfrac{\theta}{2}\right)}$ •———

> Write $e^{6i\theta}-1$ as $e^{3i\theta}(e^{3i\theta}-e^{-3i\theta})$

$=\dfrac{e^{\frac{5i\theta}{2}}(2i\sin3\theta)}{2i\sin\left(\dfrac{\theta}{2}\right)}$ •———

> ④ Since $\dfrac{e^{3i\theta}-e^{3i\theta}}{2i}=\sin(3\theta)$

$=\dfrac{e^{\frac{5i\theta}{2}}\sin3\theta}{\sin\left(\dfrac{\theta}{2}\right)}$

$=\dfrac{\left(\cos\dfrac{5\theta}{2}+i\sin\dfrac{5\theta}{2}\right)\sin3\theta}{\sin\left(\dfrac{\theta}{2}\right)}$

Therefore, $\displaystyle\sum_{r=1}^{n}\sin(r\theta)=\dfrac{\sin\left(\dfrac{5\theta}{2}\right)\sin(3\theta)}{\sin\left(\dfrac{\theta}{2}\right)}$ •———

> ⑤ Select the imaginary parts of the sum.

Example 7

Show that $\cos\theta + \cos 2\theta + \cos 3\theta + \ldots + \cos 7\theta \equiv \cos(4\theta)\sin\left(\dfrac{7\theta}{2}\right)\text{cosec}\left(\dfrac{\theta}{2}\right)$

First, let $z = \cos\theta + i\sin\theta$ and consider the sum

$$S = \sum_{r=1}^{7} \cos(r\theta) + i\sin(r\theta)$$

1 Consider a sum involving $\cos(r\theta) + i\sin(r\theta)$

$$= \sum_{r=1}^{7} z^r$$

2 Use the sum of a geometric series with $a = z, r = z$

$$= \frac{z(z^7 - 1)}{z - 1}$$

3 Use the exponential form.

$$= \frac{e^{i\theta}(e^{7i\theta} - 1)}{e^{i\theta} - 1}$$

Multiply numerator and denominator by $e^{-\frac{i\theta}{2}}$

$$= \frac{e^{\frac{i\theta}{2}}(e^{7i\theta} - 1)}{e^{\frac{i\theta}{2}} - e^{-\frac{i\theta}{2}}}$$

4 Since $\dfrac{e^{\frac{i\theta}{2}} - e^{-\frac{i\theta}{2}}}{2i} = \sin\left(\dfrac{\theta}{2}\right)$

$$= \frac{e^{\frac{i\theta}{2}}(e^{7i\theta} - 1)}{2i\sin\left(\dfrac{\theta}{2}\right)}$$

Write $e^{7i\theta} - 1$ as $e^{\frac{7i\theta}{2}}\left(e^{\frac{7i\theta}{2}} - e^{-\frac{7i\theta}{2}}\right)$

$$= \frac{e^{\frac{i\theta}{2}}e^{\frac{7i\theta}{2}}\left(e^{\frac{7i\theta}{2}} - e^{-\frac{7i\theta}{2}}\right)}{2i\sin\left(\dfrac{\theta}{2}\right)}$$

4 Since $\dfrac{e^{\frac{7i\theta}{2}} - e^{\frac{-7i\theta}{2}}}{2i} = \sin\left(\dfrac{7\theta}{2}\right)$

$$= \frac{e^{4i\theta}\left(2i\sin\left(\dfrac{7\theta}{2}\right)\right)}{2i\sin\left(\dfrac{\theta}{2}\right)}$$

$$= \frac{e^{4i\theta}\sin\left(\dfrac{7\theta}{2}\right)}{\sin\left(\dfrac{\theta}{2}\right)}$$

$$= \frac{(\cos 4\theta + i\sin 4\theta)\sin\left(\dfrac{7\theta}{2}\right)}{\sin\left(\dfrac{\theta}{2}\right)}$$

5 Select the real parts of the sum.

$$\cos\theta + \cos 2\theta + \cos 3\theta + \ldots + \cos 7\theta = \frac{\cos 4\theta\sin\left(\dfrac{7\theta}{2}\right)}{\sin\left(\dfrac{\theta}{2}\right)}$$

$$= \cos(4\theta)\sin\left(\dfrac{7\theta}{2}\right)\text{cosec}\left(\dfrac{\theta}{2}\right)$$

1 Prove each of these identities.

    a $2\cos^2\theta \equiv \cos2\theta + 1$ **b** $8\sin^3\theta \equiv 6\sin\theta - 2\sin3\theta$ **c** $4\sin^4\theta \equiv \frac{1}{2}\cos4\theta - 2\cos2\theta + \frac{3}{2}$

2 **a**   Show that $\cos^5\theta \equiv A(10\cos\theta + 5\cos3\theta + \cos5\theta)$, where A is a constant to be found.

    b   Hence find $\int \cos^5\theta \, d\theta$

3 **a**   Show that $\sin^6\theta \equiv B(15\cos2\theta - 6\cos4\theta + \cos6\theta - 10)$, where B is a constant to be found.

    b   Hence find $\int \sin^6\theta \, d\theta$

4 **a**   Show that $2\sin^3\theta \equiv \frac{3}{2}\sin\theta - \frac{1}{2}\sin3\theta$

    b   Hence solve the equation $3\sin\theta - \sin3\theta = \frac{1}{2}$ for $-\pi \le \theta \le \pi$

5 **a**   Show that $5\cos^4\theta \equiv A\cos4\theta + B\cos2\theta + C$, where A, B and C are constants to be found.

    b   Hence solve the equation $\cos4\theta + 4\cos2\theta + 3 = 2$ for $-\pi \le \theta \le \pi$

6 Use de Moivre's theorem to prove the following identities.

    a $\sin2\theta \equiv 2\cos\theta\sin\theta$ **b** $\sin3\theta \equiv 3\sin\theta - 4\sin^3\theta$

    c $\cos3\theta \equiv 4\cos^3\theta - 3\cos\theta$ **d** $\sin4\theta \equiv 4\cos\theta\sin\theta - 8\cos\theta\sin^3\theta$

7 Prove these identities.

    a $\cos6\theta \equiv 32\cos^6\theta - 48\cos^4\theta + 18\cos^2\theta - 1$ **b** $\sin6\theta \equiv 2\sin\theta\cos\theta(16\sin^4\theta - 16\sin^2\theta + 3)$

8 **a**   Use de Moivre's theorem to show that $\cos5\theta \equiv 16\cos^5\theta - 20\cos^3\theta + 5\cos\theta$

    b   Hence find 3 solutions to the equation $16x^5 - 20x^3 + 5x = 1$

9 **a**   Use de Moivre's theorem to show that $\cos4\theta \equiv 8\cos^4\theta - 8\cos^2\theta + 1$

    b   Hence find 4 solutions to the equation $x^4 - x^2 = -\frac{1}{16}$

10 Use de Moivre's theorem to show that $\tan2\theta \equiv \dfrac{2\tan\theta}{1-\tan^2\theta}$

11 Use proof by induction to prove that $\left[r(\cos\theta + i\sin\theta)\right]^n = r^n(\cos n\theta + i\sin n\theta)$ for all positive integers n

12 **a**   Given that $z = \cos\theta + i\sin\theta$, use de Moivre's theorem to show that $2\cos(n\theta) = z^n + \dfrac{1}{z^n}$

    b   Hence show that $4\cos\theta\sin^2\theta \equiv \cos\theta - \cos(3\theta)$

13 **a**   Given that $z = \cos\theta + i\sin\theta$, use de Moivre's theorem to show that $2i\sin(n\theta) = z^n - \dfrac{1}{z^n}$

    b   Hence show that $16\sin^3\theta\cos^2\theta \equiv 2\sin\theta + \sin(3\theta) - \sin(5\theta)$

14 **a**   Show that, if $z = e^{\frac{i\pi}{5}}$, then $\displaystyle\sum_{r=0}^{10} z^r = 1$

    b   Hence show that one solution of $\sin\theta + \sin2\theta + \ldots + \sin10\theta = 0$ is $\theta = \dfrac{\pi}{5}$

15 a Show that $1+e^{i\theta}+e^{2i\theta}+\ldots+e^{11i\theta}=\dfrac{e^{\frac{11i\theta}{2}}\sin(6\theta)}{\sin\left(\dfrac{\theta}{2}\right)}$

b Hence show that $\displaystyle\sum_{r=0}^{11}\cos(r\theta)=\dfrac{\cos\left(\dfrac{11\theta}{2}\right)\sin(6\theta)}{\sin\left(\dfrac{\theta}{2}\right)}$ and find an similar expression for $\displaystyle\sum_{r=0}^{11}\sin(r\theta)$

16 a Show that $\displaystyle\sum_{r=1}^{3}\cos(r\theta)\equiv\dfrac{\cos(2\theta)\sin\left(\dfrac{3\theta}{2}\right)}{\sin\left(\dfrac{\theta}{2}\right)}$

b Hence, show that $\dfrac{\cos\theta+\cos2\theta+\cos3\theta}{\sin\theta+\sin2\theta+\sin3\theta}\equiv\cot2\theta$

17 If $z=\cos\left(\dfrac{\pi}{n}\right)+i\sin\left(\dfrac{\pi}{n}\right)$, show that $\displaystyle\sum_{r=0}^{n-1}z^{r}=1+i\cot\left(\dfrac{\pi}{2n}\right)$

18 a Show that $\displaystyle\sum_{r=1}^{\infty}\dfrac{\cos r\theta+i\sin r\theta}{2^{r}}=\dfrac{e^{i\theta}}{2-e^{i\theta}}$

b Hence, show that the infinite series $\dfrac{1}{2}\cos\theta+\dfrac{1}{4}\cos2\theta+\dfrac{1}{8}\cos3\theta+\ldots$ converges to $\dfrac{2\cos\theta-1}{5-4\cos\theta}$

and find a similar expression for $\displaystyle\sum_{r=1}^{\infty}\dfrac{1}{2^{r}}\sin r\theta$

19 a Show that $\displaystyle\sum_{r=1}^{n}\cos(r\theta)=\dfrac{\cos\left(\dfrac{(n+1)\theta}{2}\right)\sin\left(\dfrac{n\theta}{2}\right)}{\sin\left(\dfrac{\theta}{2}\right)}$

(Hint: use the formulae for $\cos(A\pm B)$ and $\sin(A\pm B)$)

b Write down a similar expression for $\displaystyle\sum_{r=1}^{n}\sin(r\theta)$

Fluency and skills

A fundamental rule in maths is that an equation of order n must have n solutions. Therefore the equation $z^3 = 1$ must have three roots so there must be three cube roots of 1. One of them is real (1) and two are complex. These are called **the cube roots of unity**.

This is easily seen in the case of $z^4 = 1$: square root both sides to give $z^2 = \pm 1$

If $z^2 = 1$, then $z = 1$ or $z = -1$

If $z^2 = -1$, then $z = i$ or $z = -i$

So there are four solutions because the equation has order 4

You can use the exponential form of a complex number to solve equations of the form $z^n = 1$ to find the **nth roots of unity**.

> The three solutions to $z^3 = 1$ form the vertices of an equilateral triangle.

Example 1

a Find all the solutions to the equation $z^6 = 1$, giving your answers in Cartesian form.

b Illustrate the 6th roots of unity on an Argand diagram.

a $z^6 = e^{2k\pi i}$

$z = \left(e^{2k\pi i}\right)^{\frac{1}{6}}$

$= e^{\frac{2k\pi}{6}i}$

$= e^{\frac{k\pi}{3}i}$

> Since the modulus of 1 and the argument of any positive real number are 0 or $\pm 2\pi$ or $\pm 4\pi$ and so on, in general, this could be written $2k\pi$, where k is an integer.

Consider each of the possible values of k from 0 to 5. Going beyond 5 would just repeat the same values again.

> Use index law.

$k = 0: z = e^0 = 1$

$k = 1: z = e^{\frac{\pi}{3}i} = \cos\left(\frac{\pi}{3}\right) + i\sin\left(\frac{\pi}{3}\right) = \frac{1}{2} + \frac{\sqrt{3}}{2}i$

$k = 2: z = e^{\frac{2\pi}{3}i} = \cos\left(\frac{2\pi}{3}\right) + i\sin\left(\frac{2\pi}{3}\right) = -\frac{1}{2} + \frac{\sqrt{3}}{2}i$

$k = 3: z = e^{\frac{3\pi}{3}i} = \cos(\pi) + i\sin(\pi) = -1$

$k = 4: z = e^{\frac{4\pi}{3}i} = \cos\left(\frac{4\pi}{3}\right) + i\sin\left(\frac{4\pi}{3}\right) = -\frac{1}{2} - \frac{\sqrt{3}}{2}i$

$k = 5: z = e^{\frac{5\pi}{3}i} = \cos\left(\frac{5\pi}{3}\right) + i\sin\left(\frac{5\pi}{3}\right) = \frac{1}{2} - \frac{\sqrt{3}}{2}i$

> There should be 6 solutions to the equation.

(Continued on the next page)

b

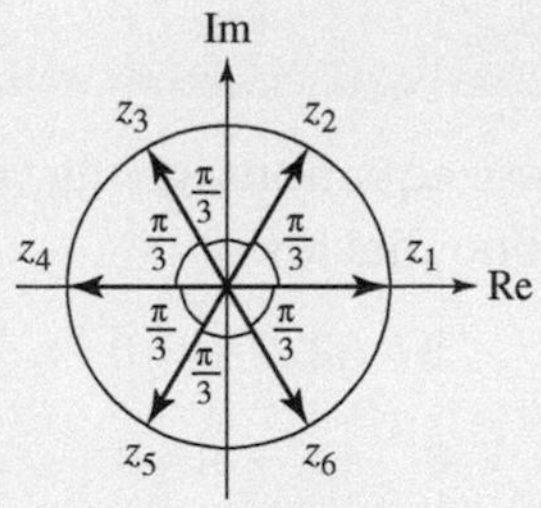

Notice how the 6 solutions split a circle into 6 equal sectors, each with angle $\frac{\pi}{3}$

The 6 solutions to $z^6 = 1$ form the vertices of a regular hexagon.

Key point

The equation $z^n = 1$ has n solutions of the form

$z = e^{\frac{2k\pi i}{n}}$, where $k = 0, 1, 2, \ldots, n$

This method can be extended to find the nth roots of any complex number.

Example 2

Solve the equation $z^4 = 2\sqrt{3} - 2i$, giving your answers in the form $re^{i\theta}$ where $r > 0$ and $-\pi < \theta \leq \pi$

$|2\sqrt{3} - 2i| = \sqrt{(2\sqrt{3})^2 + 2^2} = 4$

$\arg(2\sqrt{3} - 2i) = -\frac{\pi}{6}$

Calculate modulus and argument.

So in general argument is $-\frac{\pi}{6} + 2k\pi = \frac{(12k-1)\pi}{6}$

$z^4 = 4e^{\frac{(12k-1)\pi}{6}}$

Write a general term in exponential form.

$z = \left(4e^{\frac{(12k-1)\pi}{6}}\right)^{\frac{1}{4}}$

$= \sqrt{2}e^{\frac{(12k-1)\pi}{24}}$

Consider each of the possible values of k from 0 to 3

$k = 0: z = \sqrt{2}e^{-\frac{\pi}{24}i}$

$k = 1: z = \sqrt{2}e^{\frac{11\pi}{24}i}$

$k = 2: z = \sqrt{2}e^{\frac{23\pi}{24}i}$

$k = 3: z = \sqrt{2}e^{\frac{35\pi}{24}i}$

The final solution is not in the range $-\pi < \theta \leq \pi$

So instead write $z = \sqrt{2}e^{-\frac{13\pi i}{24}}$

Since $-\frac{13\pi}{24} = \frac{35\pi}{24}$

1 **a** Solve the equation $z^3 = 1$, giving your answers in Cartesian form.

  b Illustrate the cube roots of unity on an Argand diagram.

2 **a** Solve the equation $z^4 = 1$, giving your answers in Cartesian form.

  b Illustrate the 4th roots of unity on an Argand diagram.

3 **a** Solve the equation $z^5 = 1$, giving your answers in exponential form

  b Illustrate the 5th roots of unity on an Argand diagram.

4 Solve each of these equations, giving your solutions in Cartesian form.

  a $z^3 = 8$ **b** $z^3 = i$

  c $z^4 = -49$ **d** $z^3 = -27i$

  e $z^6 = -8$

5 Solve each of these equations, giving your solutions in exponential form.

  a $z^4 = -16i$ **b** $z^5 = 32i$

6 Solve each of these equations, giving your solutions in exponential form.

  a $z^3 = 4\sqrt{2} + 4\sqrt{2}i$ **b** $z^3 = -4\sqrt{2} + 4\sqrt{2}i$

  c $z^3 = -4\sqrt{2} - 4\sqrt{2}i$ **d** $z^3 = 4\sqrt{2} - 4\sqrt{2}i$

7 Solve each of these equations, giving your solutions in modulus–argument form with θ given to 2 decimal places.

  a $z^4 = 3\sqrt{5} + 6i$ **b** $z^4 = -3\sqrt{5} + 6i$

  c $z^4 = -6 - 3\sqrt{5}i$ **d** $z^4 = 6 - 3\sqrt{5}i$

8 Solve each of these equations, giving your solutions in exponential form.

  a $z^5 + 243 = 0$ **b** $32z^5 + 1 = 0$

  c $z^6 + 64i = 0$ **d** $4z^4 + 3i = 4$

9 Solve each of these equations, giving your solutions in Cartesian form.

  a $z^4 = 8 + 8\sqrt{3}i$ **b** $z^4 = -8 + 8\sqrt{3}i$

  c $z^4 = 8\sqrt{3} - 8i$ **d** $z^4 = -8\sqrt{3} - 8\sqrt{3}i$

10 Solve the equation $z^{\frac{3}{2}} = 2 - 2i$, giving your answers in the form $a + bi$

11 Solve the equation $z^{\frac{4}{3}} = -32\sqrt{3} + 32i$, giving your answers in the form $re^{i\theta}$ where $r > 0$ and $-\pi < \theta \le \pi$

12 **a** Solve the equation $z^4 = -32 + 32\sqrt{3}i$, giving your solutions in Cartesian form.

  b Represent the solutions on an Argand diagram.

13 Solve each of these equations, giving your solutions in the form $re^{i\theta}$ where $r > 0$ and $-\pi < \theta \le \pi$

  a $z^3 = 4\sqrt{3} + 4i$ **b** $z^4 = 2\sqrt{2} - 2\sqrt{2}i$

14 Solve each of these equations, giving your solutions in the form $r(\cos\theta + i\sin\theta)$, where $r > 0$ and $-\pi < \theta \le \pi$

  a $z^3 = \sqrt{2} - \sqrt{6}i$ **b** $z^6 = -4\sqrt{3} + 4i$

Reasoning and problem-solving

You have seen that the equation $z^n = 1$ has n solutions of the form $z = e^{\frac{2k\pi i}{n}}$, where $k = 0, 1, 2, ..., n-1$

Using index laws, you could write this as $z = \left(e^{\frac{2\pi i}{n}}\right)^k$. Therefore, if ω is a complex solution of $z^n = 1$ then $1, \omega, \omega^2, \omega^3, ..., \omega^{n-1}$ are all distinct solutions.

$$1 + \omega + \omega^2 + \omega^3 + ... + \omega^{n-1} = \left(e^{\frac{2\pi i}{n}}\right)^0 + \left(e^{\frac{2\pi i}{n}}\right)^1 + \left(e^{\frac{2\pi i}{n}}\right)^2 + \left(e^{\frac{2\pi i}{n}}\right)^3 + ... + \left(e^{\frac{2\pi i}{n}}\right)^{n-1}$$

This is a geometric series with first term 1 and common ratio $\omega = e^{\frac{2\pi i}{n}}$

So, using the formula $S_n = \frac{a(1-r^n)}{1-r}$ you can see that the sum of the series is $\dfrac{1\left(1-\left(e^{\frac{2\pi i}{n}}\right)^n\right)}{1-e^{\frac{2\pi i}{n}}}$.

This simplifies to $\dfrac{1(1-e^{2\pi i})}{1-e^{\frac{2\pi i}{n}}} = 0$ since $e^{2\pi i} = 1$

Key point

Therefore, $1 + \omega + \omega^2 + ... + \omega^{n-1} = 0$ where ω is an nth root of unity.

You can quote this result.

Strategy

To find complex solutions of an equation

1. Expand brackets or simplify fractions.
2. Use fact that $\omega^n = 1$ if ω is an nth root of unity.
3. Use $1 + \omega + \omega^2 + ... + \omega^{n-1} = 0$

Example 3

You are given that ω is a complex 4th root of unity.

a Show that $1 + \omega + \omega^2 + \omega^3 = 0$

b Evaluate $(1+\omega)(1+\omega^2) + \omega^4$

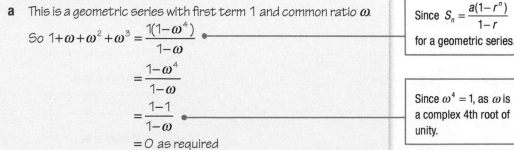

a This is a geometric series with first term 1 and common ratio ω.

So $1 + \omega + \omega^2 + \omega^3 = \dfrac{1(1-\omega^4)}{1-\omega}$ Since $S_n = \dfrac{a(1-r^n)}{1-r}$ for a geometric series.

$= \dfrac{1-\omega^4}{1-\omega}$ **(2)**

$= \dfrac{1-1}{1-\omega}$ Since $\omega^4 = 1$, as ω is a complex 4th root of unity.

$= 0$ as required

b $(1+\omega)(1+\omega^2) + \omega^4 = 1 + \omega + \omega^2 + \omega^3 + \omega^4$ **(1)**

$= 1 + \omega + \omega^2 + \omega^3 + 1$ Since $\omega^4 = 1$ **(2)**

$= 0 + 1$

$= 1$ **(3)**

19

The solutions, $z_1, z_2, ..., z_n$ of the equation $z^n = a + bi$ form the vertices of a regular n-sided polygon, also called a regular n-gon.

The vertices all lie on a circle with radius $|a + bi|^{\frac{1}{n}}$ and centre the origin.

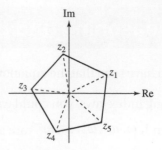

Strategy

To solve geometric problems

(1) Write the complex number representing a vertex in exponential form.

(2) Multiply by the nth root of unity to find all the vertices of a regular n-gon.

(3) Sketch a diagram.

Example 4

The equilateral triangle ABC has centre the origin. Vertex A has coordinates $(1, \sqrt{3})$

a Calculate the exact area of triangle ABC

The triangle ABC is rotated through $\dfrac{\pi}{6}$ radians anti-clockwise about the centre to triangle $A'B'C'$

b Find the coordinates of A'

a The complex number representing vertex A is $1 + \sqrt{3}i$

This has modulus $\sqrt{1^2 + (\sqrt{3})^2} = 2$

and argument $\tan^{-1}\left(\dfrac{\sqrt{3}}{1}\right) = \dfrac{\pi}{3}$

So the complex number representing vertex A is $z_1 = 2e^{\frac{\pi i}{3}}$ ● — Write in the form $re^{i\theta}$ (1)

The cube roots of unity are $1, \omega, \omega^2$ where $\omega = e^{\frac{2\pi i}{3}}$

So the vertices are at

$z_1 = 2e^{\frac{\pi i}{3}} = 1 + \sqrt{3}i$

$z_2 = 2e^{\frac{\pi i}{3}}\left(e^{\frac{2\pi i}{3}}\right) = 2e^{\pi i} = -2$ ● — Multiply z_1 by $e^{\frac{2\pi i}{3}}$ (2)

$z_3 = -2e^{\frac{2\pi i}{3}} = 1 - \sqrt{3}i$ ● — Multiply z_2 by $e^{\frac{2\pi i}{3}}$ (2)

So the vertices are at $(1, \sqrt{3}), (-2, 0), (1, -\sqrt{3})$

(Continued on the next page)

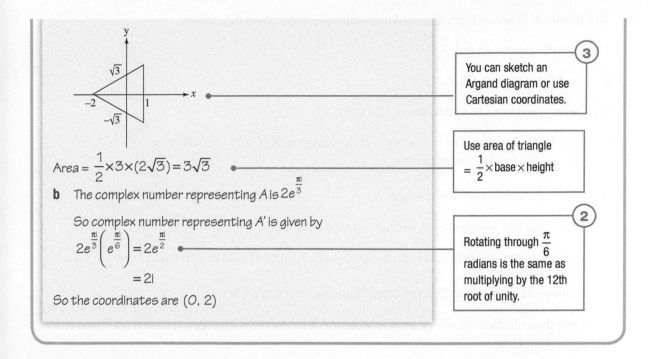

Area $= \dfrac{1}{2} \times 3 \times (2\sqrt{3}) = 3\sqrt{3}$

You can sketch an Argand diagram or use Cartesian coordinates.

Use area of triangle $= \dfrac{1}{2} \times$ base $\times$ height

b The complex number representing A is $2e^{\frac{\pi i}{3}}$

So complex number representing A' is given by

$2e^{\frac{\pi i}{3}} \left(e^{\frac{\pi i}{6}} \right) = 2e^{\frac{\pi i}{2}}$

$= 2i$

So the coordinates are $(0, 2)$

Rotating through $\dfrac{\pi}{6}$ radians is the same as multiplying by the 12th root of unity.

Exercise 6.3B Reasoning and problem-solving

1 Given that ω is a complex cube root of unity

   a Show that $1 + \omega + \omega^2 = 0$

   b Evaluate the following expressions.

      i $(1+\omega)^2 - \omega$ **ii** $(1+\omega)(1+\omega^2)$ **iii** $\omega(\omega+1)$ **iv** $\dfrac{2\omega+1}{\omega-1} + \omega$

2 Given that ω is a complex 5th root of unity

   a Show that $1 + \omega + \omega^2 + \omega^3 + \omega^4 = 0$

   b Evaluate the following expressions.

      i $\omega(1+\omega)(1+\omega^2)$ **ii** $\dfrac{\omega^2}{\omega+1} + \omega^3 + 1$ **iii** $(1+\omega)^2 + \omega(\omega^3 + \omega^2 - 1)$

3 The points A, B and C form an equilateral triangle centred on the origin.

   The coordinates of A are $\left(\dfrac{3\sqrt{3}}{2}, -\dfrac{3}{2} \right)$

   a Find the coordinates of B and C

   b Calculate the exact

      i Area,   **ii** Perimeter of triangle ABC

4 The points A, B and C form an equilateral triangle centred on the origin.

The coordinates of A are $\left(\dfrac{5}{2}, \dfrac{5\sqrt{3}}{2}\right)$

a Find the coordinates of B and C

b Calculate the exact perimeter of triangle ABC

c Find the coordinates of the vertices of the image of ABC after a rotation of $\dfrac{5\pi}{6}$ radians anti-clockwise about the centre.

5 **a** Solve the equation $z^4 = -2 + 2\sqrt{3}i$

b Show that the vertices form a square of side length 2

c Hence, find the coordinates of the vertices of a square of side length 2 and centre $(0, \sqrt{2})$

6 A regular pentagon centred on the origin has one vertex at $3 + 3i$

a Find the complex numbers representing the other vertices, giving your answers in modulus–argument form

b Show that the perimeter of the pentagon is given by $A\sin\left(\dfrac{\pi}{5}\right)$, where A is a constant to be found.

7 The points A, B, C, D, E and F are vertices of the regular hexagon shown.

Given that A, B, C, D, E and F are all solutions to $z^n = 1$

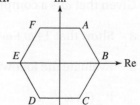

a State the value of n

b Find the coordinates of F

c Calculate the area of the hexagon.

8 A machine embroiders a logo in the shape of a regular octagon.

Relative to the centre of the octagon, the first vertex is at the point $(1, 1)$

a Find the coordinates of the other vertices of the octagon relative to the starting vertex.

b Calculate an estimate of the amount of thread required to stitch the logo given that each unit of length requires approximately 2.5 cm of thread.

Chapter summary

- A complex number $z = a + bi$ can be written
  - in modulus-argument form as $z = r(\cos\theta + i\sin\theta)$
  - in exponential form as $z = re^{i\theta}$

  where $r = |z|$ and $\theta = \arg z$, $-\pi < \theta \leq \pi$

- For complex numbers z and w,
  - $|zw| = |z||w|$
  - $\left|\dfrac{z}{w}\right| = \dfrac{|z|}{|w|}$
  - $\arg(zw) = \arg z + \arg w$
  - $\arg\left(\dfrac{z}{w}\right) = \arg z - \arg w$

- Euler's formula is $e^{i\theta} = \cos\theta + i\sin\theta$

- $\cos\theta = \dfrac{e^{i\theta} + e^{-i\theta}}{2}$ and $\sin\theta = \dfrac{e^{i\theta} - e^{-i\theta}}{2i}$

- De Moivre's theorem states that $\left[r(\cos\theta + i\sin\theta)\right]^n = r^n(\cos(n\theta) + i\sin(n\theta))$ for all integers n

- The equation $z^n = 1$ has n solutions of the form $z = e^{\frac{2k\pi i}{n}}$, where $k = 0, 1, 2, ..., n$

- Equations of the form $z^n = a + bi$ have solutions of the form $z = r^{\frac{1}{n}}e^{\frac{(\theta + 2k\pi)i}{n}}$, where $|z| = r$, $\arg(z) = \theta$ and $k = 0, 1, 2, ... n$

- $1 + \omega + \omega^2 + ... + \omega^{n-1} = 0$, where ω is an nth root of unity.

- The solutions, $z_1, z_2, ..., z_n$ of the equation $z_n = a + bi$ form the vertices of a regular n-sided polygon, also called a regular n-gon.

Check and review

You should now be able to...	Review
✓ Write complex numbers in exponential form.	1, 2
✓ Convert complex numbers from exponential to Cartesian form.	3
✓ Use rules to find the argument and modulus of products of complex numbers.	4
✓ Use rules to find the argument and modulus of quotients of complex numbers.	5
✓ Use de Moivre's theorem to simplify powers of complex numbers.	6, 7
✓ Use de Moivre's theorem to prove trigonometric identities.	8
✓ Simplify powers of trigonometric functions using a binomial expansion.	9
✓ Calculate roots of unity.	10
✓ Calculate the nth root of a complex number.	11–13

1 Write these complex numbers in exponential form.

   a $-3i$ b $1+i$

   c 5 d $-\sqrt{3}-i$

   e $\sqrt{6}-\sqrt{2}i$ f $-1+\sqrt{3}i$

2 Write each of these complex numbers in exponential form

   a $3\left(\cos\dfrac{\pi}{7}+i\sin\dfrac{\pi}{7}\right)$

   b $\sqrt{2}\left(\cos\dfrac{\pi}{9}+i\sin\dfrac{\pi}{9}\right)$

   c $\sqrt{3}\left(\cos\dfrac{\pi}{8}-i\sin\dfrac{\pi}{8}\right)$

   d $5\left(\cos\left(-\dfrac{\pi}{5}\right)-i\sin\left(-\dfrac{\pi}{5}\right)\right)$

3 Write these complex numbers in the form $a+bi$

   a $7e^{\left(\frac{\pi}{2}\right)i}$ b $6e^{\left(\frac{\pi}{3}\right)i}$

   c $\sqrt{3}e^{\left(-\frac{\pi}{6}\right)i}$ d $\sqrt{2}e^{\left(\frac{3\pi}{4}\right)i}$

   e $\sqrt{6}e^{\left(-\frac{2\pi}{3}\right)i}$ f $2\sqrt{3}e^{\left(\frac{5\pi}{6}\right)i}$

4 Find the argument and modulus of zw in each case.

   a $z=3\left(\cos\dfrac{\pi}{5}+i\sin\dfrac{\pi}{5}\right)$ and
   $w=5\left(\cos\dfrac{\pi}{7}+i\sin\dfrac{\pi}{7}\right)$

   b $z=\sqrt{2}\left(\cos\left(-\dfrac{\pi}{8}\right)+i\sin\left(-\dfrac{\pi}{8}\right)\right)$ and
   $w=\sqrt{6}\left(\cos\dfrac{\pi}{3}+i\sin\dfrac{\pi}{3}\right)$

   c $z=1+\sqrt{3}i$ and $w=3-3i$

   d $z=\dfrac{1}{2}e^{\frac{2\pi}{9}i}$ and $w=4e^{\frac{\pi}{3}i}$

5 Find the argument and modulus of $\dfrac{z}{w}$ in each case.

   a $z=5\left(\cos\dfrac{\pi}{2}+i\sin\dfrac{\pi}{2}\right)$ and
   $w=10\left(\cos\dfrac{\pi}{4}+i\sin\dfrac{\pi}{4}\right)$

   b $z=\sqrt{15}\left(\cos\left(-\dfrac{3\pi}{4}\right)+i\sin\left(-\dfrac{3\pi}{4}\right)\right)$ and
   $w=\sqrt{5}\left(\cos\left(-\dfrac{\pi}{8}\right)+i\sin\left(-\dfrac{\pi}{8}\right)\right)$

   c $z=-5+5i$ and $w=\sqrt{6}-3\sqrt{2}i$

   d $z=16e^{-\frac{2\pi}{11}i}$ and $w=\sqrt{2}e^{\frac{5\pi}{11}i}$

6 Given that $z=8\left(\cos\left(\dfrac{\pi}{2}\right)+i\sin\left(\dfrac{\pi}{2}\right)\right)$, write these powers of z in Cartesian form.

   a z^2 b z^3

7 Given that $w=-2\sqrt{3}-2i$, express the following in the form $a+bi$

   a w^2 b w^{-3}

8 a Use de Moivre's theorem to show that
   $\sin 5\theta \equiv 5\sin\theta - 20\sin^3\theta + 16\sin^5\theta$

   b Hence find 3 solutions to the equation
   $5x-20x^3+16x^5=0$
   Give your answers to 3 significant figures.

9 Prove that $\cos^3\theta \equiv A(\cos 3\theta + 3\cos\theta)$, where A is a constant to be found.

10 a Calculate the 8th roots of unity. Give your answers in exact Cartesian form.

   b Draw the roots on an Argand diagram.

11 Solve these equations, giving your answers in Cartesian form.

   a $z^8=16$ b $z^3=i$

   c $z^2=-9i$ d $z^6=-125$

12 Solve the equation $z^4=-2\sqrt{2}-2\sqrt{2}i$, giving your solutions in the form $re^{i\theta}$, where $r>0$ and $-\pi<\theta\leq\pi$

13 Solve the equation $z^{\frac{2}{5}}=-2+2\sqrt{3}i$, giving your solutions in the form $a+bi$

History

Abraham de Moivre was a French mathematician (1667–1754). He was a contemporary and friend of Isaac Newton and Edmund Halley (the astronomer after whom Halley's comet is named). As well as his work on de Moivre's formula, he wrote a major work on probability *The Doctrine of Chances*.

Halley suggested that de Moivre looked at the astronomical world, and so he also worked on the mathematical ideas associated with centripetal force.

Note

Complex numbers are widely used across many areas of engineering, particularly in electrical/electronic engineering. Engineers use 'j' rather than 'i' to represent the imaginary part of a complex number, so a complex number is written, for example, as $1 + j$

Investigation

Investigate the nth roots of unity for $n = 1, 2, 3, 4, 5, …$ Use a graph plotting package such as GeoGebra to plot these.
- What do you know about the results?
- How does it relate to what you know about geometry?
- What happens to the distance between roots as the series develops?
- What are the connections with series?
- What are the connections with calculus?

Research

The hyperbolic functions $f(x) = \sinh(x)$, $g(x) = \cosh(x)$ are complex analogues of the circular functions $f(x) = \sin(x)$ and $g(x) = \cos(x)$.

Research how the hyperbolic functions relate to the circular functions, in terms of their relationships with circles and hyperbolas.

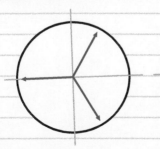
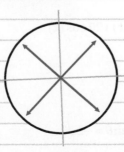

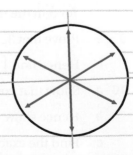

1 $f(x) = x^4 + ax^3 + bx^2 - 40x + 104$, where a and b are real constants.
Given that $z = 5 - i$ is a solution to the equation $f(x) = 0$

    a Find the value of a and the value of b **[4]**

    b Hence find all the solutions of the equation $f(x) = 0$
       You must show a correct method.   **[5]**

    c Illustrate the solutions to $f(x) = 0$ on an Argand diagram.   **[2]**

2 The complex number z is defined by $z = 3 - \sqrt{3}i$

    a Write z in the form $re^{i\theta}$ where r is given as a surd in its simplest form and
       θ is given as a multiple of π **[4]**

    b Hence work out z^4, giving your answer in modulus–argument form.
       You must show your working.   **[2]**

3 The complex number z is defined as $z = -k + k\sqrt{3}i$

    a Find the modulus and argument of z in terms of k, where appropriate.   **[4]**

    b Given that $k = 2$, use de Moivre's theorem to express $z^{\frac{1}{4}}$ in the form $a + bi$,
       where a and b are exact constants to be found.   **[4]**

4 **a** Solve the equation $z^3 = -i$, giving your answers in the form $e^{i\theta}$,
       where $-\pi < \theta \le \pi$ **[4]**

    b Sketch the solutions on an Argand diagram.   **[3]**

5 Given that $z = -2 + 2i$,

    a Write z in modulus-argument form.   **[2]**

    b Hence use de Moivre's theorem to find z^{-3} in the form $a + bi$, where a and b
       are constants to be found.   **[3]**

6 Express $(1 + \sqrt{3}i)^4$ in exact Cartesian form.
You must show your working.   **[4]**

7 Use de Moivre's theorem to prove that $(\cos\theta - i\sin\theta)^n = \cos(n\theta) - i\sin(n\theta)$ **[4]**

8 **a** Show that $\sin 5\theta \equiv 5\sin\theta - 20\sin^3\theta + 16\sin^5\theta$ **[5]**

    b Hence solve the equation $5\sin\theta - 20\sin^3\theta + 16\sin^5\theta = 0.5$
       for θ in the range $0 < \theta \le 180°$ **[4]**

9 **a** Prove that $32\cos^5 x = 2\cos(5x) + 10\cos(3x) + 20\cos(x)$ **[5]**

    b Hence find $\int \cos^5 x\, dx$ **[2]**

10 **a** Show that $\cos 3\theta \equiv \cos^3\theta - 3\cos\theta\sin^2\theta$ **[4]**

    b Hence show that $\tan 3\theta \equiv \dfrac{3\tan\theta - \tan^3\theta}{1 - 3\tan^2\theta}$ **[4]**

    c Find the exact value of $\tan 3\theta$ given that $\tan\theta = -2\sqrt{3}$ **[2]**

11 a Find the value of n such that $(\sqrt{2})^n = 8$ **[2]**

b Hence solve the equation $z^6 + 8 = 0$, giving your answers in the form

$r(\cos\theta + i\sin\theta)$, where $r > 0$ is an exact surd and $-\pi < \theta \leq \pi$ **[5]**

12 Use exponentials to prove that $\cos(2x) = 1 - 2\sin^2 x$ **[4]**

13 A complex number is given as $z = 2e^{\frac{\pi}{2}i}$

A second complex number is $w = a + bi$, where $\arg(w) = \dfrac{\pi}{4}$

Calculate the exact value of

a $\arg(zw)$ **b** $\arg\left(\dfrac{z}{w}\right)$

c $\arg(w^n)$ **d** $\arg\left(\dfrac{w}{z^2}\right)$ **[7]**

14 a Use de Moivre's theorem to show that

$\cos 7\theta = \cos\theta\,(64\cos^6\theta - 112\cos^4\theta + 56\cos^2\theta - 7)$ **[5]**

b Hence solve the equation $64x^6 - 112x^4 + 56x^2 = 7$

Give your answers to 3 significant figures. **[4]**

15 Solve the equation $z^8 + 4\sqrt{12} + 8i = 0$, giving your answers the form $re^{i\theta}$,
where $r > 0$ and $-\pi < \theta \leq \pi$ **[5]**

16 a Prove that $\sin^4\theta \equiv \dfrac{1}{8}(\cos(4\theta) - 4\cos(2\theta) + 3)$ **[4]**

b Hence find $\displaystyle\int 8\sin^4\theta\,d\theta$ **[3]**

17 a Solve the equation $z^4 = 6 - 6\sqrt{3}i$, giving your answers in modulus–argument form,
where r is given to 3 significant figures and θ is given as an exact multiple of π **[5]**

b Hence solve the equation $(z-1)^4 = 6 - 6\sqrt{3}i$, giving your answers in Cartesian
form to 2 significant figures. **[3]**

18 a Use de Moivre's theorem to show that if $z = \cos\theta + i\sin\theta$, then

$z^n - \dfrac{1}{z^n} \equiv 2i\sin(n\theta)$ **[4]**

b Hence show that $4\sin^3\theta \equiv 3\sin\theta - \sin(3\theta)$ **[4]**

c Solve the equation $2\sin(3\theta) - 6\sin\theta = 1$ for θ in the range $0 < \theta \leq 2\pi$ **[4]**

19 a Illustrate the solutions to $z^5 = 1$ on an Argand diagram. **[2]**

b Given that ω is a complex 5th root of unity, evaluate

i $\omega(1+\omega)(1+\omega^2)$ **ii** $(1+\omega^5)^2$ **[5]**

20 a Use de Moivre's theorem to show that $\cos 5\theta \equiv 16\cos^5\theta - 20\cos^3\theta + 5\cos\theta$ **[5]**

b Hence find the general form of all the solutions to the equation
$\cos 5\theta = 5\cos\theta$ **[4]**

21 a Show that $\omega = \cos\left(\dfrac{2\pi}{7}\right) + i\sin\left(\dfrac{2\pi}{7}\right)$ is one of the seventh roots of unity. **[3]**

b Write down the other non-real roots of the equation $z^7 = 1$ in terms of ω **[1]**

c State the value of $\displaystyle\sum_{r=1}^{6}\omega^r$ **[1]**

22 This question refers to the equation $(z-1)^3 = 1$

a Verify that $z_1 = \frac{1}{2} + \frac{\sqrt{3}}{2}$i is a root of the equation. **[3]**

b Find the other two roots of the equation. **[2]**

c Produce an Argand diagram showing the three roots of the equation. **[2]**

d State the centre and radius of the circle on which the three roots lie. **[2]**

23 a Solve the equation $z^4 + 16 = 0$

Give your answers in the form $a + b$i, where a and b are real numbers. **[5]**

b The points A, B, C and D in the complex plane represent these four roots.

Find the area of the square $ABCD$ **[2]**

24 a Prove that $\sin 5\theta = 16\sin^5\theta - 20\sin^3\theta + 5\sin\theta$ **[5]**

b Use the result in part **a** to solve the equation $16x^5 - 20x^3 + 5x = 1$ **[4]**

Give your solutions in the form $x = \sin k\pi$, where $0 \le k < 2$

25 a Solve the equation $(z-1)^3 = 8$

Give your solutions in the form $a + b$i where a and b are real numbers. **[4]**

The points A, B, C in the complex plane represent these three roots.

b Draw the points A, B and C on an Argand diagram. **[2]**

c Calculate the area of the triangle ABC **[3]**

The triangle ABC is rotated $\frac{\pi}{3}$ radians anticlockwise.

d Find the complex numbers that represent the vertices of the triangle after this rotation. **[4]**

26 a Use de Moivre's theorem to show that $z = \frac{1}{2}(1+i)$ is a solution to the

equation $\left(\dfrac{z-1}{z}\right)^6 = -1$ **[5]**

b Find the other 4 solutions to the equation.

Give your answers in the form $a + b$i, where a and b are real numbers. **[6]**

c The root $z = \frac{1}{2}(1+i)$ is represented by the point A in the complex plane.

Find the coordinates of the image of A following a clockwise rotation of $\frac{\pi}{6}$ radians. **[3]**

27 a Solve the equation $z^3 = 4 - 4\sqrt{3}$i

Give your answers in the form $re^{i\theta}$, where $r > 0$ and $-\pi < \theta \le \pi$ **[6]**

The roots of the equation $z^3 = 4 - 4\sqrt{3}$i are represented by the points A, B and C on an Argand diagram.

b Calculate the exact area of the triangle ABC **[3]**

c Show that $\sin\left(\dfrac{\pi}{9}\right) - \sin\left(\dfrac{5\pi}{9}\right) + \sin\left(\dfrac{7\pi}{9}\right) = 0$ **[3]**

28 Use proof by induction to show that de Moivre's theorem $(\cos\theta + i\sin\theta)^n = \cos(n\theta) + i\sin(n\theta)$ holds for all integers n

(Hint: consider separately the cases when n is positive or negative.) **[11]**

7 Series

It is important that aircraft pilots and air traffic controllers are able to locate the position of aircraft precisely, as well as their direction of travel and speed. To do so, they use a modified system of **polar coordinates**. The system takes the direction of magnetic north as 360°, and measures angles clockwise from this. Just like in Cartesian coordinates, in polar coordinates two values can be used to determine the precise location of a point accurately: distance from the origin and the angle measured from magnetic north.

Polar coordinates are useful in many situations when working with phenomena on or near the Earth's surface. For example, meteorologists use them to model existing weather systems and predict future weather, pilots of ocean-going tankers use them for navigation, and cartographers use them to map the Earth.

Orientation

What you need to know	What you will learn	What this leads to
Ch2 Algebra and series • Quadratic functions. • Cartesian coordinates.	• How to sum series and use the method of differences. • How to use polar coordinates. • How to use hyperbolic functions.	**Careers** • Air traffic control. • Meteorology.

Fluency and skills

You can sum a series using standard formulae for the sums of integers, squares and cubes. Another way to sum a series is to use the **method of differences**.

You can use this method if the general term of a function can be expressed as $f(r+1) - f(r)$. When you sum the terms in a series like this, lots of the terms cancel each other out and you can collect up and simplify the remaining terms to find an expression for the sum of the series.

Example 1

Find the sum to n terms of the series $\dfrac{1}{r} - \dfrac{1}{r+1}$

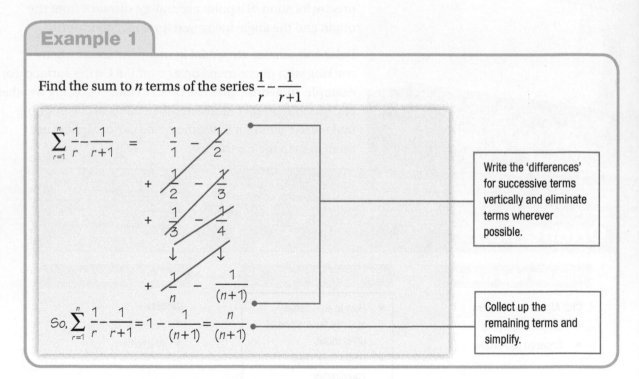

Write the 'differences' for successive terms vertically and eliminate terms wherever possible.

Collect up the remaining terms and simplify.

In order to express a function as $f(r+1) - f(r)$ you may need to use partial fractions.

You already know how to decompose some functions into their partial fractions.

Start by checking that the degree of the numerator is smaller than the degree of the denominator.

Then split into partial fractions in one of the following ways.

Key point

$$\frac{px+q}{(x-a)(x-b)} \equiv \frac{A}{(x-a)} + \frac{B}{(x-b)}$$

$$\frac{px+q}{(x-a)(x-b)^2} \equiv \frac{A}{(x-a)} + \frac{B}{(x-b)} + \frac{C}{(x-b)^2}$$

If a factor is repeated n times, $(x-a)^n$, you need n partial fractions to cover each power in the repeated factor.

You can work out the constants *A*, *B*, etc. using substitution or by comparing coefficients (or a combination of these methods).

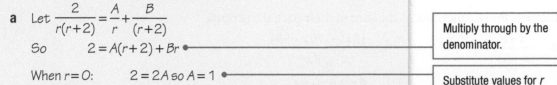

Example 2

a Find the partial fractions of $\dfrac{2}{r(r+2)}$

b Use your answer to find $\displaystyle\sum_{r=1}^{n} \dfrac{2}{r(r+2)}$

a Let $\dfrac{2}{r(r+2)} = \dfrac{A}{r} + \dfrac{B}{(r+2)}$

So $\qquad 2 = A(r+2) + Br$ •————

> Multiply through by the denominator.

When $r=0$: $\qquad 2 = 2A$ so $A = 1$ •————

When $r=-2$: $\qquad 2 = -2B$ so $B = -1$

> Substitute values for *r* to work out *A* and *B*

Hence $\dfrac{2}{r(r+2)} = \dfrac{1}{r} - \dfrac{1}{(r+2)}$

b $\displaystyle\sum_{r=1}^{n} \dfrac{2}{r(r+2)} = \sum_{r=1}^{n}\left(\dfrac{1}{r} - \dfrac{1}{(r+2)} \right)$

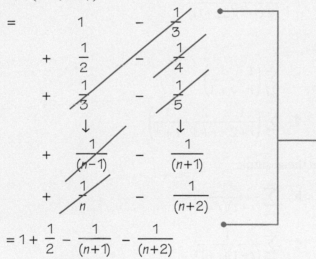

$= 1 \quad - \quad \dfrac{1}{3}$

$\quad + \dfrac{1}{2} \quad - \dfrac{1}{4}$

$\quad + \dfrac{1}{3} \quad - \dfrac{1}{5}$

$\qquad\qquad\downarrow \qquad\qquad\quad\downarrow$

$\quad + \dfrac{1}{(n-1)} \quad - \dfrac{1}{(n+1)}$

$\quad + \dfrac{1}{n} \quad - \dfrac{1}{(n+2)}$

> Write the 'differences' for successive terms vertically and eliminate where possible.
>
> This time the terms that cancel are two rows apart.

$= 1 + \dfrac{1}{2} - \dfrac{1}{(n+1)} - \dfrac{1}{(n+2)}$

$= \dfrac{3}{2} - \dfrac{(n+2)+(n+1)}{(n+1)(n+2)}$

$= \dfrac{n(3n+5)}{2(n+1)(n+2)}$ •————

> The expression can be factorised and written as a single fraction

1 Express these functions as the sum of their partial fractions.

a $\dfrac{6x+10}{(x-5)(x+5)}$ **b** $\dfrac{-4x}{(x-3)(x-7)}$ **c** $\dfrac{7x+56}{(1-x)(x+6)}$

d $\dfrac{34-9x}{(2x-5)(3x+4)}$ **e** $\dfrac{15x^2-85x+112}{(x-2)(x-3)(x-4)}$ **f** $\dfrac{27x+46}{(x-2)(x+2)(x+3)}$

g $\dfrac{6x^2-44x+196}{(2-x)(x+9)(5-x)}$ **h** $\dfrac{69x^2-11x-94}{(3x-1)(4x-3)(3x+5)}$

2 Express these functions as the sum of their partial fractions.

a $\dfrac{x+8}{(x+5)^2}$ **b** $\dfrac{12x^2-170x+594}{(x-3)(x-7)^2}$

c $\dfrac{7x^2+43x+67}{(x+4)(x+3)^2}$ **d** $\dfrac{6x^2+6x-99}{(x-5)(x+4)^2}$

e $\dfrac{x^2-45x+311}{(x+7)(x-8)^2}$ **f** $\dfrac{-2x^2+93x+870}{(x-4)(x+7)^2}$

g $\dfrac{x^2-51x+102}{(x+11)(3-x)^2}$ **h** $\dfrac{69x^2-44x-30}{(2x-7)(3x+1)^2}$

3 Find the sum of each series.

a $\displaystyle\sum_{r=1}^{n}\left(\dfrac{1}{r}-\dfrac{1}{r+1}\right)$ **b** $\displaystyle\sum_{r=1}^{n}\left(\dfrac{1}{r}-\dfrac{1}{r+3}\right)$

c $\displaystyle\sum_{r=1}^{n}\left(\dfrac{1}{2r-1}-\dfrac{1}{2r+1}\right)$ **d** $\displaystyle\sum_{r=1}^{n}\left(\dfrac{1}{r+2}-\dfrac{2}{r+3}+\dfrac{1}{r+4}\right)$

4 Use partial fractions to find these sums.

a $\displaystyle\sum_{r=2}^{n}\dfrac{1}{r(r-1)}$ **b** $\displaystyle\sum_{r=1}^{n}\dfrac{1}{(r+1)(r+2)}$

c $\displaystyle\sum_{r=1}^{n}\dfrac{1}{(r+1)(r+3)}$ **d** $\displaystyle\sum_{r=2}^{n}\dfrac{1}{(r-1)(r+1)}$

Reasoning and problem-solving

Strategy

To sum series using the method of differences

1. Write the function using partial fractions if necessary.
2. Add successive terms, cancelling where possible.
3. Simplify the resulting expression for the sum.

Example 3

Using partial fractions and the method of differences, work out $\displaystyle\sum_{r=1}^{\infty}\frac{2}{r(r+1)(r+2)}$

Let $\dfrac{2}{r(r+1)(r+2)} = \dfrac{A}{r} + \dfrac{B}{r+1} + \dfrac{C}{r+2}$

$2 = A(r+1)(r+2) + Br(r+2) + Cr(r+1)$

When $r = 0$: $\qquad 2 = 2A \qquad$ so, $A = 1$

When $r = -1$: $\qquad 2 = -B \qquad$ so $B = -2$

When $r = -2$: $\qquad 2 = 2C \qquad$ so $C = 1$

$$\frac{2}{r(r+1)(r+2)} = \frac{1}{r} - \frac{2}{r+1} + \frac{1}{r+2}$$

> **1** Multiply through and substitute to express the sum in its partial fractions.

So, $\displaystyle\sum_{r=1}^{r}\frac{2}{r(r+1)(r+2)} =$

$$\begin{array}{ccccc}
& \dfrac{1}{1} & - & \dfrac{2}{2} & + & \dfrac{1}{3} \\[6pt]
+ & \dfrac{1}{2} & - & \dfrac{2}{3} & + & \dfrac{1}{4} \\[6pt]
+ & \dfrac{1}{3} & - & \dfrac{2}{4} & + & \dfrac{1}{5} \\[6pt]
& \downarrow & & \downarrow & & \downarrow \\[6pt]
+ & \dfrac{1}{n-2} & - & \dfrac{2}{n-1} & + & \dfrac{1}{n} \\[6pt]
+ & \dfrac{1}{n-1} & - & \dfrac{2}{n} & + & \dfrac{1}{n+1} \\[6pt]
+ & \dfrac{1}{n} & - & \dfrac{2}{n+1} & + & \dfrac{1}{n+2}
\end{array}$$

> **2** This time the sum is made of three partial fractions and terms cancel across three rows.

Hence $\displaystyle\sum_{r=1}^{r}\frac{2}{r(r+1)(r+2)} = \frac{1}{1} - \frac{2}{2} + \frac{1}{2} + \frac{1}{n+1} - \frac{2}{n+1} + \frac{1}{n+2}$

$$= \frac{1}{2} - \frac{1}{n+1} + \frac{1}{n+2}$$

> **3** Collect up the remaining terms and simplify.

As $n \to \infty$, $\dfrac{1}{(n+1)} \to 0$ and $\dfrac{1}{(n+2)} \to 0$ So $\displaystyle\sum_{r=1}^{\infty}\frac{2}{r(r+1)(r+2)} = \frac{1}{2}$

Exercise 7.1B Reasoning and problem-solving

1 **a** Show that $r(r+1) - r(r-1) \equiv 2r$

    b Use this information to show that $\displaystyle\sum_{r=1}^{n}r \equiv \frac{n(n+1)}{2}$

2 **a** Simplify the expression $(2r+1)^3 - (2r-1)^3$

    b Hence show that $\displaystyle\sum_{r=1}^{n}r^2 \equiv \frac{n(n+1)(2n+1)}{6}$

3 a Simplify the expression $r^2(r+1)^2 - r^2(r-1)^2$

b Use your answer to find a formula for $\displaystyle\sum_{r=1}^{n} r^3$

4 a Use partial fractions to find $\displaystyle\sum_{r=2}^{n} \frac{4}{r^2-1}$

b Use your result to find the maximum value of $\displaystyle\sum_{r=2}^{n} \frac{1}{r^2-1}$ for $n \geq 2$

5 a Find the partial fractions of $\dfrac{1}{4r^2-1}$

b Hence find $\displaystyle\sum_{r=1}^{n} \frac{1}{4r^2-1}$

c Find the sum to infinity of this series.

6 Find $\displaystyle\sum_{r=1}^{n} \left(\frac{1}{r^2} - \frac{1}{(r+1)^2} \right)$ and hence find the sum to infinity.

7 The rth term of a series, S, is $\dfrac{2r-1}{r(r+1)(r+2)}$ Show that $\displaystyle\sum_{1}^{\infty} S = \frac{3}{4}$

8 Use partial fractions and the method of differences to find $\displaystyle\sum_{1}^{n} \frac{4(x-2)}{x(x+2)(x+4)}$

9 Use partial fractions to find the sum to infinity of the series $\dfrac{1}{1\times3} + \dfrac{1}{2\times4} + \dfrac{1}{3\times5} + \dots$

10 a Show that $\dfrac{x-1}{x} - \dfrac{x-2}{x-1} \equiv \dfrac{1}{x(x-1)}$

b Use your answer to find a formula for $\displaystyle\sum_{3}^{n} \frac{1}{x(x-1)}$

c Hence evaluate $\displaystyle\sum_{3}^{\infty} \frac{1}{x(x-1)}$

11 Find a formula for $\displaystyle\sum_{3}^{n} \ln\left(\frac{k}{(r-1)} \right)$

12 a Show that $\dfrac{1}{(x-1)!} - \dfrac{1}{x!} \equiv \dfrac{x-1}{x!}$

b Use your answer to find a formula for $\displaystyle\sum_{1}^{n} \frac{x-1}{x!}$

c Hence evaluate $\displaystyle\sum_{1}^{\infty} \frac{x-1}{x!}$

Fluency and skills

A point can be defined by its (x, y) coordinates, but you could also define it by a distance from the origin and an angle. For example, the distance of P from the origin is 2, and the angle it makes with the positive x axis is $\frac{\pi}{3}$. These two numbers describe the point perfectly, and they are called the polar coordinates of the point P, written as $\left(2, \frac{\pi}{3}\right)$

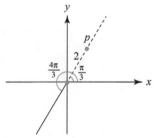

However, the angle is not unique: there are an infinite number of ways to represent a point in polar coordinates. For example,

$$\left(2, \frac{\pi}{3}\right) = \left(2, -\frac{5\pi}{3}\right) = \left(2, \frac{7\pi}{3}\right) = \left(2, -\frac{11\pi}{3}\right)$$

The letters r and θ are usual for the distance and the angle.

It is possible to use a negative value of r, by going backwards from the origin:

$$\left(2, \frac{\pi}{3}\right) = \left(-2, \frac{4\pi}{3}\right) = \left(-2, -\frac{2\pi}{3}\right)$$

Every point, however, can only be written in one way if you always choose $r \geq 0$ and $-\pi < \theta \leq \pi$ or $0 \leq \theta < 2\pi$

The point $(0, 0)$ is called the pole, and the positive x axis, which is where the angle is measured from, is called the initial line.

Key point

r is the distance from the pole to the point.

θ is the angle between the initial line and the line segment from the pole to the point, measured anticlockwise.

You can convert polar (r, θ) coordinates into Cartesian (x, y) coordinates . In the example above, the x-coordinate is $2\cos\frac{\pi}{3}$ and the y-coordinate is $2\sin\frac{\pi}{3}$

The same method will always work.

To convert from polar to Cartesian, use

$$x = r\cos\theta$$
$$y = r\sin\theta$$

To convert from Cartesian to polar, use

$$r^2 = x^2 + y^2$$
$$\tan\theta = \frac{y}{x}$$

but you should draw a diagram to ensure that you choose the right value of θ

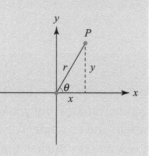

Example 1

Convert the polar coordinates $\left(5, -\frac{3\pi}{4}\right)$ into Cartesian coordinates.

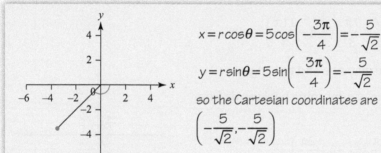

$$x = r\cos\theta = 5\cos\left(-\frac{3\pi}{4}\right) = -\frac{5}{\sqrt{2}}$$

$$y = r\sin\theta = 5\sin\left(-\frac{3\pi}{4}\right) = -\frac{5}{\sqrt{2}}$$

so the Cartesian coordinates are

$$\left(-\frac{5}{\sqrt{2}}, -\frac{5}{\sqrt{2}}\right)$$

Example 2

Convert the Cartesian coordinates $(-2\sqrt{3}, 2)$ into polar coordinates.

First, draw a sketch.

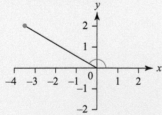

$$r^2 = x^2 + y^2 = 12 + 4 = 16$$
$$\Rightarrow r = 4$$
$$\tan\theta = \frac{y}{x} = -\frac{1}{\sqrt{3}}$$
$$\Rightarrow \theta = -\frac{\pi}{6}, \frac{5\pi}{6}$$

but the sketch shows which value to use, so the polar coordinates are $\left(4, \frac{5\pi}{6}\right)$

An equation involving x and y can be represented as a graph. So can an equation involving r and θ

Example 3

Draw the graph

$r = 3(1 - \sin\theta), \ 0 \le \theta < 2\pi$

This means that, for any value of θ in the range 0 to 2π, you can calculate the value of r and plot that point. It is a good idea to make a table of values using easy values of θ. Then you can join the points up to make a smooth curve.

θ	0	$\dfrac{\pi}{6}$	$\dfrac{\pi}{3}$	$\dfrac{\pi}{2}$	$\dfrac{2\pi}{3}$	$\dfrac{5\pi}{6}$	π	$\dfrac{7\pi}{6}$	$\dfrac{4\pi}{3}$	$\dfrac{3\pi}{2}$	$\dfrac{5\pi}{3}$	$\dfrac{11\pi}{6}$	2π
r	3	1.5	0.40	0	0.40	1.5	3	4.5	5.6	6	5.6	4.5	3

Polar graph paper can be useful: the lines represent points with the same θ coordinate, and the circles represent points with the same r coordinate.

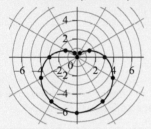

Example 4

a Draw the graph
   $r = 5\sin 3\theta, \ 0 \le \theta < 2\pi$

b Draw the graph
   $r = 5\sin 3\theta, \ 0 \le \theta < 2\pi, \ r \ge 0$

a You could draw up a table of values as before, but you do not usually do this for (x, y) graphs, and you don't always need to for (r, θ) graphs.

First, notice that $-1 \le \sin 3\theta \le 1$, so $-5 \le r \le 5$

It is useful to find the values of θ that give r these maximal and minimal values:

$r = 5 \Rightarrow \sin 3\theta = 1 \Rightarrow 3\theta = \dfrac{\pi}{2}, \dfrac{5\pi}{2}, \dfrac{9\pi}{2}, \dots$

$\Rightarrow \theta = \dfrac{\pi}{6}, \dfrac{5\pi}{6}, \dfrac{3\pi}{2} \dots$

(*Continued on the next page*)

Also,

$$r = -5 \Rightarrow \sin 3\theta = -1 \Rightarrow 3\theta = \frac{3\pi}{2}, \frac{7\pi}{2}, \frac{11\pi}{2}, \dots$$

$$\Rightarrow \theta = \frac{\pi}{2}, \frac{7\pi}{6}, \frac{11\pi}{6}, \dots$$

And it would also be a good idea to find where r is zero:

$$r = 0 \Rightarrow \sin 3\theta = 0 \Rightarrow 3\theta = 0, \pi, 2\pi, 3\pi, 4\pi, 5\pi, 6\pi, \dots$$

$$\Rightarrow \theta = 0, \frac{\pi}{3}, \frac{2\pi}{3}, \pi, \frac{4\pi}{3}, \frac{5\pi}{3}, 2\pi, \dots$$

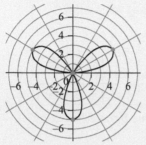

As θ increases from 0, r increases from 0 until it reaches 5, when $\theta = \frac{\pi}{6}$

Then r decreases until it reaches 0 when $\theta = \frac{\pi}{3}$

This makes the top right loop on the curve.

When θ is between $\frac{\pi}{3}$ and $\frac{2\pi}{3}$, r is negative, going from 0 to -5 and back to 0 again. This makes the bottom loop.

Then when θ is between $\frac{2\pi}{3}$ and π, r goes from 0 to 5 and back to 0 again, making the top left loop.

When θ goes above π, the curve traces over the same path that it has already completed.

b Here, you are only interested in the positive values of r (or zero), so you only need values of θ for which $5\sin 3\theta$ is positive or zero. This gives you the top right loop when

$0 \le \theta \le \frac{\pi}{3}$, the top left loop when $\frac{2\pi}{3} \le \theta \le \pi$, and the bottom loop when $\frac{4\pi}{3} \le \theta \le \frac{5\pi}{3}$

However, if the equation had been

$$r = 5\sin 3\theta, \ 0 \le \theta < \pi, \ r \ge 0$$

with θ only going as far as π, then the bottom loop would be missing.

Key point

Pay careful attention to the range of values of θ and r specified in the question.

Example 5

a Draw the graph
$r = 1 - 2\sin\theta,\ 0 \le \theta < 2\pi$

b Draw the graph
$r = 1 - 2\sin\theta,\ 0 \le \theta < 2\pi,\ r \ge 0$

a First notice that

$-1 \le 1 - 2\sin\theta \le 3$

$\Rightarrow -1 \le r \le 3$

When $r = 0$, $1 - 2\sin\theta = 0 \Rightarrow \theta = \dfrac{\pi}{6}, \dfrac{5\pi}{6}$

When $r = -1$, $1 - 2\sin\theta = -1 \Rightarrow \theta = \dfrac{\pi}{2}$

When $r = 3$, $1 - 2\sin\theta = 3 \Rightarrow \theta = \dfrac{3\pi}{2}$

When $r = 1$, $1 - 2\sin\theta = 1 \Rightarrow \theta = 0, \pi$

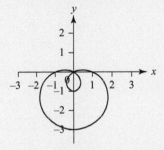

When $0 < \theta < \dfrac{\pi}{6}$, r decreases from 1 to 0

When $\dfrac{\pi}{6} < \theta < \dfrac{5\pi}{6}$, r decreases from 0 to -1 and increases back to 0

When $\dfrac{5\pi}{6} < \theta < 2\pi$, r increases from 0 to 3 and decreases back to 1

b As in example 4b, you are not interested in the values of θ for which r is negative, so the smaller inner loop is missing.

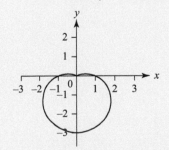

Example 6

Draw these graphs.

a $r = 3$

b $\theta = \dfrac{5\pi}{6}$

c $\theta = \dfrac{5\pi}{6},\ r \geq 0$

d $r = \theta,\ r \geq 0$

a This graph includes all points that are 3 units away from the pole, so it is a circle of radius 3 with the centre at the pole.

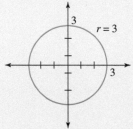

b This graph includes all points on the line that forms an angle $\dfrac{5\pi}{6}$ with the positive x axis, so it is the line $y = -\dfrac{\sqrt{3}}{3}x$

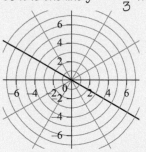

c In this case, r cannot take negative values, so this graph is only the half line.

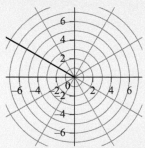

d As θ gets bigger, so does r

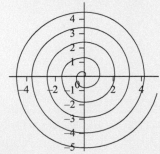

1 Draw a polar chart similar to the one shown and plot the given points.

Convert each set of polar coordinates to their Cartesian equivalent.

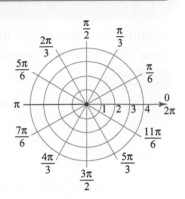

a $\left(2, \dfrac{\pi}{3}\right)$ **b** $\left(4, \dfrac{4\pi}{3}\right)$ **c** $\left(2, \dfrac{5\pi}{3}\right)$

d $\left(3, \dfrac{2\pi}{3}\right)$ **e** $\left(3, -\dfrac{5\pi}{6}\right)$ **f** $\left(3, -\dfrac{2\pi}{3}\right)$

2 Convert these Cartesian coordinates to polar form (r, θ), where $0 < \theta \le 2\pi$. Give your answers to 3 sf where appropriate.

  a $(2\sqrt{3}, 2)$ **b** $(-3, 3\sqrt{3})$ **c** $(4, -4\sqrt{3})$ **d** $(-5, -12)$

3 Convert these Cartesian coordinates to polar form (r, θ), where $-\pi < \theta \le \pi$

Give r as a surd and angles to 3 significant figures where appropriate.

  a $(-2\sqrt{3}, -2)$ **b** $(5, -5\sqrt{3})$ **c** $(-3, 4)$ **d** $(1, 6)$

4 State the polar equation of the curve $x^2 + y^2 = 5$ and describe the curve.

5 Describe the curve $r = 4$ and state its Cartesian equation.

6 Sketch the graphs for each of these polar equations and find the Cartesian equation of the line on which they lie.

  a $\theta = \dfrac{\pi}{4}, r \ge 0$ **b** $\theta = \dfrac{\pi}{2}, r \ge 0$ **c** $\theta = \dfrac{3\pi}{4}, r \ge 0$ **d** $\theta = -\dfrac{\pi}{6}, r \ge 0$

7 **a** Plot the curve $r = \theta$ in the domain $0 \le \theta \le 6\pi$ using step-sizes of $\dfrac{\pi}{4}$

  b Plot $r = \dfrac{10}{2 + \cos\theta}$ in the domain $0 \le \theta \le 2\pi$

8 Sketch these polar curves for $0 \le \theta \le 2\pi$ and state the maximum and minimum value of r in each case.

  a **i** $r = \cos 2\theta, r \ge 0$ **ii** $r = \cos 2\theta, r \in \mathbb{R}$

  b $r = \sin 4\theta$

  c **i** $r = 2\cos 3\theta, r \ge 0$ **ii** $r = 2\cos 3\theta, r \in \mathbb{R}$

  d **i** $r = 4\sin 2\theta, r \ge 0$ **ii** $r = 4\sin 2\theta, r \in \mathbb{R}$

  e $r = 1 + \cos\theta$

  f $r = 4 + \sin\theta$

  g $r = 3 - 2\cos\theta$

  h **i** $r = 3 - 5\cos\theta, r \ge 0$ **ii** $r = 3 - 5\cos\theta, r \in \mathbb{R}$

  i $r = 2\theta$

  j **i** $r^2 = 4\sin 2\theta, r \ge 0$ **ii** $r^2 = 4\sin 2\theta, r \in \mathbb{R}$

Reasoning and problem-solving

An equation that is simple in (r, θ) coordinates can be complicated in (x, y) coordinates, and vice versa. You can always convert one into the other, though.

Key point

To convert from polar to Cartesian, use

$$r^2 = x^2 + y^2$$

$$\tan\theta = \frac{y}{x}$$

$$\cos\theta = \frac{x}{r} = \pm\frac{x}{\sqrt{x^2 + y^2}}$$

$$\sin\theta = \frac{y}{r} = \pm\frac{y}{\sqrt{x^2 + y^2}}$$

[Choose + or − depending on whether r is positive or negative.]

To convert from Cartesian to polar, use

$$x = r\cos\theta$$

$$y = r\sin\theta$$

The difficulty sometimes lies in simplifying the answer.

Example 7

Convert the equation $\dfrac{x^2}{3} + \dfrac{y^2}{4} = 1$ to a polar equation.

$$\frac{(r\cos\theta)^2}{3} + \frac{(r\sin\theta)^2}{4} = 1$$

$$\Rightarrow 4r^2\cos^2\theta + 3r^2\sin^2\theta = 12$$

$$\Rightarrow r^2 = \frac{12}{4\cos^2\theta + 3\sin^2\theta}$$

This is a perfectly good answer, although you could also write

$$r^2 = \frac{12}{3(\cos^2\theta + \sin^2\theta) + \cos^2\theta}$$

$$\Rightarrow r^2 = \frac{12}{3 + \cos^2\theta}$$

Notice that this implies that r could be positive or negative. As long as the range of values of θ is big enough, however, it makes no difference whether you include the negative values of r or not.

Example 8

Convert the equation $r = \dfrac{\tan\theta}{\cos\theta}$ to a Cartesian equation.

When $r > 0$

$$\sqrt{x^2 + y^2} = \frac{y}{x} \div \frac{x}{\sqrt{x^2 + y^2}}$$

$$= \frac{y\sqrt{x^2 + y^2}}{x^2}$$

$$\Rightarrow 1 = \frac{y}{x^2}$$

$$\Rightarrow y = x^2$$

When $r < 0$

$$-\sqrt{x^2 + y^2} = \frac{y}{x} \div \frac{x}{-\sqrt{x^2 + y^2}}$$

$$\Rightarrow y = x^2$$

So this is the Cartesian equation of the graph.

You can also find the points of intersection of two curves given their polar equations without having to convert into Cartesian equations.

Example 9

a Find the points of intersection of the curves

$r = \sin 2\theta \ \ r \geq 0$ and $r = \cos 2\theta \ \ r \geq 0$

b Find the points of intersection of the curves $r = \sin 2\theta$ and $r = \cos 2\theta$, where r can be any real number.

a In this question, only the values of θ for which r is positive are considered for each curve.

At the points of intersection, $\sin 2\theta = \cos 2\theta$

$$\Rightarrow \tan 2\theta = 1$$

$$\Rightarrow 2\theta = \frac{\pi}{4}, \frac{5\pi}{4}, \frac{9\pi}{4}, \frac{13\pi}{4} \ldots$$

$$\Rightarrow \theta = \frac{\pi}{8}, \frac{5\pi}{8}, \frac{9\pi}{8}, \frac{13\pi}{8} \ldots$$

However, r is not positive on both curves for all these values of θ

When $\theta = \dfrac{\pi}{8}$ or $\theta = \dfrac{9\pi}{8}$, $r = \sin 2\theta = \cos 2\theta = \dfrac{\sqrt{2}}{2}$

When $\theta = \dfrac{5\pi}{8}$ or $\theta = \dfrac{13\pi}{8}$, $r = \sin 2\theta = \cos 2\theta = -\dfrac{\sqrt{2}}{2} < 0$

So the only solutions are

$$\left(\frac{\sqrt{2}}{2}, \frac{\pi}{8} \right), \left(\frac{\sqrt{2}}{2}, \frac{9\pi}{8} \right)$$

(*Continued on next page*)

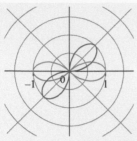

Drawing the curves will help you to visualise what is going on, but you need to choose values of θ that make r positive on each curve:

$$0 \le \theta \le \frac{\pi}{2} \ or \ \pi \le \theta \le \frac{3\pi}{2} \Rightarrow r = \sin 2\theta \ge 0$$

$$-\frac{\pi}{4} \le \theta \le \frac{\pi}{4} \ or \ \frac{3\pi}{4} \le \theta \le \frac{5\pi}{4} \Rightarrow r = \cos 2\theta \ge 0$$

b At the points of intersection, $\sin 2\theta = \cos 2\theta$

$$\Rightarrow \tan 2\theta = 1$$

$$\Rightarrow 2\theta = \frac{\pi}{4}, \frac{5\pi}{4}, \frac{9\pi}{4}, \frac{13\pi}{4} \dots$$

$$\Rightarrow \theta = \frac{\pi}{8}, \frac{5\pi}{8}, \frac{9\pi}{8}, \frac{13\pi}{8} \dots$$

This gives 4 points of intersection. However, drawing the curves shows 8 points of intersection. Where do the other 4 come from? The problem is that polar coordinates are not unique. In this case, notice that the point $\left(\frac{\sqrt{2}}{2}, \frac{3\pi}{8} \right)$ is on the curve $r = \sin 2\theta$. But this point is the same as $\left(-\frac{\sqrt{2}}{2}, \frac{11\pi}{8} \right)$, which is on the curve $r = \cos 2\theta$, and at this point, $\tan 2\theta = -1$

This means you also need to find the solutions to
$\tan 2\theta = -1$

$$\Rightarrow 2\theta = \frac{3\pi}{4}, \frac{7\pi}{4}, \frac{11\pi}{4}, \frac{15\pi}{4} \dots$$

$$\Rightarrow \theta = \frac{3\pi}{8}, \frac{7\pi}{8}, \frac{11\pi}{8}, \frac{15\pi}{8} \dots$$

giving 8 points of intersection:

$$\left(\frac{\sqrt{2}}{2}, \frac{\pi}{8} \right), \left(\frac{\sqrt{2}}{2}, \frac{3\pi}{8} \right), \left(\frac{\sqrt{2}}{2}, \frac{5\pi}{8} \right), \left(\frac{\sqrt{2}}{2}, \frac{7\pi}{8} \right), \left(\frac{\sqrt{2}}{2}, \frac{9\pi}{8} \right), \left(\frac{\sqrt{2}}{2}, \frac{11\pi}{8} \right), \left(\frac{\sqrt{2}}{2}, \frac{13\pi}{8} \right), \left(\frac{\sqrt{2}}{2}, \frac{15\pi}{8} \right)$$

Perhaps the easiest way to deal with this is to sketch the two curves, and use the idea of symmetry to see where all the solutions lie.

You can find the equation of the tangent to a polar curve at the pole (the origin) by finding the values of θ for which $r = 0$ and confirming that $\dfrac{dr}{d\theta} \neq 0$ for these values of θ

Example 10

Find the polar and Cartesian equations of the tangents at the pole to the curve
$r = \cos(3\theta)$

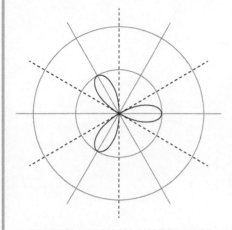

When $r = 0$,
$r = \cos(3\theta) = 0$
$$\Rightarrow 3\theta = \frac{\pi}{2}, \frac{3\pi}{2}, \frac{5\pi}{2}, \frac{7\pi}{2}, \frac{9\pi}{2}, \frac{11\pi}{2} \ldots$$
$$\Rightarrow \theta = \frac{\pi}{6}, \frac{\pi}{2}, \frac{5\pi}{6}, \frac{7\pi}{6}, \frac{3\pi}{2}, \frac{11\pi}{6} \ldots$$
and it is clear from the diagram that these are the polar equations of the tangents.

In Cartesian form, the equations of the tangents are therefore
$$y = \left(\tan\frac{\pi}{6} \right)x, \ y = \left(\tan\frac{5\pi}{6} \right)x, \ x = 0$$
or
$$y = \frac{\sqrt{3}}{3}x, \ y = -\frac{\sqrt{3}}{3}x, \ x = 0$$

Key point

If $r = 0$ when θ takes a particular value α, then the line $y = x\tan\alpha$ is a tangent to the curve at the origin.

Example 11

What are the polar and Cartesian equations of tangents at the pole on the curve

$$r = 2 - \frac{3}{2 + \cos 2\theta}$$

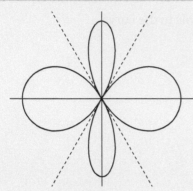

$$r = 2 - \frac{3}{2 + \cos 2\theta} = 0$$

$$\Rightarrow 3 = 4 + 2\cos 2\theta$$

$$\Rightarrow \cos 2\theta = -\frac{1}{2}$$

$$\Rightarrow 2\theta = \frac{2\pi}{3}, \frac{4\pi}{3} \ldots$$

$$\Rightarrow \theta = \frac{\pi}{3}, \frac{2\pi}{3} \ldots$$

These are the polar equations of the tangents.

The Cartesian equations of the tangents are

$$y = x\tan\frac{\pi}{3}, \; y = x\tan\frac{2\pi}{3}$$

$$\Rightarrow y = \sqrt{3}x, \; y = -\sqrt{3}x$$

Exercise 7.2B Reasoning and problem-solving

1 Find the polar equation for

a The line passing through the origin with gradient 1

b The line passing through the origin with gradient 2

c The positive y-axis

d The positive x-axis

e $y = 5$

f $x = 3$

g $y = x + 1$

h $y = mx + c$

i The circle centre the origin, radius 5

j The circle with centre $(5, 12)$ and radius 13

k The circle with centre $(3, 0)$ passing through the origin

l The circle with centre $(0, 4)$ passing through the origin

m The circle with centre $(5, 5)$ passing through the origin

n $y = x^2$

o $x^2 + y^2 + 2xy = 5$

2 Convert these polar equations into Cartesian form and describe the graphs.

a $r = \dfrac{1}{\sin\theta + \cos\theta}$ **b** $r = \dfrac{5}{\sin\theta - 2\cos\theta}$

c $r = \dfrac{2}{\cos\theta}$ **d** $r = \dfrac{-2}{3\sin\theta}$

e $r = 4\sin\theta$ **f** $r = 8\cos\theta + 6\sin\theta$

3 Find the points of intersection between these pairs of curves for $-\pi < \theta \le \pi$ and $r \ge 0$ Give angles in terms of π

a $r = 2$, $r = 3 - 2\sin\theta$

b $r = \cos 4\theta$, $r = \dfrac{1}{2}$

c $r = \sin 2\theta$, $r = \cos 2\theta$

4 Show algebraically that any spiral of the form $r = a\theta$ will intersect any circle centred on the origin precisely once.

5 For each pair of equations, the domain is $0 \le \theta \le 2\pi$, $r \ge 0$ unless otherwise stated.

i Sketch the curves on the same graph,

ii Find their points of intersection.

a $r = \sqrt{3}\sin\theta$, $r = \cos\theta$

b $r = \theta$, $r = \theta - \sin\theta$

c $r = 2 + \cos\theta$, $r = \cos^2\theta$

d $r = \sin^2\theta$, $r = \cos\theta$

e $r = -\sin 3\theta$, $r = \sqrt{3}\cos 3\theta$, $r \ge 0$

f $r = -\sin 3\theta$, $r = \sqrt{3}\cos 3\theta$, $r \in \mathbb{R}$

6 The family of curves $r = a + 4\cos\theta$, $-\pi < 0 \le \pi$, $r \in \mathbb{R}$ produces different shapes depending on the choice of value for a

a Show that if a is greater than 4 then the curve never passes through the pole.

b Prove that the curve is symmetrical about the initial line for any value of a

c Hence, sketch the curve when

i $a = 0$ **ii** $0 < a \le 4$

iii $4 < a < 8$ **iv** $a \ge 8$

d State, in terms of a, the maximum and minimum values of r for the curves in part **c**.

7 Show that the graphs of $r = \sec^2\theta$ and $r = 2\tan\theta$ intersect only once whether or not negative values of r are included, and find their point of intersection.

8 Find the points of intersection between these pairs of curves for $-\pi < \theta \le \pi$ and $r \ge 0$ Give your coordinates to 3 significant figures.

a $r = 2\cos 2\theta$, $r = 1 + \sin\theta$

b $r^2 = 4\cos\theta$, $r = 2 - \cos\theta$

What is the difference if r can take any value, including negatives?

9 Show that the points of intersection between the curves $r = 2\sin^2\theta$ and $r = \cos 2\theta$ form a rectangle and find its area.

10 Find the polar and Cartesian equations of the tangents at the pole to the following curves.

a $r = 1 - \sin 3\theta$

b $r = \cos^2\dfrac{3\theta}{2}$

11 Find the polar equations of the tangents at the pole to the curves

a $r = 1 + 2\sin 4\theta$

b $r = 1 - 2\cos^2\dfrac{5\theta}{3}$

Hyperbolic functions

Fluency and skills

If you plot all the points of the form $(\sin\theta, \cos\theta)$, where $0 \le \theta \le 2\pi$, you produce a circle (shown by the red curve).

There exists another pair of functions, denoted by $\sinh\theta$ and $\cosh\theta$. If you plot the points $(\sinh\theta, \cosh\theta)$, you produce the curve shown in blue.

This curve is called a hyperbola and so the functions are referred to as **hyperbolic** functions and defined using exponentials.

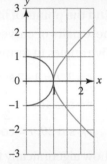

Key point

$$\sinh x = \frac{e^x - e^{-x}}{2} \qquad \cosh x = \frac{e^x + e^{-x}}{2}$$

$$\tanh x = \frac{\sinh x}{\cosh x} = \frac{e^x - e^{-x}}{e^x + e^{-x}}$$

These functions are commonly read as 'shine', 'cosh' and 'tanch' or 'than'.

Example 1

Find the exact value of

a $\tanh 0$ **b** $\sinh(\ln 3)$

Use the definition of $\tanh x$

a $\tanh 0 = \dfrac{e^0 - e^{-0}}{e^0 + e^{-0}}$

$= \dfrac{1-1}{1+1} = 0$

b $\sinh(\ln 3) = \dfrac{e^{\ln 3} - e^{-\ln 3}}{2}$

$= \dfrac{e^{\ln 3} - e^{\ln 3^{-1}}}{2}$

Use the fact that $a \ln b = \ln b^a$

$= \dfrac{3 - \dfrac{1}{3}}{2}$

$= \dfrac{4}{3}$

Calculator

Try it on your calculator

You can use a calculator to evaluate hyperbolic functions and inverse hyperbolic functions.

sinh (ln5)

$\dfrac{12}{5}$

Activity

Find out how to evaluate sinh (ln5) on your calculator

The graph $y = \cosh x$

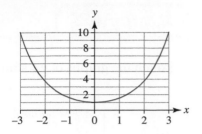

The domain of the function is $\mathbb{R}$ $(-\infty, \infty)$

The range of the function is $\mathbb{R}$ $[1, \infty)$

The curve has a minimum point at $(0, 1)$

So $y \geq 1$

The curve is symmetrical about the y-axis.

The graph $y = \sinh x$

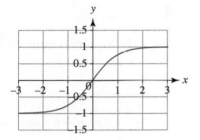

The domain of the function is $\mathbb{R}$ $(-\infty, \infty)$

The range of the function is $\mathbb{R}$ $(-\infty, \infty)$

The curve has rotational symmetry about the origin.

The graph $y = \tanh x$

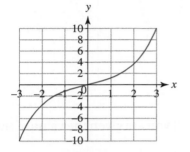

As $x \to +\infty \Rightarrow e^{-x} \to 0 \Rightarrow \tanh x \to \pm 1$

The domain of the function is $\mathbb{R}$ $(-\infty, \infty)$

The range of the function is $\mathbb{R}$ $(-1, 1)$

The curve has asymptotes at $y = 1$ and $y = -1$

So $-1 < y < 1$

The curve has rotational symmetry about the origin.

Key point

Graphs of hyperbolic functions can be transformed
in the same way as other functions, so for $y = f(x)$

- $y = af(x)$ is $y = f(x)$ stretched vertically by scale factor a
- $y = f(ax)$ is $y = f(x)$ stretch horizontally by scale factor $\dfrac{1}{a}$
- $y = a + f(x)$ is $y = f(x)$ translated vertically by a units
- $y = f(x+a)$ is $y = f(x)$ translated horizontally by $-a$ units.

Example 2

Sketch the graph of $y = 2 + \cosh x$

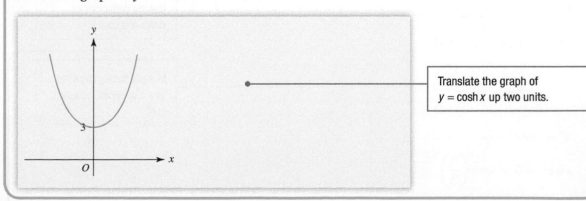

Translate the graph of
$y = \cosh x$ up two units.

Sketch the graphs of $y = \sinh x$ and $y = \sinh \dfrac{x}{2}$ on the same axes.

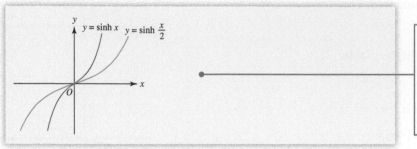

The graph of $y = \sinh \dfrac{x}{2}$ is $y = \sinh x$ stretched horizontally by scale factor 2 Ensure that you label each curve.

If you are given the value of $\sinh x$, $\cosh x$ or $\tanh x$ you can use their definitions to find the value of x

Solve the equation $\tanh x = \dfrac{1}{2}$

$$\frac{e^x - e^{-x}}{e^x + e^{-x}} = \frac{1}{2}$$

Use the definition of $\tanh x$

$$2(e^x - e^{-x}) = e^x + e^{-x} \Rightarrow e^x = 3e^{-x}$$

$$e^{2x} = 3$$

Multiply both sides by e^x

$$2x = \ln 3 \Rightarrow x = \frac{1}{2}\ln 3$$

Take logarithms on both sides.

This example could also be done on a calculator using $\tanh^{-1} \dfrac{1}{2}$. However, your calculator will give a decimal not an exact answer.

Solve the equation $\sinh x + 2 \cosh x = 2$

$$\frac{e^x - e^{-x}}{2} + \frac{2(e^x + e^{-x})}{2} = 2$$

Use the definitions of $\sinh x$ and $\cosh x$

$$e^x - e^{-x} + 2e^x + 2e^{-x} = 4$$

$$3e^x - 4 + e^{-x} = 0$$

$$3e^{2x} - 4e^x + 1 = 0$$

Multiply through by e^x to give a quadratic equation in e^x

$$(3e^x - 1)(e^x - 1) = 0$$

$$e^x = 1 \text{ or } e^x = \frac{1}{3}$$

$$x = \ln 1 = 0 \text{ or } x = \ln\left(\frac{1}{3}\right)$$

1 Work out the value of

   a $\sinh 0$ **b** $\cosh(-1)$ **c** $\sinh(\ln 2)$

   d $\cosh(-\ln 3)$ **e** $\cosh 0$ **f** $\tanh(\ln 2)$

2 Sketch these graphs.

   a $y = \sinh(x-2)$ **b** $y = 10 + \sinh x$ **c** $y = \tanh(x-1)$

   d $y = \cosh(x+2)$ **e** $y = 2 + \tanh(x)$ **f** $y = 1 - \cosh(x)$

3 The function f is defined as $f(x) = \sinh(x)$

   a   Sketch the graphs of $y = f(x)$ and $y = f(2x)$ on the same set of axes.

   b   Solve the equation $f(2x) = 2$, giving your answer to 3 significant figures.

4 Sketch the graphs of $y = \cosh\left(\dfrac{x}{3}\right)$ and $y = \cosh(x)$ on the same set of axes.

5 The function f is defined as $f(x) = \tanh(x)$

   a   Sketch the graph $y = -2f(x)$

   b   Solve the equation $-2f(x) = 1$, giving your answer to 3 significant figures.

6 Given that $f(x) = \tanh(x+a)$, $a > 0$

   a   Sketch the graph of $y = f(x)$

   b   Write down the equations of the asymptotes.

7 Given that $f(x) = \tanh x$, $a > 0$

   a   Sketch the graph of $y = a + f(x)$

   b   Write down the equations of the asymptotes.

8 Given that $f(x) = \cosh x$, $x \in \mathbb{R}$, sketch the graphs of

   a $y = f(x-a)$, for $a > 0$ **b** $y = f(x) + a$, for $a > 0$ **c** $y = af(x)$, for $a < 0$

9 Solve each of these equations, giving your answers to 3 significant figures.

   a $\sinh x = 5$ **b** $\cosh x = 2$ **c** $\tanh x = -\dfrac{1}{2}$

   d $\cosh(x+1) = 3$ **e** $\sinh(3x) = 4$ **f** $2\tanh(x) + 1 = 2$

10 Solve these equations, leaving your answer as a logarithm in its simplest form where appropriate. Show your working.

   a $\sinh x = 0$ **b** $\cosh x = 1$ **c** $\sinh x = \dfrac{3}{4}$

   d $\cosh x = \dfrac{17}{8}$ **e** $\tanh x = \dfrac{3}{5}$ **f** $\tanh x = \dfrac{40}{41}$

11 Solve these equations, leaving your answer as a logarithm in its simplest form where appropriate.

   a $\cosh x - \sinh x = 2$ **b** $2\sinh x + 3\cosh x = 3$

   c $3 - \sinh x = 2\cosh x$ **d** $\dfrac{1}{\sinh x} - \dfrac{1}{\cosh x} = 4e^x$

Reasoning and problem-solving

Hyperbolic functions have identities which are very similar to the trigonometric ones you already know. You can prove these using the definitions.

> Notice how the sign is different from the trigonometric identity
> $\cos^2 x + \sin^2 x \equiv 1$

Key point

Learn this important identity: $\cosh^2 x - \sinh^2 x \equiv 1$

Strategy 1

To solve problems using hyperbolic functions

1. Use the definitions of $\sinh x$, $\cosh x$ and $\tanh x$
2. Use laws of indices and logarithms, as necessary, to simplify expressions.

Example 6

Prove that $\cosh(A+B) \equiv \cosh(A)\cosh(B) + \sinh(A)\sinh(B)$

$$\cosh(A)\cosh(B) + \sinh(A)\sinh(B)$$
$$\equiv \left(\frac{e^A + e^{-A}}{2}\right)\left(\frac{e^B + e^{-B}}{2}\right) + \left(\frac{e^A - e^{-A}}{2}\right)\left(\frac{e^B - e^{-B}}{2}\right)$$

1 Use the definitions of $\cosh x$ and $\sinh x$

$$\equiv \frac{e^{A+B} + e^{(A-B)} + e^{-(A-B)} + e^{-(A+B)}}{4} + \frac{e^{A+B} - e^{(A-B)} - e^{-(A-B)} + e^{-(A+B)}}{4}$$

2 Expand the brackets, using laws of indices.

$$\equiv \frac{2e^{A+B} + 2e^{-(A+B)}}{4}$$
$$\equiv \frac{e^{A+B} + e^{-(A+B)}}{2}$$
$$\equiv \cosh(A+B), \text{ as required}$$

Example 7

Prove that $\cosh^2 x - \sinh^2 x \equiv 1$

$$\cosh^2 x - \sinh^2 x \equiv \left(\frac{e^x + e^{-x}}{2}\right)^2 - \left(\frac{e^x - e^{-x}}{2}\right)^2$$

1 Use the exponential definitions.

$$\equiv \frac{e^{2x} + 2 + e^{-2x}}{4} - \frac{e^{2x} - 2 + e^{-2x}}{4}$$

2 Simplify the numerator.

$$\equiv \frac{4}{4}$$
$$\equiv 1, \text{ as required.}$$

Example 8

Since cosh is a many-to-one function, you need to restrict its domain for an inverse function to exist.

Derive the logarithmic form of the inverse function $\cosh^{-1} x$

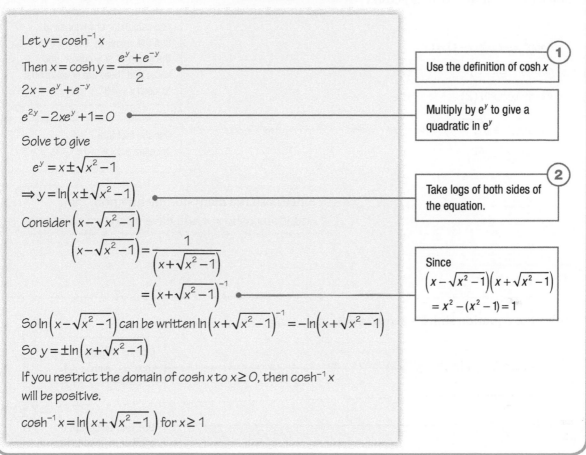

Let $y = \cosh^{-1} x$

Then $x = \cosh y = \dfrac{e^y + e^{-y}}{2}$

$2x = e^y + e^{-y}$

$e^{2y} - 2xe^y + 1 = 0$

Solve to give

$e^y = x \pm \sqrt{x^2 - 1}$

$\Rightarrow y = \ln\left(x \pm \sqrt{x^2 - 1}\right)$

Consider $\left(x - \sqrt{x^2 - 1}\right)$

$\left(x - \sqrt{x^2 - 1}\right) = \dfrac{1}{\left(x + \sqrt{x^2 - 1}\right)}$

$= \left(x + \sqrt{x^2 - 1}\right)^{-1}$

So $\ln\left(x - \sqrt{x^2 - 1}\right)$ can be written $\ln\left(x + \sqrt{x^2 - 1}\right)^{-1} = -\ln\left(x + \sqrt{x^2 - 1}\right)$

So $y = \pm\ln\left(x + \sqrt{x^2 - 1}\right)$

If you restrict the domain of $\cosh x$ to $x \geq 0$, then $\cosh^{-1} x$ will be positive.

$\cosh^{-1} x = \ln\left(x + \sqrt{x^2 - 1}\right)$ for $x \geq 1$

1 Use the definition of $\cosh x$

Multiply by e^y to give a quadratic in e^y

2 Take logs of both sides of the equation.

Since
$\left(x - \sqrt{x^2 - 1}\right)\left(x + \sqrt{x^2 - 1}\right)$
$= x^2 - (x^2 - 1) = 1$

The other inverse hyperbolic functions can be derived in a similar way.

Key point

$\sinh^{-1} x = \ln\left(x + \sqrt{x^2 + 1}\right)$

$\cosh^{-1} x = \ln\left(x + \sqrt{x^2 - 1}\right), \ x \geq 1$

$\tanh^{-1} x = \dfrac{1}{2}\ln\left(\dfrac{1+x}{1-x}\right), \ -1 < x < 1$

You can also refer to the inverse functions as $\operatorname{arsinh} x$, $\operatorname{arcosh} x$ and $\operatorname{artanh} x$

Strategy 2

To solve quadratic equations involving hyperbolic functions

1 Use identities to write the equation in terms of a single hyperbolic function.

2 Solve the quadratic equation to find the value of $\sinh x$, $\cosh x$ or $\tanh x$

3 Use the definitions of the inverse hyperbolic functions to find the exact values of x

Example 9

Solve the equation $\cosh^2 x + \cosh x = 6$

$\cosh^2 x + \cosh x - 6 = 0$

$\cosh x = 2, (-3)$ ●————————

$x = \pm \text{arccosh}\,(2)$

Use calculator or factorise to find the values of $\cosh x$; from the graph of $y = \cosh x$ you know that there are no solutions to $\cosh x = -3$

$= \pm \ln(2 + \sqrt{2^2 - 1})$ ●————————

Use the fact that
$\text{arcosh}\,x = \ln(x + \sqrt{x^2 - 1}),\ x \geq 1$

$= \pm \ln(2 + \sqrt{3})$

Since cosh is a many-to-one function, the equation $\cosh x = a$ will have up to two solutions.

Example 10

Solve the equation $\sinh x + 2\cosh^2 x = 3$

$\sinh x + 2(1 + \sinh^2 x) = 3$ ●————————

Use the identity
$\cosh^2 x - \sinh^2 x \equiv 1$

$2\sinh^2 x + \sinh x - 1 = 0$

$\sinh x = \dfrac{1}{2}, -1$ ●————————

Use your calculator or another method to find values of $\sinh x$

$x = \text{arsinh}\left(\dfrac{1}{2}\right), \text{arsinh}(-1)$

$\text{arsinh}\left(\dfrac{1}{2}\right) = \ln\left(\dfrac{1}{2} + \sqrt{\left(\dfrac{1}{2}\right)^2 + 1}\right)$ ●————————

Use the fact that
$\text{arsinh}\,x = \ln(x + \sqrt{x^2 + 1})$

$= \ln\left(\dfrac{1 + \sqrt{5}}{2}\right)$

$\text{arsinh}(-1) = \ln(-1 + \sqrt{(-1)^2 + 1})$

$= \ln(-1 + \sqrt{2})$

So $x = \ln\left(\dfrac{1 + \sqrt{5}}{2}\right), \ln(-1 + \sqrt{2})$

sinh is a one-to-one function, so the equation $\sinh x = a$ will always have one solution.

1 Solve the equations, giving your answers as exact logarithms where appropriate.

 a $\sinh^2 x + \sinh x = 6$

 b $2\cosh^2 x - 5\cosh x + 3 = 0$

 c $3\tanh^2 x - 2 = 0$

 d $\cosh^2(2x) - 3\cosh(2x) = 0$

2 Solve each of the equations. Give your answers as exact logarithms where appropriate.

 a $\sinh^2 x + \cosh x = 1$

 b $\cosh^2 x + 3\sinh x = 5$

 c $2\sinh x + 3\cosh^2 x = 3$

 d $2\tanh^2 x + 3\tanh x = 2$ **e** $\sinh 2x = 3$

 f $\cosh 3x = 4$ **g** $2\tanh 3x = 1$

3 Use the definitions of $\sinh x$ and $\cosh x$ to prove that $\cosh^2 x - \sinh^2 x \equiv 1$

4 Use the definitions of $\cosh x$ and $\sinh x$ to prove these identities.

 a $\sinh(A+B) = \sinh(A)\cosh(B)$
 $+ \sinh(B)\cosh(A)$

 b $\sinh(A-B) = \sinh(A)\cosh(B)$
 $- \sinh(B)\cosh(A)$

 c $\sinh(2x) = 2\sinh(x)\cosh(x)$

5 **a** Use the definitions of $\sinh x$ and $\cosh x$
 to prove that $\tanh x = \dfrac{e^{2x}-1}{e^{2x}+1}$

 b Hence show that $\tanh(2x) = \dfrac{2\tanh x}{1+\tanh^2 x}$

6 **a** Use the definition of $\cosh x$ to prove that $\cosh(2x) \equiv 2\cosh^2 x - 1$

 b Hence find the exact solutions to the equation $\cosh(2x) + \cosh x = 5$

7 Use a similar method to that used in Example 8 to prove that
$\operatorname{arsinh} x = \ln\left(x + \sqrt{x^2+1}\right)$

Explain clearly why $\operatorname{arsinh} x \neq \ln\left(x - \sqrt{x^2+1}\right)$

8 **a** Prove that $\operatorname{artanh} x = \dfrac{1}{2}\ln\left(\dfrac{1+x}{1-x}\right)$

 b Explain why the formula in part **a** is only valid for $-1 < x < 1$

9 Solve the equation $\cosh 2x + 7 = 7\cosh x$

10 **a** Show that $3\sinh\left(\dfrac{x}{2}\right) - \sinh(x)$
 $\equiv \sinh\left(\dfrac{x}{2}\right)\left(3 - 2\cosh\left(\dfrac{x}{2}\right)\right)$

 b Hence find the exact solutions to the equation $3\sinh\left(\dfrac{x}{2}\right) - \sinh(x) = 0$

11 **a** Show that $\dfrac{1}{\cosh(x)+1} - \dfrac{1}{\cosh(x)-1}$
 $\equiv -\dfrac{2}{\sinh^2 x}$

 b Hence solve the equation
 $\dfrac{1}{\cosh(x)+1} - \dfrac{1}{\cosh(x)-1} + 8 = 0$

12 **a** Given that $\sinh(2x) = \cosh^2 x$,
 show that $\tanh x = \dfrac{1}{2}$

 b Hence solve the equation
 $\sinh(2x) - \cosh^2 x = 0$

13 Solve the equation $\operatorname{arcosh} x^2 = 2\ln x$

14 Given that $\sinh x = 2$, $x > 0$, calculate the exact value of

 a $\cosh x$ **b** $\tanh x$

15 Given that $\cosh x = 3$, $x > 0$, calculate the exact value of

 a $\sinh x$ **b** $\tanh x$

16 Given that $\tanh x = \dfrac{1}{2}$, $x > 0$, calculate the exact value of

 a $\sinh x$ **b** $\cosh x$

17 **a** Use the definitions to find the derivatives of $\sinh x$ and $\cosh x$

 b Find the gradient of both curves when $x = 0$

 c Show that, for $x > 0$, the curve $y = \sinh x$ is always steeper than $y = \cosh x$

 d Find the gradient of the tangent to the curve $y = \tanh x$ at $x = 0$

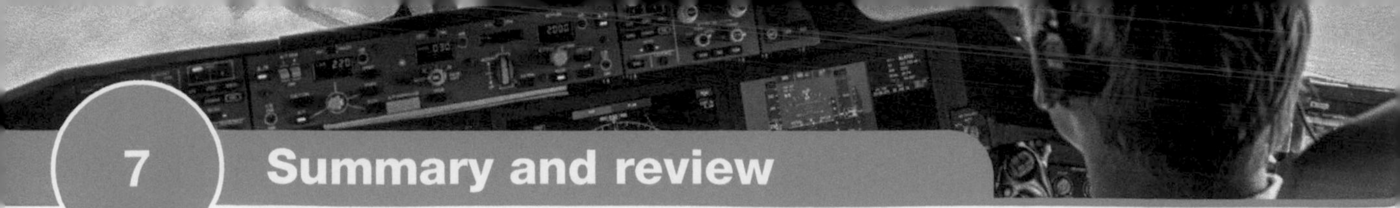

7 Summary and review

Chapter summary

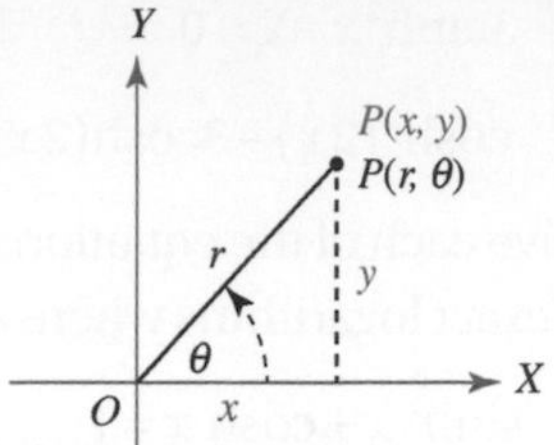

- If you can write a function f(x) such that its general term, u_r, can be expressed as $f(r+1)-f(r)$, then you can find the sum of the series using the method of differences.
- You can use partial fractions to express a function in the form $f(r+1)-f(r)$
- r is the distance from the pole to the point.
- θ is the angle between the initial line and the line segment from the pole to the point, measured anticlockwise.
- To convert from polar to Cartesian form, use

  $x = r\cos\theta$

  $y = r\sin\theta$
- To convert from Cartesian to polar form, use

  $r^2 = x^2 + y^2$

  $\tan\theta = \dfrac{y}{x}$

  but you should draw a diagram to ensure that you choose the right value of θ
- To convert from polar to Cartesian form, use

  $r^2 = x^2 + y^2$

  $\tan\theta = \dfrac{y}{x}$

  $\cos\theta = \dfrac{x}{r} = \pm\dfrac{x}{\sqrt{x^2+y^2}}$

  $\sin\theta = \dfrac{y}{r} = \pm\dfrac{y}{\sqrt{x^2+y^2}}$

  [Choose + or – depending on whether r is positive or negative.]
- To convert from Cartesian to polar form, use

  $x = r\cos\theta$

  $y = r\sin\theta$
- If $r = 0$ when θ takes a particular value α, then the line $y = x\tan\alpha$ is a tangent to the curve at origin.
- The hyperbolic functions are defined using exponentials.
  - $\sinh x = \dfrac{e^x - e^{-x}}{2}$ $\cosh x = \dfrac{e^x + e^{-x}}{2}$ $\tanh = \dfrac{e^x - e^{-x}}{e^x + e^{-x}}$
- $\cosh^2 x - \sinh^2 x \equiv 1$
- The inverse hyperbolic functions are

  $\sinh^{-1} x = \ln(x + \sqrt{x^2+1})$ $\cosh^{-1} x = \ln(x + \sqrt{x^2-1}), x \geq 1$ $\tanh^{-1} x = \dfrac{1}{2}\ln\left(\dfrac{1+x}{1-x}\right), -1 < x < 1$

Checkout

You should now be able to...	Review Questions
✔ Understand and use the method of differences.	1–3
✔ Convert between polar and Cartesian coordinates and equations.	4–7
✔ Sketch polar curves and find points of intersection.	8–11
✔ Find tangents to polar curve.	12
✔ Know the definition of $\sinh x$, $\cosh x$ and $\tanh x$ in terms of exponentials; calculate exact values and solve equations.	13, 16, 17
✔ Sketch hyperbolic graphs and their transformations and know each domain and range.	14
✔ Recall and use the identity $\cosh^2 x - \sinh^2 x \equiv 1$	15
✔ Derive the logarithmic form of the inverse hyperbolic functions and calculate exact values.	17, 18

1 Find the sums of these series to n terms.

   a $\dfrac{1}{r+1} - \dfrac{1}{r+2}$ **b** $\dfrac{1}{r+2} - \dfrac{1}{r+3}$

2 Use partial fractions and the method of differences to find the sum to n terms of the series

   $\displaystyle\sum_{r=2}^{n} \frac{1}{r(r-1)}$ and hence write down the sum to infinity.

3 The expression $\dfrac{2n+1}{n^2(n+1)^2}$ can be written as $\dfrac{1}{n^2} - \dfrac{1}{(n+1)^2}$. Use the method of differences on

   $\left(\dfrac{1}{n^2} - \dfrac{1}{(n+1)^2} \right)$ to show that $\displaystyle\sum_{1}^{k} \left(\dfrac{1}{n^2} - \dfrac{1}{(n+1)^2} \right)$ is $\dfrac{k(k+2)}{(k+1)^2}$

4 Convert these polar coordinates to Cartesian coordinates.

   a $\left(5, \dfrac{\pi}{3}\right)$ **b** $\left(5, -\dfrac{\pi}{3}\right)$

5 Convert these Cartesian coordinates to polar coordinates.

   a $(2\sqrt{3}, 2)$ **b** $(-2\sqrt{3}, -2)$

6 Express these Cartesian equations in polar form.

   a $y = 3x$

   b $y = 2x + 1$

   c $x^2 + y^2 = 16$

   d $(x-1)^2 + (y-1)^2 = 2$

7 Express these polar equations in Cartesian form.

    a The circle $r = 4\sin\theta$, $r \geq 0$, stating its centre and radius,

    b The graph $r = 2 - 4\sin\theta$, $r \geq 0$

8 For each of these polar equations

    i Sketch the line it represents,

    ii Give the Cartesian equation of the line on which it lies.

    a $\theta = -\dfrac{3\pi}{4}$

    b $\theta = \dfrac{\pi}{3}$

9 For each of these polar equations

    i Sketch the graph for $r \geq 0$ and $0 \leq \theta < 2\pi$

    ii State the maximum and minimum values of r

    a $r = 3$

    b $r = 4\cos 2\theta$

    c $r = 2 + \sin\theta$

    d $r = 3\theta$

10 Find the points of intersection between the polar curves $r = 2$ and $r = 3 - 2\cos\theta$.

11 Find the points of intersection between the polar curves $r = \sin 2\theta$ and $r = \sqrt{3}\cos 2\theta$

12 Find the polar and Cartesian equations of the tangents at the pole to the curve
$r = \sin\dfrac{3\theta}{2}$

13 Calculate the exact value of

    a $\cosh(\ln 5)$ **b** $\tanh(-\ln 2)$

14 Sketch these graphs for $a \geq 1$ and state each domain and range.

    a $y = a\cosh x$ **b** $y = \sinh(x - a)$

    c $y = a\tanh x$ **d** $y = \cosh x - a$

15 Solve $\cosh^2 x + \sinh x = 3$

16 Solve these equations, giving your answers as logarithms.

    a $\sinh x + \cosh x = 3$

    b $\cosh x + 1 = e^x$

17 Use the definition of $\cosh x$ to prove that
$\text{arcosh}(2x) = \ln\left(2x + \sqrt{4x^2 - 1}\right)$

18 Find the exact solution to these equations, assuming that $r \geq 0$

    a $\sinh x = 2$ **b** $\tanh x = -\dfrac{1}{2}$

History

Two mathematicians, the Flemish Gregoire de Saint-Vincent and the Italian Bonaventura Cavalieri, introduced the concepts of the polar coordinate system independently in the mid-seventeenth century. Cavalieri used them to solve a problem involving the area within an Archimedian spiral. The French mathematician Blaise Pascal used polar coordinates to calculate the lengths of parabolic arcs like the ones on this bridge.

Did you know?

Although many bridges are known as suspension bridges, they are actually suspended deck bridges. The cables follow a parabolic curve.

Using a photograph of a famous bridge, such as the Forth Road Bridge, together with graph plotting technology, explore how well you can model the suspension cables using quadratic and hyperbolic functions.

Research

Coordinate systems are used to locate the position of a point in space. Explore coordinate systems that can be used effectively in three-dimensions, for example, the spherical coordinate system commonly used by mathematicians. You should also research cylindrical coordinate systems. These are often used by engineers. Explore when, and why, these systems are used.

Different coordinate systems are used in other subject areas. For example, geographers use a geographical coordinate system, and space scientists and astronomers use a celestial coordinate system. Explore the advantages and disadvantages of these systems.

Assessment

7

1 **a** Show that $\dfrac{1}{x}-\dfrac{1}{x+1}=\dfrac{1}{x(x+1)}$ [1]

  b Find an expression in n for $\displaystyle\sum_{r=1}^{n}\dfrac{1}{r(r+1)}$ [3]

2 **a** Show that $(r+1)^3-(r-1)^3=6r^2+2$ [2]

  b Hence use the method of differences to prove that $\displaystyle\sum_{r=1}^{n}(3r^2+1)=\dfrac{n}{2}(2n^2+3n+3)$ [4]

3 **a** Express the Cartesian coordinates $(\sqrt{3},-3)$ as polar coordinates, giving the exact value of r and writing θ in terms of π, where $-\pi<\theta\le\pi$ [4]

  b Sketch the polar graph with equation

    i $r=9$ **ii** $\theta=\dfrac{\pi}{6}$ [2] [1]

  c Find the Cartesian equations of the curves in part **b**. [5]

4 **a** Express the polar coordinates $\left(4,-\dfrac{5\pi}{6}\right)$ as exact Cartesian coordinates. [3]

  b Find the Cartesian equation of the polar curve

    i $r=2\sin\theta$ **ii** $r=\sec\theta$ [3] [3]

  c Sketch the graphs of the curves in part **b** on the same Cartesian axes. [3]

  d State the polar coordinates of the point of intersection between $r=2\sin\theta$ and $r=\sec\theta$ [4]

5 Find the polar equations of these curves.

  a $x^2+y^2=x$ **b** $y=2$ [2] [2]

6 Sketch these polar curves.

  a $r=4\theta$ for $0\le\theta\le3\pi$ **b** $r=\cos3\theta$ [2] [2]

  c $r=1+\sin\theta$ **d** $r=8+3\cos\theta$ [2] [2]

7 Find the polar and Cartesian equations of the tangents at the pole to the curve

$r=\sin^2 4\theta$ [6]

8 Find the points of intersection between the curve $r = 1 + \sin 4\theta$, $r \geq 0$, $0 \leq \theta \leq 2\pi$, and the curve $x^2 + y^2 = 4$; $x, y \in \mathbb{R}$ **[4]**

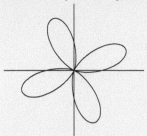

9 **a** Sketch the graph of

  i $y = 2\cosh x$ **ii** $y = 2 + \cosh x$ **[2] [2]**

  b State the range of each function in part **a**.   **[2]**

10 **a** Use the exponential definition of $\sinh x$ to show that $\sinh(\ln 2) = \dfrac{3}{4}$ **[3]**

  b Solve the equation $\sinh x = 3$, giving your solution to 1 decimal place.   **[3]**

11 **a** Sketch the graph of $y = \tanh(x+1)$ **[2]**

  b Write down the equations of the asymptotes to the curve in part **a**.   **[2]**

  c Use the exponential definitions of $\sinh x$ and $\cosh x$ to show that $\tanh x = \dfrac{e^{2x}-1}{e^{2x}+1}$ **[4]**

  d Hence solve the equation $\tanh x = \dfrac{1}{2}$. Give your answer as a logarithm.   **[3]**

12 **a** Write $\dfrac{3}{r(r+1)}$ in partial fractions.   **[3]**

  b Hence use the method of differences to prove that $\displaystyle\sum_{r=1}^{n} \dfrac{3}{r(r+1)} = \dfrac{3n}{n+1}$ **[5]**

  c Evaluate $\dfrac{1}{2} + \dfrac{1}{6} + \dfrac{1}{12} + \dfrac{1}{20} + \ldots + \dfrac{1}{930}$ **[3]**

13 **a** Write $\dfrac{2}{r^2-1}$ in partial fractions.   **[3]**

  b Hence use the method of differences to prove that $\displaystyle\sum_{r=2}^{n} \dfrac{1}{r^2-1} = \dfrac{(3n+2)(n-1)}{4n(n+1)}$ **[6]**

  c Show that $\displaystyle\sum_{r=2}^{\infty} \dfrac{1}{r^2-1} = \dfrac{3}{4}$ **[3]**

14 Show that the curve $y = \dfrac{1}{x}$ has polar equation $r^2 = 2\operatorname{cosec} 2\theta$ **[4]**

15 **a** Give the Cartesian equation of the curve with polar equation $r = 3\sin 2\theta$ **[4]**

  b Sketch the polar curve found in part **a**.   **[2]**

  c State the maximum value of $r = 3\sin 2\theta$ and the values of θ at which it occurs.   **[3]**

16 a State the maximum and minimum values of $r = 4 + 3\cos\theta$, where $0 \leq \theta < 2\pi$ [2]

b Sketch the curve $r = 4 + 3\cos\theta$ [2]

17 a Sketch the graphs of $y = \sinh x$ and $y = \sinh 2x$ on the same diagram. [3]

b If $y = \sinh x$, use the exponential definition of $\sinh x$ to show that $x = \ln(y + \sqrt{y^2 + 1})$ [5]

18 a Prove that $\text{arcosh}\, x = \pm\ln(x + \sqrt{x^2 - 1})$ [7]

b Hence find the exact solutions to $\cosh x = 3$ [3]

19 a Show that $\dfrac{1}{r!} - \dfrac{1}{(r+1)!} = \dfrac{r}{(r+1)!}$ [3]

b Hence use the method of differences to show that $\displaystyle\sum_{r=1}^{n} \frac{r}{(r+1)!} = 1 - \frac{1}{(n+1)!}$ [3]

c Find an expression for $\dfrac{n}{(n+1)!} + \dfrac{n+1}{(n+2)!} + \cdots + \dfrac{2n}{(2n+1)!}$ as a single fraction. [3]

20 Using the standard formula for $\sum 1, \sum r, \sum r^2, \sum r^3$ and by considering $\displaystyle\sum_{r=1}^{n}[(r+1)^5 - r^5]$,

find a fully factorised formula in terms of n for $\sum r^4$ [5]

21 a Sketch these curves on the same diagram.

    i $r = \sin 4\theta$ **ii** $r = 2\sin 3\theta$ [4]

b Find the points of intersections of the two curves.
Give your solutions as polar coordinates. [8]

22 a Sketch the graph of $y = 5 + \tanh x$ [2]

b Write the equations of the asymptotes to the curve. [2]

c Find the inverse of $y = 5 + \tanh x$ and state its domain. [3]

8 Integration and differentiation

Economics is an area of study and work that has become increasingly important. You can see this by the time and space that is devoted to business and the economy in various parts of the media. Many aspects of the work of economists has become more mathematical, as they develop, and use, a range of mathematical models. These models allow them to understand not only what has happened in the past, but to also predict what may happen in the future.

Aspects of calculus are important in the mathematics used by economists. Methods and techniques associated with integration and differentiation allow economists to consider how measurable quantities change with time. For example, supply and demand, in relation to (number of) sales and pricing, are quantities that can be modelled using mathematical functions. Once the model is set up, calculus can be used to work out optimal conditions.

Orientation

What you need to know	What you will learn	What this leads to
Chapter 3 Integration	<ul><li>How to work with improper integrals.</li><li>How to find mean values.</li><li>How to find volumes of revolution for parametric functions.</li><li>How to differentiate inverse trigonometric functions.</li><li>How to differentiate hyperbolic functions.</li><li>How to use partial fractions to help with integration.</li><li>How to use integration to find areas enclosed by a polar curve.</li><li>How to find and use Maclaurin series.</li></ul>	**Careers** • Economics.

8.1 Improper integrals

Fluency and skills

You know how to find the value of a definite integral, by integrating and then substituting in the limits. This value represents a finite area between the curve and the axis.

Key point

An **improper integral** is a definite integral where either:

- one or both of the limits is $\pm\infty$
- the integrand (expression to be integrated) is undefined at one of the limits of the integral
- the integrand is undefined at some point between the limits of the integral.

It is sometimes possible to calculate the value of an improper integral by replacing a limit of $\pm\infty$ with a variable and then considering what happens as that variable tends to $\pm\infty$

Key point

To evaluate an improper integral with a limit of $\pm\infty$ use

$$\int_a^\infty f(x)dx = \lim_{t\to\infty} \int_a^t f(x)dx \quad \text{or} \quad \int_{-\infty}^a f(x)dx = \lim_{t\to-\infty} \int_t^a f(x)dx$$

If the limit exists, then the improper integral is called **convergent**. If the limit does not exist, then the improper integral is called **divergent**.

For example, the integral, $\int_1^\infty \frac{1}{x^2} dx$ is an improper integral because one of the limits is infinite. To evaluate it, use

$$\int_1^\infty \frac{1}{x^2}dx = \lim_{t\to\infty}\int_1^t \frac{1}{x^2}dx$$

$$= \lim_{t\to\infty}\left(1-\frac{1}{b}\right) \quad \text{since } \int_1^t \frac{1}{x^2}dx = \left[-\frac{1}{x}\right]_1^t = 1-\frac{1}{t}$$

$$= 1 \text{ since } \frac{1}{t}\to 0 \text{ as } t\to\infty$$

Therefore, the improper integral $\int_1^\infty \frac{1}{x^2}dx$ is convergent and represents a finite area of 1

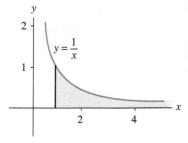

However, if you were to try to evaluate the integral $\int_0^\infty x^2 dx$ then

$$\lim_{t\to\infty}\int_0^t x^2 dx = \lim_{t\to\infty}\left(\frac{t^3}{3}\right) \quad \text{since } \int_0^t x^2 dx = \left[\frac{x^3}{3}\right]_0^t = \frac{t^3}{3}$$

$\frac{t^3}{3}\to\infty$ as $t\to\infty$; therefore the improper integral $\int_0^\infty x^2 dx$ is divergent and represents an infinite area.

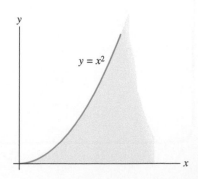

Example 1

Find the value of $\int\limits_{2}^{\infty}\dfrac{2}{x^3}\,dx$

$$\int\limits_{2}^{\infty}\frac{2}{x^3}dx = \lim_{t\to\infty}\int\limits_{2}^{t}2x^{-3}dx$$ •————————————— Replace ∞ with t

$$= \lim_{t\to\infty}\left[-x^{-2}\right]_{2}^{t}$$

$$= \lim_{t\to\infty}((-t^{-2})-(-2^{-2}))$$

$$= \lim_{t\to\infty}\left(-\frac{1}{t^2}+\frac{1}{4}\right)$$

$$= \frac{1}{4}\ \text{since}\ \frac{1}{t^2}\to 0\ \text{as}\ t\to\infty$$

If both the limits are $\pm\infty$, then the integral needs to be split into two integrals, each with one finite limit. Any point can be chosen for this limit, so just choose a convenient value.

Example 2

When splitting the integral, you need to use different variables for ∞ and $-\infty$. This is because both parts of the integral must be convergent for the original integral to exist.

a Find the value of $\int\limits_{-\infty}^{\infty}\dfrac{x}{e^{x^2}}\,dx$

b Show that the improper integral $\int\limits_{-\infty}^{\infty}e^x$ is divergent.

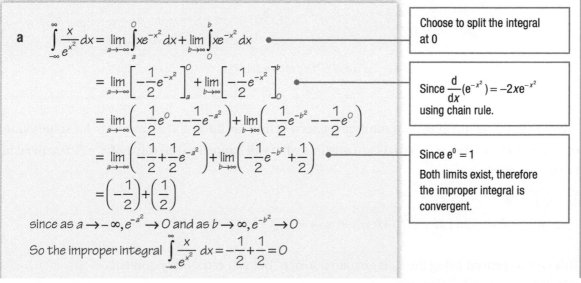

a
$$\int\limits_{-\infty}^{\infty}\frac{x}{e^{x^2}}dx = \lim_{a\to-\infty}\int\limits_{a}^{0}xe^{-x^2}dx + \lim_{b\to\infty}\int\limits_{0}^{b}xe^{-x^2}dx$$ •————— Choose to split the integral at 0

$$= \lim_{a\to-\infty}\left[-\frac{1}{2}e^{-x^2}\right]_{a}^{0} + \lim_{b\to\infty}\left[-\frac{1}{2}e^{-x^2}\right]_{0}^{b}$$ •————— Since $\dfrac{d}{dx}(e^{-x^2})=-2xe^{-x^2}$ using chain rule.

$$= \lim_{a\to-\infty}\left(-\frac{1}{2}e^{0}--\frac{1}{2}e^{-a^2}\right) + \lim_{b\to\infty}\left(-\frac{1}{2}e^{-b^2}--\frac{1}{2}e^{0}\right)$$

$$= \lim_{a\to-\infty}\left(-\frac{1}{2}+\frac{1}{2}e^{-a^2}\right) + \lim_{b\to\infty}\left(-\frac{1}{2}e^{-b^2}+\frac{1}{2}\right)$$ •————— Since $e^0=1$. Both limits exist, therefore the improper integral is convergent.

$$= \left(-\frac{1}{2}\right)+\left(\frac{1}{2}\right)$$

since as $a\to-\infty, e^{-a^2}\to 0$ and as $b\to\infty, e^{-b^2}\to 0$

So the improper integral $\int\limits_{-\infty}^{\infty}\dfrac{x}{e^{x^2}}\,dx = -\dfrac{1}{2}+\dfrac{1}{2}=0$

(*Continued on the next page*)

b $\displaystyle\int_{-\infty}^{\infty}e^x\,dx = \lim_{a\to-\infty}\int_{a}^{0}e^x\,dx + \lim_{b\to\infty}\int_{0}^{b}e^x\,dx$

$\displaystyle\qquad = \lim_{a\to-\infty}\Big[e^x\Big]_{a}^{0} + \lim_{b\to\infty}\Big[e^x\Big]_{0}^{b}$

$\displaystyle\qquad = \lim_{a\to-\infty}(e^0 - e^a) + \lim_{b\to\infty}(e^b - e^0)$

$\displaystyle\qquad = \lim_{a\to-\infty}(1 - e^a) + \lim_{b\to\infty}(e^b - 1)$

As $a\to-\infty$, $e^a\to 0$ so $\displaystyle\lim_{a\to-\infty}\int_{a}^{0}e^x\,dx = 1$

However, as $b\to\infty$, $e^b\to\infty$ so $\displaystyle\lim_{b\to\infty}\int_{0}^{b}e^x\,dx$ does not exist.

Therefore, the improper integral $\displaystyle\int_{-\infty}^{\infty}e^x\,dx$ is divergent.

> Both limits must exist for the improper integral to be convergent.

Another type of improper integral is where the integrand is undefined at one of the limits of the integral. To evaluate these integrals you replace that limit of integration with a variable as before. You then consider what happens as the variable tends to the original value of the limit of integration.

Example 3

Find the value of $\displaystyle\int_{0}^{4}\frac{1}{\sqrt{x}}\,dx$

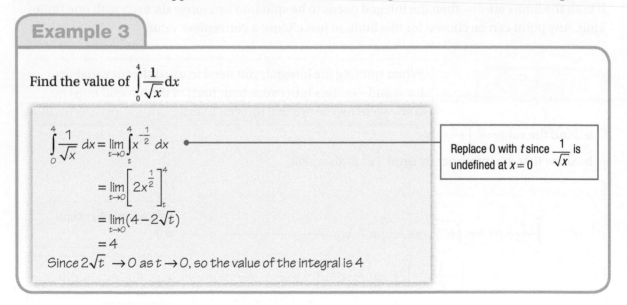

$\displaystyle\int_{0}^{4}\frac{1}{\sqrt{x}}\,dx = \lim_{t\to 0}\int_{t}^{4}x^{-\frac{1}{2}}\,dx$

> Replace 0 with t since $\dfrac{1}{\sqrt{x}}$ is undefined at $x=0$

$\displaystyle\qquad = \lim_{t\to 0}\left[2x^{\frac{1}{2}}\right]_{t}^{4}$

$\displaystyle\qquad = \lim_{t\to 0}(4 - 2\sqrt{t})$

$\displaystyle\qquad = 4$

Since $2\sqrt{t}\to 0$ as $t\to 0$, so the value of the integral is 4

When evaluating integrals, you sometimes need to find the limit of $x^k e^{-x}$ as $x\to\infty$ for some value of k. Since $x^k\to\infty$ but $e^{-x}\to 0$ it is not obvious what will happen to the value of $x^k e^{-x}$. You need to know this result.

Key point

For any real number k, $x^k e^{-x}\to 0$ when $x\to\infty$

This can be proved using the series expansion of e^x but you can simply quote it.

Another common limit is that of $x^k \ln x$ as $x\to 0$. Again, as $x^k\to 0$ and $\ln x\to\infty$ it is not obvious what the result will be, but the previous result can be used to prove the following.

For any real number k, $x^k \ln x \to 0$ when $x \to 0+$ (this means that x approaches zero from above as x must be positive for $\ln x$ to be defined).

You can also quote this result.

Example 4

Find the value of $\int_0^1 2 - \ln x \, dx$

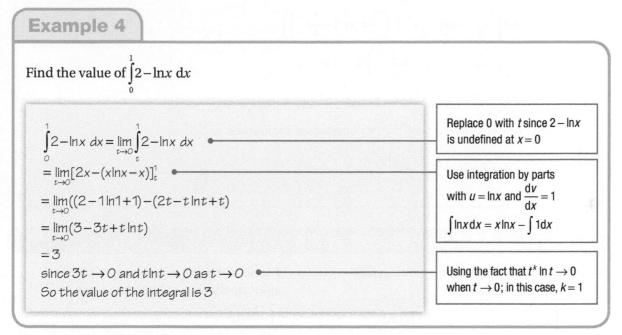

$\int_0^1 2 - \ln x \, dx = \lim_{t \to 0} \int_t^1 2 - \ln x \, dx$ — Replace 0 with t since $2 - \ln x$ is undefined at $x = 0$

$= \lim_{t \to 0} [2x - (x \ln x - x)]_t^1$ — Use integration by parts with $u = \ln x$ and $\frac{dv}{dx} = 1$

$= \lim_{t \to 0} ((2 - 1 \ln 1 + 1) - (2t - t \ln t + t))$ — $\int \ln x \, dx = x \ln x - \int 1 \, dx$

$= \lim_{t \to 0} (3 - 3t + t \ln t)$

$= 3$

since $3t \to 0$ and $t \ln t \to 0$ as $t \to 0$ — Using the fact that $t^k \ln t \to 0$ when $t \to 0$; in this case, $k = 1$

So the value of the integral is 3

To evaluate an improper integral where the integrand is undefined at a point between the limits of the integral, you need to split the integral into two parts about the point of discontinuity.

Example 5

You need to identify the point of discontinuity and split the integral at this point.

Evaluate the improper integral $\int_{-1}^{e} x \ln |x| \, dx$

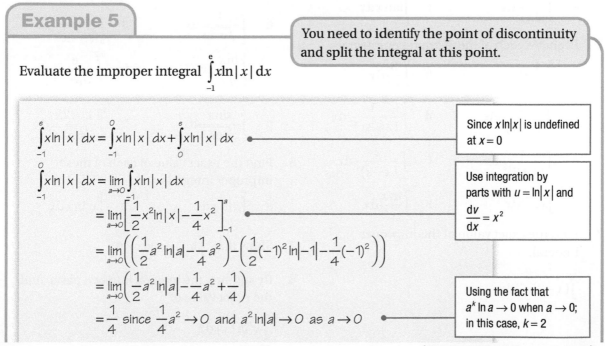

$\int_{-1}^{e} x \ln |x| \, dx = \int_{-1}^{0} x \ln |x| \, dx + \int_{0}^{e} x \ln |x| \, dx$ — Since $x \ln |x|$ is undefined at $x = 0$

$\int_{-1}^{0} x \ln |x| \, dx = \lim_{a \to 0} \int_{-1}^{a} x \ln |x| \, dx$ — Use integration by parts with $u = \ln |x|$ and $\frac{dv}{dx} = x^2$

$= \lim_{a \to 0} \left[\frac{1}{2} x^2 \ln |x| - \frac{1}{4} x^2 \right]_{-1}^{a}$

$= \lim_{a \to 0} \left(\left(\frac{1}{2} a^2 \ln |a| - \frac{1}{4} a^2 \right) - \left(\frac{1}{2} (-1)^2 \ln |-1| - \frac{1}{4} (-1)^2 \right) \right)$

$= \lim_{a \to 0} \left(\frac{1}{2} a^2 \ln |a| - \frac{1}{4} a^2 + \frac{1}{4} \right)$

$= \frac{1}{4}$ since $\frac{1}{4} a^2 \to 0$ and $a^2 \ln |a| \to 0$ as $a \to 0$ — Using the fact that $a^k \ln a \to 0$ when $a \to 0$; in this case, $k = 2$

(Continued on the next page)

$$\int_0^e x\ln|x|\,dx = \lim_{b\to 0}\int_b^e x\ln|x|\,dx$$

$$= \lim_{b\to 0}\left[\frac{1}{2}x^2\ln|x| - \frac{1}{4}x^2\right]_b^e$$

$$= \lim_{b\to 0}\left(\left(\frac{1}{2}e^2\ln|e| - \frac{1}{4}e^2\right) - \left(\frac{1}{2}b^2\ln|b| - \frac{1}{4}b^2\right)\right)$$

$$= \lim_{b\to 0}\left(\frac{1}{4}e^2 - \frac{1}{2}b^2\ln|b| + \frac{1}{4}b^2\right) \text{ since } \ln|e| = 1$$

$$= \frac{1}{2}e^2 - \frac{1}{4}e^2 \text{ since } \frac{1}{4}b^2 \to 0 \text{ and } \frac{1}{2}b^2\ln|b| \to 0 \text{ as } b \to 0$$

Therefore, since both limits exist, the improper integral converges

$$\int_{-1}^e x\ln|x|\,dx = \frac{1}{4} + \frac{1}{2}e^2 - \frac{1}{4}e^2 = \frac{1}{4}(1+e^2)$$

Exercise 8.1A Fluency and skills

1 Which of these are improper integrals? Explain your answers.

a $\int_0^5 e^{-x}dx$

b $\int_0^2 \ln x\,dx$

c $\int_1^\infty \frac{1}{x^2}dx$

d $\int_{-\infty}^\infty \sin x\,dx$

e $\int_{\frac{\pi}{4}}^{\frac{\pi}{2}} \frac{1}{\sin x}dx$

f $\int_0^\pi \tan x\,dx$

2 Evaluate each of these improper integrals.

a $\int_1^\infty \frac{1}{x^2}dx$

b $\int_2^\infty \frac{3}{x^4}dx$

c $\int_{-\infty}^0 \frac{1}{(1-x)^2}dx$

d $\int_{-\infty}^0 \frac{1}{(2-3x)^2}dx$

e $\int_0^\infty \frac{1}{(x+2)^3}dx$

f $\int_{-\infty}^1 \frac{1}{(x-2)^4}dx$

g $\int_0^\infty xe^{-2x}dx$

h $\int_1^\infty \frac{\ln x}{x^2}dx$

3 Find the exact value of this improper integral.

$$\int_3^\infty (x-3)e^{-x}dx$$

4 Evaluate this improper integral by first splitting the integral into two parts.

$$\int_{-\infty}^\infty \frac{1}{e^{|x|}}dx$$

5 Evaluate each of these improper integrals.

a $\int_0^9 \frac{1}{\sqrt{x}}dx$

b $\int_0^{27} \frac{1}{x^{\frac{1}{3}}}dx$

c $\int_2^4 \frac{1}{\sqrt{x-2}}dx$

d $\int_0^3 \frac{x}{\sqrt{9-x^2}}dx$

e $\int_0^1 \frac{\ln x}{\sqrt{x}}dx$

f $\int_0^{\ln 2} \frac{e^x}{\sqrt{e^x-1}}dx$

g $\int_0^{\frac{\pi}{2}} \frac{\sin x}{\sqrt{\cos x}}dx$

h $\int_0^{\frac{\pi}{12}} \frac{\cos 2x}{\sqrt{\sin 2x}}dx$

6 Find the exact value of each of these improper integrals.

a $\int_0^e x\ln x\,dx$

b $\int_0^e x^2\ln x\,dx$

7 Show that $\int_{1-e}^1 \ln(1-x)dx = 0$

8 By splitting the integral into two parts, find the exact value of

$$\int_{-e}^e x^2\ln|x|\,dx$$

Reasoning and problem-solving

Many improper integrals are divergent because the areas they represent are not finite. You can use algebra to show that an improper integral is divergent.

Strategy

To decide whether an improper integral exists

(1) Replace the limit where the integrand is undefined by a variable.

(2) Integrate and substitute in the limits.

(3) Consider the behaviour of the integral as the variable tends towards the original limit.

Example 6

Show that the improper integral $\int_0^5 \frac{1}{x^2} dx$ does not exist.

$$\int_a^5 \frac{1}{x^2} dx = \lim_{a \to 0} \int_a^5 x^{-2} dx$$

$\frac{1}{x^2}$ is undefined at $x = 0$ so replace 0 with a **(1)**

$$= \lim_{a \to 0} \left[-x^{-1} \right]_a^5$$

$$= \lim_{a \to 0} \left(-\frac{1}{5} - -\frac{1}{a} \right)$$

Integrate and substitute in the limits. **(2)**

$$= \lim_{a \to 0} \left(\frac{1}{a} - \frac{1}{5} \right)$$

However, $\frac{1}{a} \to \infty$ as $a \to 0$

Consider the behaviour of the function as $a \to 0$ **(3)**

Therefore $\int_a^5 \frac{1}{x^2} dx$ is divergent: the improper integral cannot be evaluated.

You may need to consider the behaviour of more complex expressions, such as those involving **rational expressions**.

You do this by dividing the numerator and the denominator by the highest power of the variable.

For example, the expression $\frac{x^2 + 2x}{3x^2 - 5}$ can be written as $\dfrac{1 + \frac{2}{x}}{3 - \frac{5}{x^2}}$ by dividing the numerator and the denominator by x^2

As $x \to \infty, \frac{2}{x} \to 0$ and $\frac{5}{x^2} \to 0$; therefore you can see that $\dfrac{1 + \frac{2}{x}}{3 - \frac{5}{x^2}} \to \frac{1}{3}$

Example 7

Is the improper integral $\int_{1}^{\infty}\left(\dfrac{2x}{x^2+1}-\dfrac{4x-1}{2x^2-x}\right)dx$ convergent or divergent?

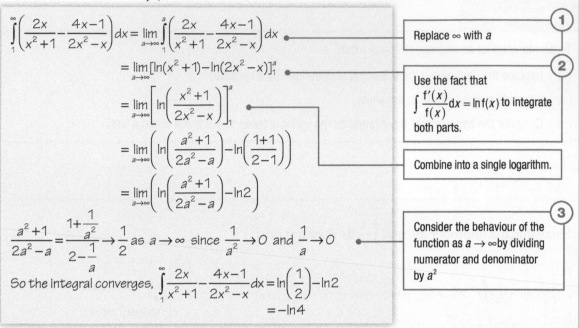

$$\int_{1}^{\infty}\left(\frac{2x}{x^2+1}-\frac{4x-1}{2x^2-x}\right)dx=\lim_{a\to\infty}\int_{1}^{a}\left(\frac{2x}{x^2+1}-\frac{4x-1}{2x^2-x}\right)dx$$

Replace ∞ with a ①

$$=\lim_{a\to\infty}[\ln(x^2+1)-\ln(2x^2-x)]_{1}^{a}$$

Use the fact that
$\int\dfrac{f'(x)}{f(x)}dx=\ln f(x)$ to integrate
both parts. ②

$$=\lim_{a\to\infty}\left[\ln\left(\frac{x^2+1}{2x^2-x}\right)\right]_{1}^{a}$$

$$=\lim_{a\to\infty}\left(\ln\left(\frac{a^2+1}{2a^2-a}\right)-\ln\left(\frac{1+1}{2-1}\right)\right)$$

Combine into a single logarithm.

$$=\lim_{a\to\infty}\left(\ln\left(\frac{a^2+1}{2a^2-a}\right)-\ln2\right)$$

$$\frac{a^2+1}{2a^2-a}=\frac{1+\dfrac{1}{a^2}}{2-\dfrac{1}{a}}\to\frac{1}{2}\text{ as }a\to\infty\text{ since }\frac{1}{a^2}\to0\text{ and }\frac{1}{a}\to0$$

Consider the behaviour of the
function as $a\to\infty$ by dividing
numerator and denominator
by a^2 ③

So the integral converges, $\displaystyle\int_{1}^{\infty}\frac{2x}{x^2+1}-\frac{4x-1}{2x^2-x}dx=\ln\left(\frac{1}{2}\right)-\ln2$

$$=-\ln4$$

Exercise 8.1B Reasoning and problem-solving

1 Decide whether or not each of these integrals converges.

If it does converge, find its value. If it diverges, explain why.

a $\displaystyle\int_{0}^{1}\frac{1}{x^4}dx$ b $\displaystyle\int_{1}^{\infty}\frac{1}{x^4}dx$

c $\displaystyle\int_{0}^{\frac{\pi}{4}}\tan x\,dx$ d $\displaystyle\int_{0}^{\frac{\pi}{2}}\tan x\,dx$

e $\displaystyle\int_{0}^{\infty}\cos x\,dx$ f $\displaystyle\int_{1}^{\infty}\frac{1}{x}dx$

g $\displaystyle\int_{-\infty}^{0}\frac{1}{3-x}dx$ h $\displaystyle\int_{0}^{7}\frac{1}{\sqrt{7-x}}dx$

i $\displaystyle\int_{0}^{7}\frac{1}{(7-x)^2}dx$ j $\displaystyle\int_{2}^{\infty}\left(\frac{1}{x-1}-\frac{2}{2x-1}\right)dx$

k $\displaystyle\int_{1}^{\infty}\left(\frac{1}{x}-\frac{2x}{x^2+1}\right)dx$ l $\displaystyle\int_{1}^{\infty}\left(\frac{x}{x^2+1}-\frac{2x}{2x^2+1}\right)dx$

2 a Show that
$$\int\left(\frac{x}{x^2+3}-\frac{2}{2x+3}\right)dx=\frac{1}{2}\ln\frac{(x^2+3)}{4x^2+12x+9}+c$$

b Hence show that $\displaystyle\int_{0}^{\infty}\left(\frac{x}{x^2+3}-\frac{2}{2x+3}\right)dx=\ln k$,

where k is a constant to be found.

3 Show that each of these integrals converges and give its value.

a $\displaystyle\int_{-\infty}^{0}\left(\frac{6}{3x-2}-\frac{2x}{x^2+4}\right)dx$

b $\displaystyle\int_{-\infty}^{0}\left(\frac{3-x}{x^2-6x+1}-\frac{8}{5-8x}\right)dx$

4 Find the range of values of p for which the improper integral $\displaystyle\int_{0}^{1}\frac{1}{x^p}dx$ converges and find its value in terms of p

5 Find the range of values of p for which the improper integral $\displaystyle\int_{1}^{\infty}\frac{1}{x^p}dx$ converges and find its value in terms of p

Mean values

Fluency and skills

You already know that you can use integration to find the area, A, enclosed by the curve $y = f(x)$, the x-axis and the lines $x = a$ and $x = b$: $A = \int_a^b f(x)\, dx$

The mean value theorem is based on the idea that for any continuous function there is always a rectangle with exactly the same area, A. The base of this rectangle will be between a and b. The height, h, of the rectangle is known as the mean value of the function f(x)

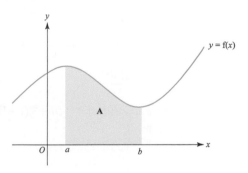

You can calculate the height of the rectangle by dividing its area, found by integration, by its width. So $h = \dfrac{1}{b-a} \int_a^b f(x)\, dx$

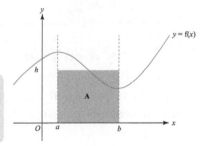

Key point

The mean value of a function $f(x)$ in the range

$a \le x \le b$ is given by $\dfrac{1}{b-a} \int_a^b f(x)\, dx$

Example 1

Calculate the mean value of the function $f(x) = 2x^3 + 6x - 1$ in the interval $[2, 5]$

$$\text{Mean value} = \frac{1}{5-2} \int_2^5 (2x^3 + 6x - 1)\, dx$$

Use the definition of the mean value as $\dfrac{1}{b-a} \int_a^b f(x)\, dx$

$$= \frac{1}{3} \left[\frac{x^4}{2} + 3x^2 - x \right]_2^5$$

$$= \frac{1}{3} \left(\frac{625}{2} + 75 - 5 \right) - \frac{1}{3}(8 + 12 - 2)$$

Substitute the limits.

$$= \frac{243}{2}$$

So the mean value of $2x^3 + 6x - 1$ in the interval $[2, 5]$ is $\dfrac{243}{2}$

(or 121.5)

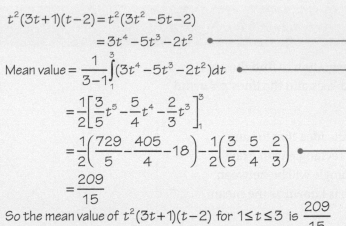

Example 2

Calculate the mean value with respect to t of the function $t^2(3t+1)(t-2)$ for $1 \le t \le 3$

$t^2(3t+1)(t-2) = t^2(3t^2 - 5t - 2)$

$\qquad\qquad\qquad\qquad = 3t^4 - 5t^3 - 2t^2$

> Expand the brackets and use index rules to simplify.

Mean value $= \dfrac{1}{3-1} \displaystyle\int_1^3 (3t^4 - 5t^3 - 2t^2)\,dt$

> Use the definition of the mean value as $\dfrac{1}{b-a}\displaystyle\int_a^b f(t)\,dt$ since our function is in terms of the variable t

$\qquad = \dfrac{1}{2}\left[\dfrac{3}{5}t^5 - \dfrac{5}{4}t^4 - \dfrac{2}{3}t^3\right]_1^3$

$\qquad = \dfrac{1}{2}\left(\dfrac{729}{5} - \dfrac{405}{4} - 18\right) - \dfrac{1}{2}\left(\dfrac{3}{5} - \dfrac{5}{4} - \dfrac{2}{3}\right)$

> Substitute the limits.

$\qquad = \dfrac{209}{15}$

So the mean value of $t^2(3t+1)(t-2)$ for $1 \le t \le 3$ is $\dfrac{209}{15}$ (or 13.9)

Calculator

Try it on your calculator

Definite integrals can be worked out on your calculator.

$$\int_{-2}^{1}\left(4x^3 - 2x\right)dx$$

-12

Activity

Find out how to work out $\displaystyle\int_{-2}^{1}\left(4x^3 - 2x\right)dx$ on your calculator.

Exercise 8.2A Fluency and skills

1 The graph of $y = x^3 - 3x^2 + 6$ is shown.

   a Calculate the area bounded by the curve, the coordinate axes and the line $x = 3$

   b Work out the mean value of $x^3 - 3x^2 + 6$ for $0 \le x \le 3$

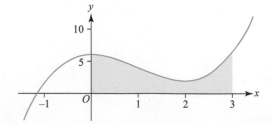

2 $f(x) = x^2 - 3x + 2$

   a Sketch the graph of $y = f(x)$, labelling where the curve crosses the coordinate axes.

   b Calculate the area bounded by the curve $y = f(x)$ and the x-axis.

   c Work out the mean value of $f(x)$ for each of these intervals.

     i $1 \le x \le 2$ **ii** $0 \le x \le 1$

3 $g(x) = x^3 + 5x^2$

   a Sketch the graph of $y = g(x)$, labelling where the curve crosses the coordinate axes.

b Calculate the area bounded by the curve $y = g(x)$ and the x-axis.

c Work out the mean value of $f(x)$ for each of these intervals.

   i $-5 \leq x \leq 0$ **ii** $0 \leq x \leq 2$

4 Calculate the mean value of the function $f(x) = 5x^4$ in each of these intervals.

a $[0, 4]$ **b** $[1, 3]$

c $[-1, 2]$ **d** $\left[-\dfrac{1}{2}, \dfrac{1}{2}\right]$

5 Calculate the mean value of $2x^2 + 3$ for each of these ranges of values of x

a $0 \leq x \leq 3$ **b** $2 \leq x \leq 6$

c $-1 \leq x \leq 2$ **d** $-2 \leq x \leq -1$

6 Calculate the mean value of $3x - 8x^3$ for x between

a 1 and 4 **b** 0 and 3

c -1 and 1 **d** -3 and -2

7 Show that the mean value of $\dfrac{1}{x^2}$ for $2 \leq x \leq 5$ is $\dfrac{1}{10}$

8 Show that the mean value of $\sqrt{x}$ for $1 \leq x \leq 4$ is $\dfrac{14}{9}$

9 Show that the mean value of $\dfrac{1}{\sqrt{x}}$ for $4 \leq x \leq 9$ is 0.4

10 Given that $f(x) = \dfrac{2\sqrt{x} + 3x}{2x}$

a Write $f(x)$ in the form $Ax^c + B$, for constants A, B and c and state their values.

b Show that the mean value of $f(x)$ in the interval $[1, 9]$ is 2

11 Given that $g(x) = \dfrac{3x - x^2}{5\sqrt{x}}$

a Write $g(x)$ in the form $Ax^c + Bx^d$ for constants A, B, c and d and state their values.

b Calculate the exact mean value of $g(x)$ for $2 \leq x \leq 4$

12 Find the mean value of each of these functions of t for the range given.

a $2t\left(5t^3 - 8t + 1\right)$ for $-1 \leq t \leq 0$

b $3t\sqrt{t} - 2t^2$ for $\dfrac{1}{4} \leq t \leq 4$

c $\dfrac{1}{2t\sqrt{t}}$ for $1 \leq t \leq 9$

d $\dfrac{(t^3)^{\frac{1}{2}}}{t}$ for $4 \leq t \leq 64$

13 Find an expression for the mean value of $(1 - 2t)^2$ across the interval

a $0 \leq t \leq T$ **b** $T \leq t \leq T + 2$

Give your answers as polynomials in T

14 Calculate the mean value of the function $\dfrac{4 + x}{\sqrt{x}}$ for $2 \leq x \leq 8$

Give your answer in the form $A\sqrt{2}$

15 Derive an expression in terms of a for the mean value of the function $3x^3 - \dfrac{1}{x^3}$ for x in the interval $[a, 2a]$. Give your answer in its simplest form.

16 Find an expression in terms of X for the mean value of the function $\dfrac{1}{2x^4}$, for x in the interval $\left[0, \dfrac{1}{X}\right]$. Simplify your answer.

Reasoning and problem-solving

Strategy

To solve problems on mean values

(**1**) Use the formula for the mean value of a function.

(**2**) Form one or more equations using the information provided.

(**3**) Solve quadratic or simultaneous equations.

Example 3

The velocity of a particle after t seconds is given by $v = \dfrac{1}{8}t\left(9 - \dfrac{4}{\sqrt{t}}\right)$ m s^{-1}.

a Show that the mean velocity for $1 \leq t \leq 4$ is 2.03 m s^{-1} to 3 significant figures.

b Calculate the mean acceleration of the particle over the same time period.

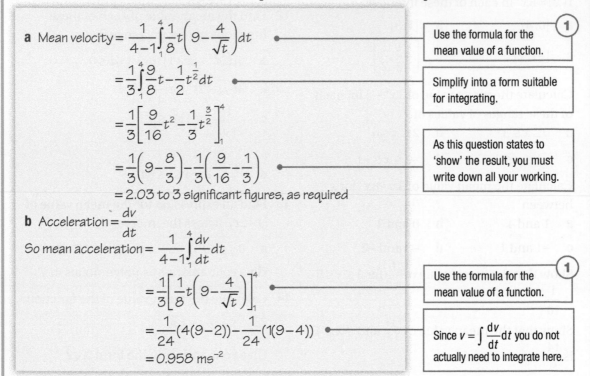

a Mean velocity $= \dfrac{1}{4-1}\displaystyle\int_1^4 \dfrac{1}{8}t\left(9 - \dfrac{4}{\sqrt{t}}\right)dt$

⟶ Use the formula for the mean value of a function. ①

$= \dfrac{1}{3}\displaystyle\int_1^4 \dfrac{9}{8}t - \dfrac{1}{2}t^{\frac{1}{2}}\,dt$

⟶ Simplify into a form suitable for integrating.

$= \dfrac{1}{3}\left[\dfrac{9}{16}t^2 - \dfrac{1}{3}t^{\frac{3}{2}}\right]_1^4$

$= \dfrac{1}{3}\left(9 - \dfrac{8}{3}\right) - \dfrac{1}{3}\left(\dfrac{9}{16} - \dfrac{1}{3}\right)$

⟶ As this question states to 'show' the result, you must write down all your working.

$= 2.03$ to 3 significant figures, as required

b Acceleration $= \dfrac{dv}{dt}$

So mean acceleration $= \dfrac{1}{4-1}\displaystyle\int_1^4 \dfrac{dv}{dt}\,dt$

$= \dfrac{1}{3}\left[\dfrac{1}{8}t\left(9 - \dfrac{4}{\sqrt{t}}\right)\right]_1^4$

⟶ Use the formula for the mean value of a function. ①

$= \dfrac{1}{24}(4(9-2)) - \dfrac{1}{24}(1(9-4))$

⟶ Since $v = \displaystyle\int \dfrac{dv}{dt}\,dt$ you do not actually need to integrate here.

$= 0.958$ ms^{-2}

See Maths Ch7

For a reminder on the link between differentiation and kinematics

Example 4

The mean value of the function $4x + 7$ in the interval $[a, b]$ is 2 and in the interval $[a, 2b]$ is 3. Calculate the values of a and b

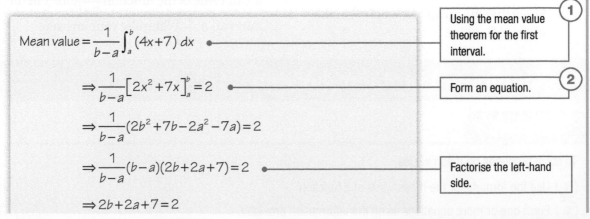

Mean value $= \dfrac{1}{b-a}\displaystyle\int_a^b (4x+7)\,dx$

⟶ Using the mean value theorem for the first interval. ①

$\Rightarrow \dfrac{1}{b-a}\left[2x^2 + 7x\right]_a^b = 2$

⟶ Form an equation. ②

$\Rightarrow \dfrac{1}{b-a}(2b^2 + 7b - 2a^2 - 7a) = 2$

$\Rightarrow \dfrac{1}{b-a}(b-a)(2b + 2a + 7) = 2$

⟶ Factorise the left-hand side.

$\Rightarrow 2b + 2a + 7 = 2$

(*Continued on the next page*)

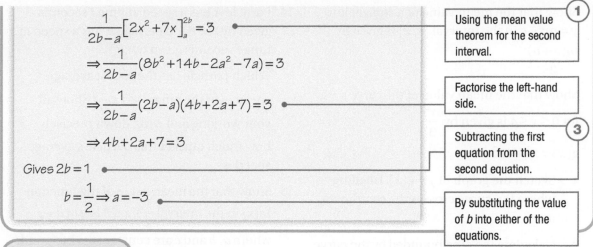

$$\frac{1}{2b-a}\left[2x^2+7x\right]_a^{2b}=3$$

Using the mean value theorem for the second interval. ①

$$\Rightarrow \frac{1}{2b-a}(8b^2+14b-2a^2-7a)=3$$

$$\Rightarrow \frac{1}{2b-a}(2b-a)(4b+2a+7)=3$$

Factorise the left-hand side.

$$\Rightarrow 4b+2a+7=3$$

Subtracting the first equation from the second equation. ③

Gives $2b=1$

$$b=\frac{1}{2} \Rightarrow a=-3$$

By substituting the value of b into either of the equations.

Example 5

Given that $f(x)=\dfrac{2x+1}{\sqrt{x}}$, show that the mean value of $f(x)$ for x in the interval $2 \le x \le 8$ is $\dfrac{31}{9}\sqrt{2}$

$$f(x)=\frac{2x+1}{x^{\frac{1}{2}}}$$

Write in simplified index form.

$$=2x^{\frac{1}{2}}+x^{-\frac{1}{2}}$$

$$\text{Mean value} = \frac{1}{8-2}\int_2^8\left(2x^{\frac{1}{2}}+x^{-\frac{1}{2}}\right)dx$$

Use the definition of the mean value.

$$=\frac{1}{6}\left[\frac{4}{3}x^{\frac{3}{2}}+2x^{\frac{1}{2}}\right]_2^8$$

$$=\frac{1}{6}\left(\frac{64}{3}\sqrt{2}+4\sqrt{2}\right)-\frac{1}{6}\left(\frac{8}{3}\sqrt{2}+2\sqrt{2}\right)$$

$$=\frac{31}{9}\sqrt{2}, \text{ as required}$$

Exercise 8.2B Reasoning and problem-solving

1 The velocity of a particle after t seconds is given by $v=\dfrac{2}{5}t^3-\dfrac{1}{5}t$ (m s⁻¹).

   a Calculate the mean velocity for $1 \le t \le 3$
   b Show that the mean acceleration for $1 \le t \le 3$ is 5 m s⁻².

2 The velocity of a particle after t seconds is given by $v=\dfrac{3t^2-5}{5}$ ms⁻¹.

   a Show that the mean velocity over the first 5 seconds is 4 m s⁻¹.
   b Calculate the mean acceleration over the first 5 seconds.

3 The displacement of a particle after t seconds is given by $s=2t\sqrt{t}$
   Show that the mean acceleration over the range $\dfrac{1}{2} \le t \le 1$ is $6-3\sqrt{2}$ m s⁻².

4 The acceleration of a particle after t seconds is given by $a=\dfrac{t}{10}-\dfrac{10}{t^3}$ for $t>0$

   a Show that the mean acceleration for $1 \le t \le 3$ is $-\dfrac{91}{45}$ m s⁻².
   b Given that after 1 second the particle is travelling at 5 m s⁻¹, calculate the mean velocity for $1 \le t \le 3$

5 Show that the mean value of a straight line $y = mx + c$ in the interval $[a, b]$ is given by $\dfrac{m(a+b)}{2} + c$

6 Show that the mean value of the curve $y = x^2$ for $0 \le x \le a$ is given by $\dfrac{a^2}{3}$

7 $g(x) = x^3 + 4x^2 - 5x$

   a Sketch the graph of $y = g(x)$, labelling where the curve crosses the coordinate axes.

   b Calculate the area bounded by the curve $y = g(x)$ and the x-axis.

   c Find the mean value of $g(x)$ for values of x in each of these intervals.

      i $[-5, 0]$ **ii** $[0, 1]$

8 $g(x) = \sqrt{x}$

The mean value of $g(x)$ in the interval $[1, 4]$ is double the mean value of $g(x)$ in the interval $[0, k]$ for $k > 0$

Find the exact value of k

9. $f(x) = 2x^2 + a$

The mean value of $f(x)$ in the interval $[-3, 3]$ is -2. Calculate the value of a

10 The mean value of the function $f(x) = x - 5$ for $0 \le x \le a$ is -1.5

Calculate the value of a

11 The mean value of the function $g(x) = x^2 + 2x - 1$ for $0 \le x \le X$ is 17

Calculate the possible values of X

12 $h(x) = 2x + 1$

The mean value of $h(x)$ for $a \le x \le b$ is 4 and the mean value of $h(x)$ for $a \le x \le 2b$ is 8

Calculate the values of a and b

13 Is the mean value of the function $x^4 - 2x^3 + 3x - 5$ greater in the range $[1, 3]$ or the range $[-2, -1]$? Show your working and state how much bigger the greater mean value is.

14 Particle A has a speed at time t seconds given by $t^2 + t$ and particle B has a speed at time t seconds given by $\sqrt{t}$

Which particle has the fastest average speed over the range $0 \le t \le \dfrac{1}{2}$? Show all your working and write down precisely how much quicker this particle's average speed is.

15 Show that the mean value of the function $\ln(x)$ in the interval $\dfrac{1}{2} \le x \le 3$ is $a \ln b + c$ where a, b and c are constants to be found.

16 $f(x) = \dfrac{x}{4x^2 - 3}$

Find an expression for the mean value of $f(x)$ in the interval $[k, k+1]$. Give your answer as a single logarithm.

17 The graph of $y = A + \dfrac{1}{x - B}$ is shown.

Calculate the mean value of the function for $5 \le x \le 7$

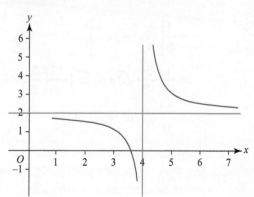

18 Calculate the mean value of each of these functions for the interval given.

   a $\sin x$ for $\left[0, \dfrac{\pi}{3}\right]$ **b** $\cos 2x$ for $\left[\dfrac{\pi}{6}, \dfrac{\pi}{3}\right]$

   c e^{3x} for $[0, 1]$ **d** $2xe^x$ for $[0, 3]$

Volumes of revolution for parametric functions

Fluency and skills

You can find the volume of revolution when a curve is rotated around the x-axis or the y-axis.

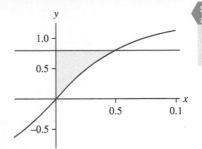

See CP1 Ch3.1 and 3.2 for a reminder on volumes of revolution

Key point

- The volume of the solid formed by rotating the curve $y = f(x)$ between $x = a$ and $x = b$ through a full turn around the x-axis is given by $V = \pi \int_a^b y^2 dx$

- The volume of the solid formed by rotating the curve $x = f(y)$ between $y = a$ and $y = b$ through a full turn around the y-axis is given by $V = \pi \int_a^b x^2 dy$

You can apply this to any function that you are able to integrate.

Example 1

Find the volume of the solid formed when the area enclosed by the curve with equation $y = 2\sin x$ and the x-axis between $x = \dfrac{\pi}{2}$ and $x = \dfrac{3\pi}{2}$ is rotated through 2π radians around the x-axis.

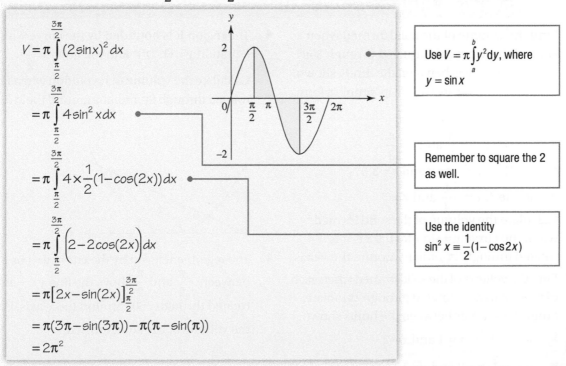

$$V = \pi \int_{\frac{\pi}{2}}^{\frac{3\pi}{2}} (2\sin x)^2 dx$$

Use $V = \pi \int_a^b y^2 dy$, where $y = \sin x$

$$= \pi \int_{\frac{\pi}{2}}^{\frac{3\pi}{2}} 4\sin^2 x \, dx$$

Remember to square the 2 as well.

$$= \pi \int_{\frac{\pi}{2}}^{\frac{3\pi}{2}} 4 \times \frac{1}{2}(1 - \cos(2x)) dx$$

Use the identity $\sin^2 x \equiv \dfrac{1}{2}(1 - \cos 2x)$

$$= \pi \int_{\frac{\pi}{2}}^{\frac{3\pi}{2}} \left(2 - 2\cos(2x)\right) dx$$

$$= \pi \left[2x - \sin(2x)\right]_{\frac{\pi}{2}}^{\frac{3\pi}{2}}$$

$$= \pi(3\pi - \sin(3\pi)) - \pi(\pi - \sin(\pi))$$

$$= 2\pi^2$$

Example 2

The shaded region is bounded by the curve with equation

$y = \dfrac{1}{x^2} - 3$, $x > 0$, the y-axis and the lines $y = 1$ and $y = 5$

Calculate the volume of revolution when the shaded region is rotated through $360°$ around the y-axis.

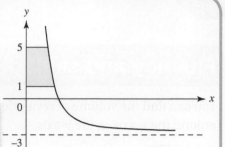

$y = \dfrac{1}{x^2} - 3 \Rightarrow \dfrac{1}{x^2} = y + 3$

$\Rightarrow x^2 = \dfrac{1}{y+3}$

$\Rightarrow x = \dfrac{1}{\sqrt{y+3}}$

First rearrange to make x the subject of the equation.

$V = \pi \displaystyle\int_{1}^{5} \left(\dfrac{1}{\sqrt{y+3}} \right)^2 dy$

Use $V = \pi \displaystyle\int_{a}^{b} x^2 \, dx$, where

$x = \dfrac{1}{\sqrt{y+3}}$

$= \pi \displaystyle\int_{1}^{5} \dfrac{1}{y+3} \, dy$

$= \pi \left[\ln(y+3) \right]_{1}^{5}$

$= \pi(\ln(8) - \ln(4))$

$= \pi \ln(2)$

Use a law of logarithms to simplify.

Exercise 8.3A Fluency and skills

1   Find the volume of the solid formed when each of these curves is rotated through $360°$ around the x-axis between the limits shown. Give each of your answers in its simplest form.

   a $y = \dfrac{1}{2\sqrt{x}}$, $x = 2$ and $x = 5$

   b $y = \sec x$, $x = \dfrac{\pi}{6}$ and $x = \dfrac{\pi}{3}$

   c $y = (2x - 5)^3$, $x = 3$ and $x = 3.5$

   d $y = 3e^{2x}$, $x = -\dfrac{1}{4}$ and $x = 0$

2   Calculate the volume of the solid formed when the curve $y = \sin x$ for $0 \le x \le \pi$ is rotated through 2π radians around the x-axis.

3   Find the volume of the solid formed when each of these curves is rotated through π radians around the x-axis between the limits shown.

   a $y = x\sqrt{\ln x}$, $x = 1$ and $x = 2$

   b $y = xe^{\frac{x}{2}}$, $x = 0$ and $x = 1$

4   The region R is bounded by the curve with equation $y = (1-x)^2$ and the line $y = \dfrac{x}{2}$

Calculate the volume of revolution when R is rotated through 2π radians around the x-axis.

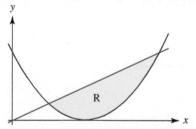

5   The region enclosed by the curve $y = \cos x$ between $-\dfrac{\pi}{2}$ and $\dfrac{\pi}{2}$ above the line $y = \dfrac{1}{2}$ is rotated through $360°$ around the x-axis. Find the volume of the solid formed.

6 The shaded region is bounded by the curve with equation $y = x^3 + 8$ and the coordinate axes.

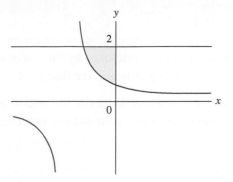

Find the volume of the solid formed when the shaded region is rotated through π radians around

  a The x-axis,

  b The y-axis.

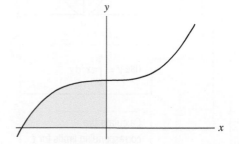

7 The shaded region is bounded by the curve with equation $y = \dfrac{1}{x+2}$, the y-axis and the line $y = 2$

The shaded region is rotated through $360°$ around the y-axis. Find an expression involving $\ln 4$ for the volume of revolution.

8 The curve C has equation $y = 2\ln x$

Find the volume of the solid formed when the section of C between $x = 1$ and $x = 2$ is rotated through $360°$ around the y-axis.

9 The section of the curve with equation $x = \sqrt{\tan y}$ between $y = \dfrac{\pi}{6}$ and $y = \dfrac{\pi}{3}$ is rotated through $180°$ around the y-axis.

Show that the volume of revolution is $A\pi \ln 3$, where A is a constant to be found.

10 The region enclosed between the curve with equation $y = \text{arcsec}\, 2x$, the coordinate axes and the line $y = \dfrac{\pi}{3}$ is rotated through 2π radians around the y-axis. Find the volume of the solid formed.

Reasoning and problem-solving

You can also find volumes when curves are defined by parametric equations.

Suppose a curve defined by $y = f(t)$ and $x = g(t)$ is rotated around the x-axis.

Then the formula for the volume of revolution, $V = \displaystyle\int_a^b \pi y^2 \, dx$, becomes $\displaystyle\int_{t_1}^{t_2} \pi \left[f(t) \right]^2 \dfrac{dx}{dt} \, dt$

See Maths Ch12.3 for a reminder on parametric equations

Strategy

To find the volume of revolution when the equation of the curve is given parametrically

  ① Chose the correct formula for the volume of revolution, $V = \displaystyle\int_a^b \pi y^2 \, dx$ or $V = \displaystyle\int_a^b \pi x^2 \, dy$

  ② Work out the corresponding limits for t

  ③ Substitute for x (or y for rotation around x-axis) in terms of t

  ④ Replace dy by $\dfrac{dy}{dt} \, dt$ (or dx by $\dfrac{dx}{dt} \, dt$ for rotation around x-axis).

  ⑤ Calculate the integral.

Example 3

The shaded region is enclosed by the curve with parametric equations $x = 2t$, $y = 4t^2$, the y-axis and the line $y = 4$

Calculate the area when the shaded region is rotated through $360°$ around the y-axis. Give your answer in terms of π

$V = \int_0^4 \pi x^2 \, dy$

When $y = 0$, $t = 0$ and when $y = 4$, $t = 1$ (or -1)

$V = \int_0^1 \pi (2t)^2 \, dy$

$= \pi \int_0^1 4t^2 (8t) \, dt$

$= 32\pi \int_0^1 t^3 \, dt$

$= 32\pi \left[\dfrac{t^4}{4} \right]_0^1$

$= 32\pi \left(\dfrac{1}{4} - 0 \right)$

$= 8\pi$

You could check this answer by writing x in terms of y

(1) Use $V = \int_a^b \pi x^2 \, dy$

(2) Calculate the corresponding limits for t

(3) Substitute $2t$ in place of x

(4) Since $\dfrac{dy}{dt} = 8t \Rightarrow dy = 8t \, dt$

(5) Give your answer in terms of π

Key point

The volume of revolution for a curve defined by parametric equations $y = f(t)$ and $x = g(t)$ is

- $V = \int_{t_1}^{t_2} \pi y^2 \dfrac{dx}{dt} \, dt$ for rotation around the x-axis

- $V = \int_{t_1}^{t_2} \pi x^2 \dfrac{dy}{dt} \, dt$ for rotation around the y-axis.

Exercise 8.3B Reasoning and problem-solving

1 The curve C has parametric equations $y = 6t^2$, $x = 3t$. Find the volume of revolution when the section of C between $t = 0$ and $t = \dfrac{1}{2}$ is rotated $360°$ around

   a The x-axis,   **b** The y-axis.

2 The curve C has parametric equations $y = 5t$, $x = \dfrac{5}{2} t^2$, $t > 0$. Find the volume of revolution when the section of C between $x = 5$ and $x = 10$ is rotated $180°$ around

   a The x-axis,   **b** The y-axis.

3 The region R is bounded by the curve
$$y = t^3, \quad x = \frac{t^2}{3}, \quad t > 0,$$ the x-axis and the line
$$x = \frac{1}{3}$$
Find the volume of the solid formed when R is rotated 2π radians around

  a The x-axis, **b** The y-axis.

4 **a** Show that $\sin^3\theta = \sin\theta - \sin\theta\cos^2\theta$

The curve C is defined by the parametric equations $x = \cos\theta$, $y = \sin\theta$

  b Find the volume of revolution when the section of C in the range $0 \le \theta \le \dfrac{\pi}{2}$ is rotated 2π radians around the x-axis.

5 The curve C has parametric equations
$$x = 3\cos t, \quad y = \cos(2t)$$

The shaded region is bounded by C and the x-axis.

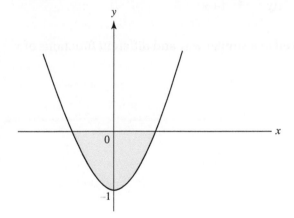

a Use parametric integration to calculate the volume of the solid formed when the shaded region is rotated $180°$ around

  i The x-axis, **ii** The y-axis.

b By first finding the Cartesian equation of C, verify your answers to part **a**.

6 The curve C has parametric equations
$$x = 3\sin\theta, \quad y = \sqrt{3}\cos\theta, \quad 0 \le \theta \le \frac{\pi}{2}$$

The shaded region is bounded by C, the x-axis and the line $y = x$

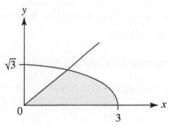

a Show that $\displaystyle\int \cos^3\theta\, d\theta = \sin x - \frac{1}{3}\sin^3 x + c$

b Hence calculate the volume of revolution when the shaded region is rotated 2π radians around the x-axis.

Inverse trigonometric functions

Fluency and skills

To differentiate inverse trigonometric functions you need to use the relationship $\dfrac{dy}{dx} = \dfrac{1}{\frac{dx}{dy}}$

For example, to differentiate $y = \arcsin x$, first rearrange to give $x = \sin y$

Then you know that $\dfrac{dx}{dy} = \cos y$ so $\dfrac{dy}{dx} = \dfrac{1}{\cos y}$

Now rewrite this in terms of x, using the fact that
$\cos y = \pm\sqrt{1-\sin^2 y} = \pm\sqrt{1-x^2}$

The gradient of $y = \arcsin x$ is always positive, so substituting the positive square root gives the result $\dfrac{dy}{dx} = \dfrac{1}{\sqrt{1-x^2}}$

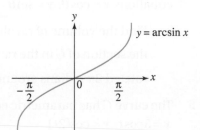

Key point

$$\frac{d(\arcsin x)}{dx} = \frac{1}{\sqrt{1-x^2}}, \quad \frac{d(\arccos x)}{dx} = -\frac{1}{\sqrt{1-x^2}}, \quad \frac{d(\arctan x)}{dx} = \frac{1}{1+x^2}$$

The derivatives of $\arccos x$ and $\arctan x$ can be derived in a similar way and different functions of x can be used.

Example 1

Differentiate $y = \arctan 3x$ with respect to x

$3x = \tan y \Rightarrow x = \dfrac{1}{3}\tan y$

$\dfrac{dx}{dy} = \dfrac{1}{3}\sec^2 y$

$\dfrac{dy}{dx} = \dfrac{1}{\frac{1}{3}\sec^2 y}$ Use $\dfrac{dy}{dx} = \dfrac{1}{\frac{dx}{dy}}$

$= \dfrac{3}{1+\tan^2 y}$ Use $1 + \tan^2 y = \sec^2 y$

$= \dfrac{3}{1+(3x)^2}$ Write in terms of x, using the fact that $3x = \tan y$

$= \dfrac{3}{1+9x^2}$

Using the fundamental theorem of calculus, you can use these derivatives to obtain the following integration results.

$$\int \frac{1}{\sqrt{1-x^2}}\,dx = \arcsin x + c, \quad \int \frac{1}{\sqrt{1-x^2}}\,dx = -\arccos x + c, \quad \int \frac{1}{1+x^2}\,dx = \arctan x + c$$

These results can also be derived using a suitable substitution.

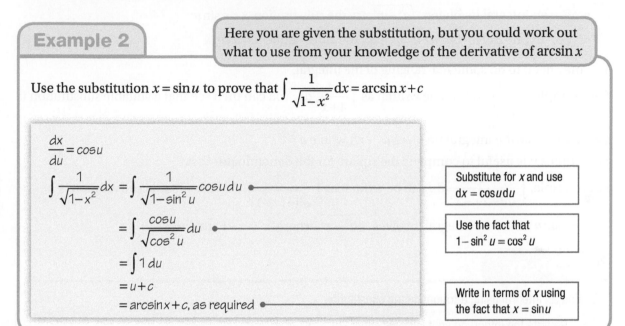

Example 2

Here you are given the substitution, but you could work out what to use from your knowledge of the derivative of $\arcsin x$

Use the substitution $x = \sin u$ to prove that $\int \frac{1}{\sqrt{1-x^2}}\,dx = \arcsin x + c$

$$\frac{dx}{du} = \cos u$$

$$\int \frac{1}{\sqrt{1-x^2}}\,dx = \int \frac{1}{\sqrt{1-\sin^2 u}}\cos u\,du$$

Substitute for x and use $dx = \cos u\,du$

$$= \int \frac{\cos u}{\sqrt{\cos^2 u}}\,du$$

Use the fact that $1-\sin^2 u = \cos^2 u$

$$= \int 1\,du$$

$$= u + c$$

$$= \arcsin x + c, \text{ as required}$$

Write in terms of x using the fact that $x = \sin u$

Exercise 8.4A Fluency and skills

1 Use the technique in the example above to differentiate these expressions with respect to x

  a $\arccos x$ **b** $\arctan x$

  c $\arcsin 2x$ **d** $\arccos 5x$

  e $\arctan(x-1)$ **f** $2\arcsin x$

  g $3\arccos\left(\dfrac{x}{3}\right)$ **h** $3\arcsin(2-x)$

  i $\arccos x^2$ **j** $x\arcsin x$

2 Prove that $\dfrac{d(\operatorname{arcsec} x)}{dx} = \dfrac{1}{x\sqrt{x^2-1}}$

3 Prove that $\dfrac{d(\operatorname{arccosec} x)}{dx} = -\dfrac{1}{x\sqrt{x^2-1}}$

4 Prove that $\dfrac{d(\operatorname{arccot} x)}{dx} = -\dfrac{1}{x^2+1}$

5 Use the derivatives of $\arccos x$, $\arcsin x$ and $\arctan x$ to find $\dfrac{dy}{dx}$ in each case.

  a $y = e^x \arctan x$ **b** $y = \arccos(3x^2-1)$

  c $y = \sin x \arccos 2x$ **d** $y = (\arcsin x)^2$

  e $y = \arcsin(e^x)$ **f** $e^{\arctan 2x}$

6 Use the substitution $x = \cos u$ to show that

$$\int \frac{1}{\sqrt{1-x^2}}\,dx = -\arccos x + c$$

7 Use the substitution $x = \tan u$ to show that

$$\int \frac{1}{1+x^2}\,dx = \arctan x + c$$

8 Use the substitution $x = \dfrac{1}{3}\sin u$ to integrate

$$\int \frac{1}{\sqrt{1-9x^2}}\,dx$$

9 Use the substitution $x = 5\sin u$ to integrate

$$\int \frac{5}{\sqrt{25-x^2}}\,dx$$

10 Use the substitution $x = 3\tan u$ to integrate

$$\int \frac{1}{9+x^2}\,dx$$

Reasoning and problem-solving

You will not always be told what substitution to use so you will need to choose a suitable one. From the questions in Exercise 8.4A, you can see that these substitutions are often suitable.

You may need to do some rearranging of the integral.

For example, $\int \dfrac{1}{1 + \dfrac{x^2}{4}} dx$ can be written as $\int \dfrac{4}{4 + x^2} dx$. You can then see that a suitable substitution is

$x = 2\tan u$ since the integral involves $a^2 + x^2$, where $a = 2$

Sometimes it is useful to complete the square for the denominator first.

For example, $\int \dfrac{1}{\sqrt{8 + 2x - x^2}} dx$ can be written as $\int \dfrac{1}{\sqrt{9 - (x-1)^2}} dx$, so a suitable substitution is

$x - 1 = 3\sin u$ since the integral involves $\sqrt{a^2 - x^2}$, where $`x` = x - 1$ and $a = 3$

Strategy

To find an integral using a trigonometric substitution

1. Rewrite the integrand so it involves either $\sqrt{a^2 - x^2}$ or $a^2 + x^2$

2. Choose the correct substitution, either $x = a\sin u$ or $x = a\tan u$

3. Use integration by substitution to work out the integral and then write the answer in terms of x

Example 3

Work out each of these integrals.

a $\displaystyle\int \dfrac{3}{\sqrt{1 - 9x^2}} dx$ **b** $\displaystyle\int \dfrac{9}{x^2 + 6x + 25} dx$

a $\displaystyle\int \dfrac{3}{\sqrt{1 - 9x^2}} dx = \int \dfrac{1}{\sqrt{\dfrac{1}{9} - x^2}} dx$ — Divide the numerator and denominator by 3 ①

Let $x = \dfrac{1}{3}\sin u$

Then $\dfrac{dx}{du} = \dfrac{1}{3}\cos u$ — $a^2 = \dfrac{1}{9}$, so $a = \dfrac{1}{3}$ ②

$\displaystyle\int \dfrac{1}{\sqrt{\dfrac{1}{9} - \left(\dfrac{1}{3}\sin u\right)^2}} \dfrac{1}{3}\cos u \, du = \dfrac{1}{3}\int \dfrac{\cos u}{\sqrt{\dfrac{1}{9} - \dfrac{1}{9}\sin^2 u}} du$

$= \dfrac{1}{3}\int \dfrac{\cos u}{\sqrt{\dfrac{1}{9}\cos^2 u}} du$ — Use $1 - \sin^2 u = \cos^2 u$

(*Continued on the next page*)

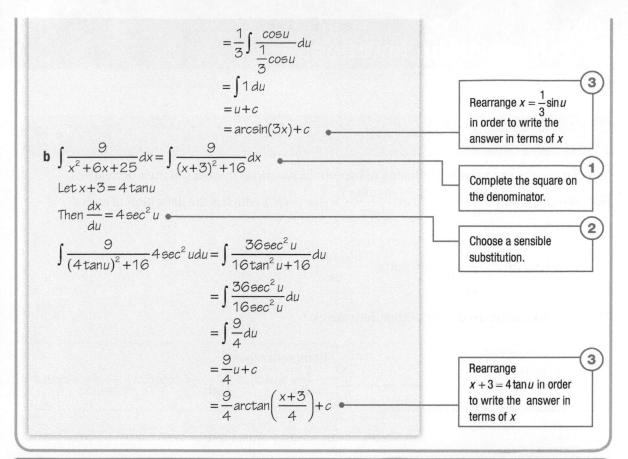

$$= \frac{1}{3}\int \frac{cosu}{\frac{1}{3}cosu}du$$

$$= \int 1\,du$$

$$= u + c$$

$$= arcsin(3x) + c$$

3 Rearrange $x = \frac{1}{3}\sin u$ in order to write the answer in terms of x

b $\int \frac{9}{x^2+6x+25}dx = \int \frac{9}{(x+3)^2+16}dx$

1 Complete the square on the denominator.

Let $x + 3 = 4\tan u$

Then $\frac{dx}{du} = 4\sec^2 u$

2 Choose a sensible substitution.

$$\int \frac{9}{(4\tan u)^2+16}4\sec^2 u\,du = \int \frac{36\sec^2 u}{16\tan^2 u+16}du$$

$$= \int \frac{36\sec^2 u}{16\sec^2 u}du$$

$$= \int \frac{9}{4}du$$

$$= \frac{9}{4}u + c$$

$$= \frac{9}{4}arctan\left(\frac{x+3}{4}\right) + c$$

3 Rearrange $x + 3 = 4\tan u$ in order to write the answer in terms of x

Exercise 8.4B Reasoning and problem-solving

1 Work out each of these integrals by choosing a suitable substitution.

a $\int \frac{1}{\sqrt{9-x^2}}dx$

b $\int \frac{1}{\sqrt{100-x^2}}dx$

c $-\int \frac{2}{\sqrt{36-x^2}}dx$

d $\int \frac{-4}{\sqrt{8-x^2}}dx$

e $\int \frac{1}{25+x^2}dx$

f $\int \frac{1}{49+x^2}dx$

g $\int \frac{-3}{x^2+2}dx$

h $\int \frac{6}{x^2+6}dx$

i $\int \sqrt{1-x^2}dx$

j $\int \sqrt{16-x^2}dx$

2 Work out each of these integrals by first rearranging the integrand and then choosing a suitable substitution.

a $\int \frac{1}{1+\frac{x^2}{64}}dx$

b $\int \frac{25}{1+25x^2}dx$

c $\int \frac{2}{\frac{x^2}{2}+4}dx$

d $\int \frac{4}{2x^2+6}dx$

e $\int \frac{1}{\sqrt{1-\frac{x^2}{9}}}dx$

f $\int \frac{1}{\sqrt{16-4x^2}}dx$

g $-\int \frac{1}{\sqrt{1-\frac{x^2}{36}}}dx$

h $\int \frac{3}{\sqrt{18-9x^2}}dx$

i $\int \sqrt{1-\frac{x^2}{81}}dx$

j $\int \sqrt{36-9x^2}dx$

3 Work out each of these integrals.

a $\int \frac{1}{x^2+2x+5}dx$

b $\int \frac{1}{x^2-6x+10}dx$

c $\int \frac{1}{x^2-14x+53}dx$

d $\int \frac{2}{x^2+12x+38}dx$

e $\int \frac{1}{\sqrt{20-8x-x^2}}dx$

f $\int \frac{1}{\sqrt{1+2x-x^2}}dx$

g $-\int \frac{2}{\sqrt{8x-x^2}}dx$

h $\int \frac{1}{\sqrt{4-4x-x^2}}dx$

i $\int \sqrt{12+4x-x^2}dx$

j $\int \sqrt{16-6x-x^2}$

4 Calculate the exact value of the integral

$$\int_{-\frac{1}{4}}^{\frac{1}{2}} \frac{1}{\sqrt{8+4x-4x^2}}dx$$

Hyperbolic functions

Fluency and skills

See Maths
Chapter 15.5
for a
reminder of
the chain
rule

You can use the exponential definition of hyperbolic functions to work out their derivatives.

For example, $\sinh x = \dfrac{1}{2}(e^x - e^{-x})$ so $\dfrac{d(\sinh x)}{dx} = \dfrac{1}{2}(e^x + e^{-x})$, which is the definition of $\cosh x$

See Chapter
7.3
for a
reminder
about
Hyperbolic
functions

Key point

$$\frac{d(\sinh x)}{dx} = \cosh x, \quad \frac{d(\cosh x)}{dx} = \sinh x, \quad \frac{d(\tanh x)}{dx} = \text{sech}^2 x$$

These results can be used to find other derivatives.

Example 1

Remember that

$$\frac{1}{\cosh x} = \text{sech}\, x, \quad \frac{1}{\sinh x} = \text{cosech}\, x, \quad \frac{1}{\tanh x} = \coth x$$

Differentiate $\text{sech}\, x$ with respect to x

$$\text{sech}\, x = (\cosh x)^{-1}$$

$$\frac{d(\text{sech}\, x)}{dx} = -(\cosh x)^{-2}(\sinh x)$$ •————— Use the chain rule.

$$= -\frac{\sinh x}{\cosh^2 x}$$

$$= -\tanh x\, \text{sech}\, x$$ •————— Notice the difference in sign compared with the derivative of $\sec x$ which is $\tan x \sec x$

You can also use these derivatives:

Key point

$$\frac{d(\text{sech}\, x)}{dx} = -\text{sech}\, x \tanh x, \quad \frac{d(\text{cosech}\, x)}{dx} = -\text{cosech}\, x \coth x, \quad \frac{d(\coth x)}{dx} = -\text{cosech}^2 x$$

You can integrate $\sinh x$ and $\cosh x$ using the fundamental theorem of calculus:

Key point

$$\int \sinh x = \cosh x + c, \quad \int \cosh x = \sinh x + c$$

To find the integral of $\tanh x$, use the fact that $\tanh x = \dfrac{\sinh x}{\cosh x}$. Then

$$\int \tanh x\, dx = \int \frac{\sinh x}{\cosh x} dx$$

$$= \ln \cosh x + c \quad \text{since } \int \frac{f'(x)}{f(x)} dx = \ln f(x) + c''$$

Example 2

Use the identity $\cosh(2x) \equiv 1 + 2\sinh^2 x$

Work out the exact value of $\displaystyle\int_0^{\ln 3} \sinh^2 x\,dx$

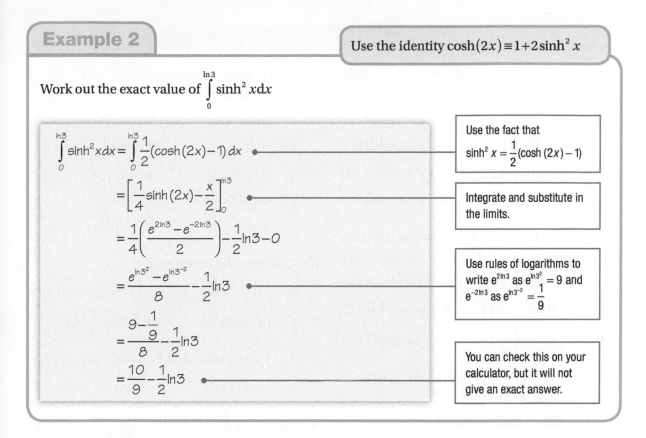

$$\int_0^{\ln 3} \sinh^2 x\,dx = \int_0^{\ln 3} \frac{1}{2}(\cosh(2x) - 1)\,dx$$

Use the fact that $\sinh^2 x = \dfrac{1}{2}(\cosh(2x) - 1)$

$$= \left[\frac{1}{4}\sinh(2x) - \frac{x}{2}\right]_0^{\ln 3}$$

Integrate and substitute in the limits.

$$= \frac{1}{4}\left(\frac{e^{2\ln 3} - e^{-2\ln 3}}{2}\right) - \frac{1}{2}\ln 3 - 0$$

$$= \frac{e^{\ln 3^2} - e^{\ln 3^{-2}}}{8} - \frac{1}{2}\ln 3$$

Use rules of logarithms to write $e^{2\ln 3}$ as $e^{\ln 3^2} = 9$ and $e^{-2\ln 3}$ as $e^{\ln 3^{-2}} = \dfrac{1}{9}$

$$= \frac{9 - \frac{1}{9}}{8} - \frac{1}{2}\ln 3$$

$$= \frac{10}{9} - \frac{1}{2}\ln 3$$

You can check this on your calculator, but it will not give an exact answer.

You can differentiate the inverse hyperbolic functions using a similar method to that used for inverse trigonometric functions in Section 8.4

Example 3

Use the identity $\cosh^2 x - \sinh^2 x \equiv 1$

Differentiate $\operatorname{arcosh} x$ with respect to x

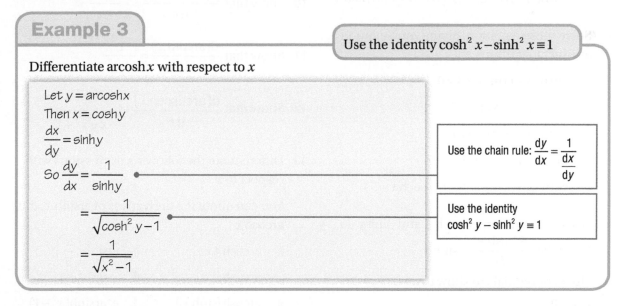

Let $y = \operatorname{arcosh} x$

Then $x = \cosh y$

$$\frac{dx}{dy} = \sinh y$$

So $\dfrac{dy}{dx} = \dfrac{1}{\sinh y}$

Use the chain rule: $\dfrac{dy}{dx} = \dfrac{1}{\dfrac{dx}{dy}}$

$$= \frac{1}{\sqrt{\cosh^2 y - 1}}$$

Use the identity $\cosh^2 y - \sinh^2 y \equiv 1$

$$= \frac{1}{\sqrt{x^2 - 1}}$$

The derivative of $\operatorname{arsinh} x$ can be found in a similar way.

Key point

$$\frac{d(\operatorname{arcosh} x)}{dx} = \frac{1}{\sqrt{x^2 - 1}}, \quad \frac{d(\operatorname{arsinh} x)}{dx} = \frac{1}{\sqrt{x^2 + 1}}$$

1 Sketch each of these and state their domain and range.

   a $y = \text{cosech}(x-1)$ **b** $y = 3\text{sech}(x)$

   c $y = 2\coth(x)$ **d** $y = \text{cosech}(-x)$

   e $y = \text{sech}(x+2)$ **f** $y = \coth(x+1)$

   g $y = 2+\text{cosech}(x)$ **h** $y = 1+\coth(x)$

2 Sketch each of these and state their domain and range.

   a $y = \text{arsinh}(x+2)$ **b** $y = 3\,\text{arcosh}(x)$

   c $y = \text{artanh}(x+1)$ **d** $y = \text{arcosh}(2x)$

   e $y = \text{artanh}\left(\dfrac{x}{4}\right)$ **f** $y = \text{arcosh}(x+1)$

3 Use the definitions of $\cosh x$ and $\sinh x$ to prove these identities.

   a $\cosh 2x \equiv \cosh^2 x + \sinh^2 x$

   b $\sinh(A+B) \equiv \sinh A \cosh B + \sinh B \cosh A$

   c $\cosh(A-B) \equiv \cosh A \cosh B - \sinh B \sinh A$

4 Solve these equation, giving your solutions as exact logarithms.

   a $\cosh^2 x - \cosh x - 6 = 0$

   b $\sinh^2 x - 3\cosh x = 3$

   c $\tanh^2 x - \text{sech}^2 x = 0$

   d $2\text{cosech}^2 x - 3\coth x = 0$

   e $\coth^2 x + 2\text{cosech}^2 x + \text{cosech} x = 5$

5 Use the definition of $\cosh x$ and $\sinh x$ to show that $\dfrac{d(\cosh x)}{dx} = \sinh x$

6 Use exponentials to show that $\dfrac{d(\tanh x)}{dx} = \text{sech}^2 x$

7 Differentiate these expressions with respect to x

   a $\text{cosech} x$ **b** $\coth x$

   c $\sinh 2x$ **d** $\cosh\left(\dfrac{x}{3}\right)$

   e $\tanh(x-2)$ **f** $\sinh x^2$

   g $\cosh(2x-3)$ **h** $\sinh^2 x$

   i $x\tanh x$ **j** $x\sqrt{\text{sech} x}$

8 Calculate these integrals.

   a $\displaystyle\int \text{sech}^2 x \, dx$ **b** $\displaystyle\int \coth x \,\text{cosech} x \, dx$

   c $\displaystyle\int \cosh 2x \, dx$ **d** $\displaystyle\int \sinh\left(\dfrac{x}{3}\right) dx$

   e $\displaystyle\int \cosh x \sinh^2 x \, dx$ **f** $\displaystyle\int \cosh^2 x \, dx$

   g $\displaystyle\int \coth 4x \, dx$ **h** $\displaystyle\int \text{cosech}^2 5x \, dx$

9 Show that $\dfrac{d(\text{arsinh} x)}{dx} = \dfrac{1}{\sqrt{x^2+1}}$

10 Show that $\dfrac{d(\text{artanh} 2x)}{dx} = \dfrac{2}{1-4x^2}$

11 Show that $\dfrac{d(\text{arcoth} x)}{dx} = \dfrac{1}{1-x^2}$

12 Show that $\dfrac{d(\text{arcosech} x)}{dx} = \dfrac{1}{-x\sqrt{x^2+1}}$

13 Differentiate the following expressions with respect to x

You can quote the derivatives of $\text{arsinh} x$ and $\text{arcosh} x$

   a $\text{arcosh} 4x$ **b** $\text{arsinh}\left(\dfrac{x}{5}\right)$

   c $\text{arcosh}(x^2)$ **d** $\text{arsinh}(e^x)$

   e $\text{arcosh}(\sinh x)$ **f** $e^x \text{arsinh}(x^2-1)$

Reasoning and problem-solving

You can use the derivatives of $\operatorname{arsinh} x$ and $\operatorname{arcosh} x$ to choose a suitable substitution to use when integrating.

You may need to rearrange the integral first, using the same techniques that you saw in Section 8.4

Strategy

To find an integral using a hyperbolic substitution

(1) Rewrite the integrand so that it involves either $\sqrt{x^2 + a^2}$ or $\sqrt{x^2 - a^2}$

(2) Choose the correct substitution, either $x = a \sinh u$ or $x = a \cosh u$

(3) Use integration by substitution to work out the integral and then write the answer in terms of x

Example 4

Work out each of these integrals.

a $\displaystyle\int \frac{6}{\sqrt{36 + 4x^2}}\, dx$ **b** $\displaystyle\int \frac{1}{\sqrt{x^2 - 8x - 20}}\, dx$

a $\displaystyle\int \frac{6}{\sqrt{36 + 4x^2}}\, dx = \int \frac{3}{\sqrt{9 + x^2}}\, dx$

> **(1)** Divide the numerator and denominator by 2

Let $x = 3 \sinh u$

Then $\dfrac{dx}{du} = 3 \cosh u$

> **(2)** $a^2 = 9$ so $a = 3$

$\displaystyle\int \frac{3}{\sqrt{9 + (3\sinh u)^2}}\, 3\cosh u\, du = \int \frac{9\cosh u}{\sqrt{9 + 9\sinh^2 u}}\, du$

$\displaystyle = \int \frac{9\cosh u}{\sqrt{9\cosh^2 u}}\, du$

> Use $1 + \sinh^2 u = \cosh^2 u$

$\displaystyle = \int \frac{9\cosh u}{3\cosh u}\, du$

$\displaystyle = \int 3\, du$

$= 3u + c$

> **(3)** Rearrange $x = 3\sinh u$ in order to write the answer in terms of x

$= 3 \operatorname{arsinh}\left(\dfrac{x}{3}\right) + c$

(Continued on the next page)

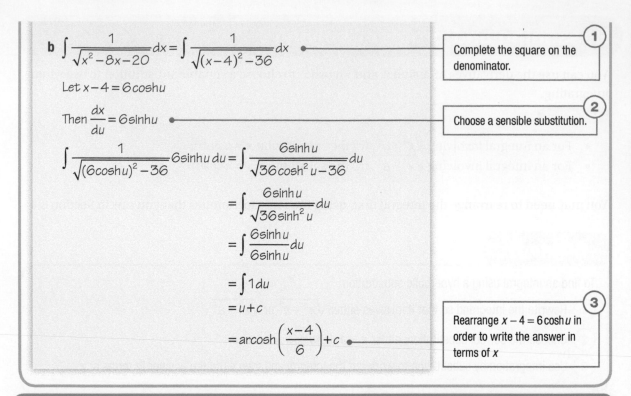

b $\displaystyle\int \frac{1}{\sqrt{x^2-8x-20}}\,dx = \int \frac{1}{\sqrt{(x-4)^2-36}}\,dx$

> **1** Complete the square on the denominator.

Let $x-4 = 6\cosh u$

Then $\dfrac{dx}{du} = 6\sinh u$

> **2** Choose a sensible substitution.

$\displaystyle\int \frac{1}{\sqrt{(6\cosh u)^2-36}}\, 6\sinh u\,du = \int \frac{6\sinh u}{\sqrt{36\cosh^2 u - 36}}\,du$

$\displaystyle = \int \frac{6\sinh u}{\sqrt{36\sinh^2 u}}\,du$

$\displaystyle = \int \frac{6\sinh u}{6\sinh u}\,du$

$\displaystyle = \int 1\,du$

$= u + c$

$= \operatorname{arcosh}\left(\dfrac{x-4}{6}\right) + c$

> **3** Rearrange $x-4 = 6\cosh u$ in order to write the answer in terms of x

Exercise 8.5B Reasoning and problem-solving

1 Work out each of these integrals.

a $\displaystyle\int \frac{1}{\sqrt{x^2+49}}\,dx$

b $\displaystyle\int \frac{1}{\sqrt{x^2-81}}\,dx$

c $\displaystyle\int \frac{2}{\sqrt{64+4x^2}}\,dx$

d $\displaystyle\int \frac{-1}{\sqrt{\dfrac{x^2}{9}-1}}\,dx$

e $\displaystyle\int \frac{1}{\sqrt{x^2+6x+25}}\,dx$

f $\displaystyle\int \frac{1}{\sqrt{x^2-10x+26}}\,dx$

g $\displaystyle\int \frac{1}{\sqrt{x^2+14x+24}}\,dx$

h $\displaystyle\int \frac{3}{\sqrt{x^2-24x+44}}\,dx$

2 Evaluate these integrals, giving your answers to 3 significant figures.

a $\displaystyle\int_0^1 \frac{1}{\sqrt{6+x^2}}\,dx$

b $\displaystyle\int_6^{12} \frac{\sqrt{27}}{\sqrt{\dfrac{x^2}{12}-3}}\,dx$

3 Calculate the exact values of these integrals.

a $\displaystyle\int_0^3 \sqrt{\frac{3}{27+3x^2}}\,dx$

b $\displaystyle\int_4^5 \frac{1}{\sqrt{3x^2-48}}\,dx$

4 Integrate each of these expressions with respect to x

a $\dfrac{x+1}{\sqrt{16+9x^2}}$

b $\dfrac{3-x}{\sqrt{\dfrac{x^2}{4}-3}}$

c $\dfrac{x+4}{\sqrt{16-2x^2}}$

d $\dfrac{2x-5}{\dfrac{x^2}{7}+7}$

5 Use integration by parts to show that

$\displaystyle\int \operatorname{arsinh} x\,dx = x\operatorname{arsinh} x - \sqrt{x^2+1} + c$

6 Work out each of these integrals.

a $\displaystyle\int \operatorname{arcosh} x\,dx$

b $\displaystyle\int \operatorname{arcoth} x\,dx$

7 One of these improper integrals exists for $a>0$ and the other does not.

A: $\displaystyle\int_a^{2a} \frac{1}{\sqrt{x^2-a^2}}\,dx$

B: $\displaystyle\int_{2a}^{\infty} \frac{1}{\sqrt{x^2-a^2}}\,dx$

Explain why one doesn't exist and find the exact value of the integral that does exist.

Partial fractions

Fluency and skills

Rational functions which have linear factors in the denominator can be split into partial fractions to help you to integrate the function.

If the degree of the numerator is the same as or greater than the degree of the denominator, then this is an **improper** fraction.

See Maths Ch12.5
For a reminder on partial fractions

> **Key point**
>
> A fraction $\dfrac{f(x)}{g(x)}$ where degree of $f(x) \geq$ degree of $g(x)$ is called an **improper** fraction and can be written in the form $P(x) + \dfrac{Q(x)}{g(x)}$

It may then be possible to write $\dfrac{Q(x)}{g(x)}$ in partial fractions.

You can use long division to find $P(x)$ and $Q(x)$. However, if you prefer, you can consider the degree of the numerator and the denominator to decide on a general form for the quotient.

For example, if you have $\dfrac{f(x)}{g(x)}$ where $f(x)$ is a quartic (polynomial of degree 4) and $g(x)$ is a quadratic, then you know that the quotient is of the form $Ax^2 + Bx + C$

> **Example 1**
>
> **a** Write $\dfrac{30x^3 - 13x^2 + 6x + 6}{15x^2 + x - 6}$ in partial fractions. **b** Hence work out $\displaystyle\int \dfrac{30x^3 - 13x^2 + 6x + 6}{15x^2 + x - 6}\,\mathrm{d}x$
>
> **a** Dividing a cubic by a quadratic will give a linear quotient.
>
> $\dfrac{30x^3 - 13x^2 + 6x + 6}{15x^2 + x - 6} = Ax + B + \dfrac{C}{3x + 2} + \dfrac{D}{5x - 3}$
>
> > Alternatively, you could use long division.
>
> $30x^3 - 13x^2 + 6x + 6 = (Ax + B)(15x^2 + x - 6) + C(5x - 3) + D(3x + 2)$
>
> > Multiply both sides by $(x - 2)(3x - 5)$
>
> Equating coefficients
>
> $x^3: 30 = 15A \Rightarrow A = 2$
>
> $x^2: -13 = A + 15B \Rightarrow B = -1$
>
> $x: 6 = -6A + B + 5C + 3D \Rightarrow 5C + 3D = 19$
>
> $1: 6 = -6B - 3C + 2D \Rightarrow 3C - 2D = 0$
>
> Solve simultaneously to give $C = 2, D = 3$
>
> $\dfrac{30x^3 - 13x^2 + 6x + 6}{15x^2 + x - 6} = 2x - 1 + \dfrac{2}{3x + 2} + \dfrac{3}{5x - 3}$
>
> **b** $\displaystyle\int \dfrac{30x^3 - 13x^2 + 6x + 6}{15x^2 + x - 6}\,\mathrm{d}x = \int\left(2x - 1 + \dfrac{2}{3x + 2} + \dfrac{3}{5x - 3}\right)\mathrm{d}x$
>
> > Use the answer from part **a**.
>
> $= x^2 - x + \dfrac{2}{3}\ln(3x + 2) + \dfrac{3}{5}\ln(5x - 3) + c$

You can also find partial fractions when the denominator includes a quadratic factor which cannot be factorised, for example x^2+5

In these cases, you should use a linear expression $Ax+B$ as the numerator.

Key point

$\dfrac{f(x)}{(\alpha x^2+\beta)(\gamma x+\delta)}$ can be split into partial fractions of the form $\dfrac{Ax+B}{\alpha x^2+\beta}+\dfrac{C}{\gamma x+\delta}$

Example 2

In some questions you may have to factorise the denominator yourself.

Work out $\displaystyle\int \frac{2x+12}{(x+1)(x^2+9)}\,dx$

$\dfrac{2x+12}{(x+1)(x^2+9)}=\dfrac{A}{x+1}+\dfrac{Bx+C}{x^2+9}$

$2x+12=A(x^2+9)+(Bx+C)(x+1)$ — Multiply both sides by $(x+1)(x^2+9)$

$=Ax^2+9A+Bx^2+Bx+Cx+C$

Equating coefficients — Or use an alternative method to find the values of A, B and C

$x^2: 0=A+B$ so $A=-B$

$x: 2=B+C$ (equation 1)

$1: 12=9A+C$ (equation 2)

Subtract equation 1 from equation 2 to give $9A-B=10$ — Solve the three equations simultaneously.

$A=-B\Rightarrow-9B-B=10\Rightarrow B=-1$

So $A=1$ and $C=3$

$\displaystyle\int\frac{2x+12}{(x+1)(x^2+9)}dx=\int\left(\frac{1}{x+1}+\frac{3-x}{x^2+9}\right)dx$

$\phantom{\int\frac{2x+12}{(x+1)(x^2+9)}dx}=\displaystyle\int\left(\frac{1}{x+1}+\frac{3}{x^2+9}-\frac{x}{x^2+9}\right)dx$ — Split the numerator of the second fraction.

$\phantom{\int\frac{2x+12}{(x+1)(x^2+9)}dx}=\ln(x+1)+\arctan\left(\dfrac{x}{3}\right)-\dfrac{1}{2}\ln\left(x^2+9\right)+c$ — Use the substitution $x=3\tan u$ to integrate the second fraction.

which can be written as $\ln\left(\dfrac{x+1}{\sqrt{x^2+9}}\right)+\arctan\left(\dfrac{x}{3}\right)+c$

Exercise 8.6A Fluency and skills

1   Work out each of these integrals by first expressing the integrand in partial fractions.

a $\displaystyle\int\frac{5x^2+14x-42}{(x-2)(x+4)}dx$

b $\displaystyle\int\frac{2+11x-3x^2}{x^2-4x-5}dx$

c $\displaystyle\int\frac{5x^3+x^2-46x-24}{x^2-9}dx$

d $\displaystyle\int\frac{8x^3+92x^2+243x+72}{x^3+11x^2+24x}dx$

e $\displaystyle\int\frac{x^4-7x^3-20x^2+14x+46}{(x-9)(x+1)}dx$

f $\displaystyle\int\frac{9x^4}{(x+1)(x-2)^2}dx$

g $\displaystyle\int\frac{12x^4+7x^3-22x-9}{12-7x-12x^2}dx$

2 Work out each of these integrals by first expressing the integrand in partial fractions.

a $\int \dfrac{9-7x}{(x+3)(x^2+1)}\,\mathrm{d}x$ **b** $\int \dfrac{10+8x}{(x-3)(x^2+25)}\,\mathrm{d}x$

c $\int \dfrac{x+66}{(x^2+36)(6-x)}\,\mathrm{d}x$ **d** $\int \dfrac{7x-1}{(1-x)(x^2+2)}\,\mathrm{d}x$

e $\int \dfrac{36x+6}{(x+1)(x-2)(x^2+9)}\,\mathrm{d}x$

f $\int \dfrac{79-28x}{(x-1)^2(x^2+16)}\,\mathrm{d}x$

3 Work out each of these integrals by first expressing the integrand in partial fractions.

a $\int \dfrac{3x+49}{x^3+49x}\,\mathrm{d}x$ **b** $\int \dfrac{x-192}{x^3+64x}\,\mathrm{d}x$

c $\int \dfrac{10x+9}{3x^3-x^2+12x-4}\,\mathrm{d}x$

4 a Given that x^2+5 is a factor of $x^4-6x^3+14x^2-30x+45$, write the fraction $\dfrac{20x-46}{x^4-6x^3+14x^2-30x+45}$ in partial fractions.

b Hence work out $\int \dfrac{20x-46}{x^4-6x^3+14x^2-30x+45}\,\mathrm{d}x$

Reasoning and problem-solving

You can also use partial fractions when evaluating improper integrals. The integral will only exist if all parts of it exist.

Strategy

To decide if an improper integral exists when the integrand is a rational function

(1) Factorise the denominator as far as possible.

(2) Split into partial fractions and establish any points of discontinuity for each fraction.

(3) Integrate and consider the behaviour of the integral as the variable tends towards the original limit.

Example 3

a Show algebraically that the improper integral $\displaystyle\int_{\frac{1}{2}}^{1} \dfrac{3x+7}{2x^3-x^2+8x-4}\,\mathrm{d}x$ does not exist.

b Explain whether the improper integral $\displaystyle\int_{1}^{\infty} \dfrac{3x+7}{2x^3-x^2+8x-4}\,\mathrm{d}x$ exists.

a $\dfrac{3x+7}{2x^3-x^2+8x-4} = \dfrac{3x+7}{(2x-1)(x^2+4)}$

This can be written in partial fractions: $\dfrac{3x+7}{(2x-1)(x^2+4)} = \dfrac{A}{2x-1}+\dfrac{Bx+C}{x^2+4}$

x^2+4 cannot be factorised further so will be a quadratic factor.

$3x+7 = A(x^2+4)+(Bx+C)(2x-1)$

Consider coefficients

$x^2: 0 = A+2B$

$x: 3 = -B+2C$

$1: 7 = 4A-C$

(Continued on the next page)

Solve simultaneously to give $A=2, B=-1, C=1$

So $\dfrac{3x+7}{(2x-1)(x^2+4)} = \dfrac{2}{2x-1} - \dfrac{x}{x^2+4} + \dfrac{1}{x^2+4}$

②

Consider any points of discontinuity for either fraction within the limits given.

$\dfrac{2}{2x-1}$ has a point of discontinuity at $x = \dfrac{1}{2}$ so replace limit with a variable:

$\displaystyle\int_a^1 \left(\dfrac{2}{2x-1} - \dfrac{x}{x^2+4} + \dfrac{1}{x^2+4} \right) dx = \left[\ln(2x-1) - \dfrac{1}{2}\ln(x^2+4) + \dfrac{1}{2}\arctan\left(\dfrac{x}{2}\right) \right]_a^1$

$\ln(2x-1) \to \infty$ as $a \to \dfrac{1}{2}$ since $2a-1 \to 0$

Therefore, the integral does not exist.

b Replace the top limit with a

$\left[\ln(2x-1) - \dfrac{1}{2}\ln(x^2+4) + \dfrac{1}{2}\arctan\left(\dfrac{x}{2}\right) \right]_1^a = \left[\dfrac{1}{2}\ln\left(\dfrac{(2x-1)^2}{x^2+4}\right) + \dfrac{1}{2}\arctan\left(\dfrac{x}{2}\right) \right]_1^a$

$= \left(\dfrac{1}{2}\ln\left(\dfrac{(2a-1)^2}{a^2+4}\right) + \dfrac{1}{2}\arctan\left(\dfrac{a}{2}\right) \right)$

$- \left(\dfrac{1}{2}\ln\left(\dfrac{1}{5}\right) + \dfrac{1}{2}\arctan\left(\dfrac{1}{2}\right) \right)$

$\dfrac{(2a-1)^2}{a^2+4} = \dfrac{4a^2 - 4a + 1}{a^2+4}$

$= \dfrac{4 - \dfrac{4}{a} + \dfrac{1}{a^2}}{1 + \dfrac{4}{a^2}} \to 4$ as $a \to \infty$ since $\dfrac{4}{a}, \dfrac{1}{a^2} \to 0$

So $\dfrac{1}{2}\ln\left(\dfrac{(2a-1)^2}{a^2+4}\right) \to \dfrac{1}{2}\ln 4$

③

So this part of the integral does exist.

However, $\arctan\left(\dfrac{a}{2}\right) \to \infty$ as $a \to \infty$

Therefore the integral does not exist.

Exercise 8.6B Reasoning and problem-solving

1 Find the exact value of these improper integrals.

a $\displaystyle\int_0^\infty \dfrac{1}{x^2+7x+12} dx$ **b** $\displaystyle\int_1^\infty \dfrac{10}{3-8x-3x^2} dx$

c $\displaystyle\int_0^\infty \dfrac{7x+26}{x^3+9x^2+24x+20} dx$

d $\displaystyle\int_{-\infty}^0 \dfrac{9}{8x^2-22x+5} dx$

2 One of these improper integrals converges and the other does not. Explain which does not exist and find the value of the integral that does exist.

A: $\displaystyle\int_1^\infty \dfrac{2x+12}{2x^3-x^2+6x-3} dx$ **B:** $\displaystyle\int_0^1 \dfrac{2x+12}{2x^3-x^2+6x-3} dx$

3 a Express $\dfrac{2x+a}{x^3+ax}$ in partial fractions.

b Hence show that the improper integral $\displaystyle\int_0^a \dfrac{2x+a}{x^3+ax} dx$ does not exist.

4 For what range of values of a (if any) does each improper integral exist? Explain your answers.

a $\displaystyle\int_{-\infty}^0 \dfrac{3-a}{(x+a)(x+3)} dx$

b $\displaystyle\int_0^\infty \dfrac{3-a}{(x+a)(x+3)} dx$

8.7 Polar graphs and areas

Fluency and skills

You can use integration to calculate the area enclosed by a polar curve between two half-lines.

Split the area into n slices, each with an angle of $\delta\theta$

The area of each slice can be approximated as the area of a sector of a circle. The formula for the area of a sector of angle θ from a circle of radius r is $A = \dfrac{1}{2}r^2\theta$

Therefore, the area of the slice shown is $\delta A \approx \dfrac{1}{2}\left[f(\theta_i)\right]^2 \delta\theta$

Taking the limit as $n \to \infty$ of the sum of these areas gives the integral $\int_a^b \dfrac{1}{2}\left[f(\theta)\right]^2 d\theta$, which can be written as $\dfrac{1}{2}\int_a^b r^2 d\theta$

> **Remember that θ must be measured in radians.**

Key point

> The area of a sector of a polar curve $r = f(\theta)$ between the half-lines $\theta = a$ and $\theta = b$ is given by $A = \dfrac{1}{2}\int_a^b r^2 d\theta$

Example 1

Calculate the area enclosed by the curve $r = \sin(2\theta)$

$\sin(2\theta) = 0 \Rightarrow 2\theta = 0, \pi, 2\pi, 3\pi, 4\pi$

$\Rightarrow \theta = 0, \dfrac{\pi}{2}, \pi, \dfrac{3\pi}{2}, 2\pi$

So r is positive for $0 < \theta < \dfrac{\pi}{2}, \pi < \theta < \dfrac{3\pi}{2}$

Consider the graph of $y = \sin(2\theta)$

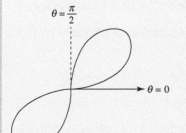

$\theta = \dfrac{\pi}{2}$

$\theta = 0$

> A sketch of $r = \sin(2\theta)$ will help you to see the required areas.

(Continued on the next page)

Calculate the area of one 'loop':

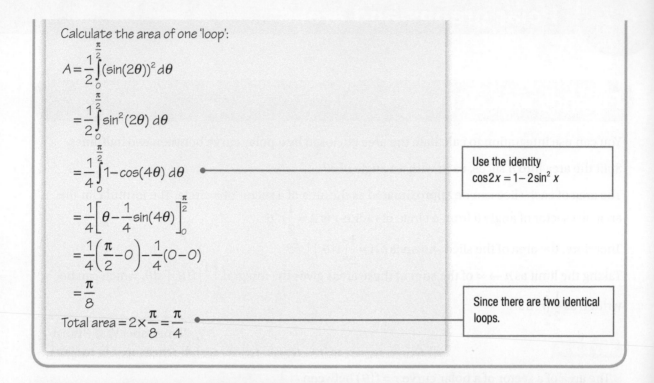

$$A = \frac{1}{2}\int_0^{\frac{\pi}{2}} (\sin(2\theta))^2 \, d\theta$$

$$= \frac{1}{2}\int_0^{\frac{\pi}{2}} \sin^2(2\theta) \, d\theta$$

$$= \frac{1}{4}\int_0^{\frac{\pi}{2}} 1 - \cos(4\theta) \, d\theta$$

Use the identity
$\cos 2x = 1 - 2\sin^2 x$

$$= \frac{1}{4}\left[\theta - \frac{1}{4}\sin(4\theta) \right]_0^{\frac{\pi}{2}}$$

$$= \frac{1}{4}\left(\frac{\pi}{2} - 0 \right) - \frac{1}{4}(0 - 0)$$

$$= \frac{\pi}{8}$$

Total area $= 2 \times \dfrac{\pi}{8} = \dfrac{\pi}{4}$

Since there are two identical loops.

Exercise 8.7A Fluency and skills

1 For each of these polar equations

i State the maximum and the minimum values of r

ii Sketch the curve.

    a $r = 1 - \cos\theta$ **b** $r = 2 + \sin\theta$

    c $r = 2(1 - \sin\theta)$ **d** $r = 6 + 3\cos\theta$

    e $r = 5 + \sin\theta$ **f** $r = \cos(2\theta)$

    g $r = a\sin(3\theta)$ **h** $r = b\cos(4\theta)$

    i $r^2 = c^2 \sin\theta$

2 Calculate the area bounded by the curve with equation $r = 4\cos\theta$ and the half-lines $\theta = 0$ and $\theta = \dfrac{\pi}{3}$

3 Calculate the area bounded by the curve with equation $r = \theta$ and the half-lines $\theta = 0$ and $\theta = \dfrac{\pi}{2}$

4 Calculate the area bounded by the curve with equation $r = 2\sin\theta$ and the half-lines $\theta = \dfrac{\pi}{12}$ and $\theta = \dfrac{5\pi}{12}$

5 Calculate the area bounded by the curve with equation $r^2 = 4\sin\theta$ and the half-lines $\theta = \dfrac{\pi}{4}$ and $\theta = \dfrac{3\pi}{4}$

6 Calculate the area enclosed by each of these cardioids.

    a $r = 3 + \cos\theta$ **b** $r = 5 + 2\sin\theta$

    c $r = 1 + \sin\theta$ **d** $r = 3 - 2\cos\theta$

7 Calculate the total area enclosed by these curves.

    a $r = \cos 2\theta$ **b** $r = \sin 4\theta$

    c $r = 4\cos 3\theta$ **d** $r = 5\sin 3\theta$

    e $r^2 = 2\sin 2\theta$ **f** $r^2 = (5 + 2\sin\theta)$

Reasoning and problem-solving

To find the area between two polar curves, you first need to work out where they intersect. As with Cartesian equations, you do this by solving the equations simultaneously.

You can then calculate the area between the two curves by adding the areas of the different parts.

Strategy

To calculate the area between two polar curves

1. Sketch the curves and solve simultaneously to find their point of intersection.

2. Calculate the area of a sector for each curve using $\dfrac{1}{2}\displaystyle\int_a^b r^2 d\theta$

3. Add the areas together to find the required area.

Example 2

Find the area bounded by the curves $r = 3\cos\theta$ and $r = 1 + \cos\theta$

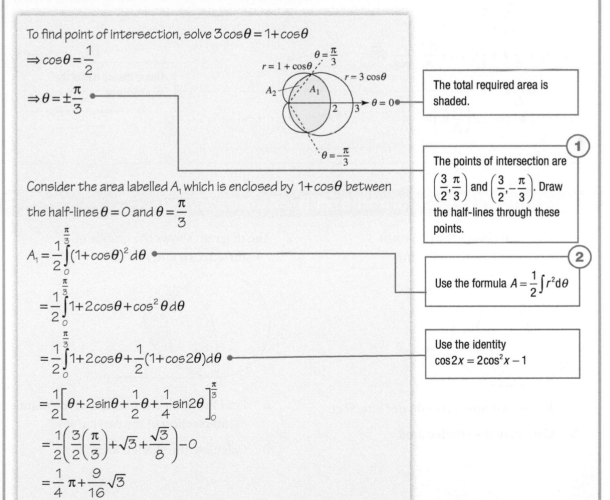

To find point of intersection, solve $3\cos\theta = 1 + \cos\theta$

$\Rightarrow \cos\theta = \dfrac{1}{2}$

$\Rightarrow \theta = \pm\dfrac{\pi}{3}$

The total required area is shaded.

The points of intersection are $\left(\dfrac{3}{2}, \dfrac{\pi}{3}\right)$ and $\left(\dfrac{3}{2}, -\dfrac{\pi}{3}\right)$. Draw the half-lines through these points. **(1)**

Consider the area labelled A_1 which is enclosed by $1 + \cos\theta$ between the half-lines $\theta = 0$ and $\theta = \dfrac{\pi}{3}$

$A_1 = \dfrac{1}{2}\displaystyle\int_0^{\frac{\pi}{3}} (1 + \cos\theta)^2 d\theta$

Use the formula $A = \dfrac{1}{2}\displaystyle\int r^2 d\theta$ **(2)**

$= \dfrac{1}{2}\displaystyle\int_0^{\frac{\pi}{3}} 1 + 2\cos\theta + \cos^2\theta \, d\theta$

$= \dfrac{1}{2}\displaystyle\int_0^{\frac{\pi}{3}} 1 + 2\cos\theta + \dfrac{1}{2}(1 + \cos 2\theta) d\theta$

Use the identity $\cos 2x = 2\cos^2 x - 1$

$= \dfrac{1}{2}\left[\theta + 2\sin\theta + \dfrac{1}{2}\theta + \dfrac{1}{4}\sin 2\theta\right]_0^{\frac{\pi}{3}}$

$= \dfrac{1}{2}\left(\dfrac{3}{2}\left(\dfrac{\pi}{3}\right) + \sqrt{3} + \dfrac{\sqrt{3}}{8}\right) - 0$

$= \dfrac{1}{4}\pi + \dfrac{9}{16}\sqrt{3}$

(*Continued on the next page*)

Consider the area labelled A_2 which is enclosed by $3\cos\theta$ between

the half-lines $\theta = \dfrac{\pi}{3}$ and $\theta = \dfrac{\pi}{2}$

$$A_2 = \frac{1}{2}\int_{\frac{\pi}{3}}^{\frac{\pi}{2}} (3\cos\theta)^2\, d\theta$$

$$= \frac{1}{2}\int_{\frac{\pi}{3}}^{\frac{\pi}{2}} 9\cos^2\theta\, d\theta$$

$$= \frac{9}{4}\int_{\frac{\pi}{3}}^{\frac{\pi}{2}} (1+\cos2\theta)\, d\theta$$

$$= \frac{9}{4}\left[\theta + \frac{1}{2}\sin2\theta\right]_{\frac{\pi}{3}}^{\frac{\pi}{2}}$$

$$= \frac{9}{4}\left(\frac{\pi}{2} + 0\right) - \frac{9}{4}\left(\frac{\pi}{3} + \frac{1}{4}\sqrt{3}\right)$$

$$= \frac{3\pi}{8} - \frac{9}{16}\sqrt{3}$$

$$A_1 + A_2 = \frac{1}{4}\pi + \frac{9}{16}\sqrt{3} + \frac{3\pi}{8} - \frac{9}{16}\sqrt{3}$$

$$= \frac{5}{8}\pi$$

This is the top half of the required area.

3

So total area $= 2 \times \left(\dfrac{5}{8}\pi\right) = \dfrac{5}{4}\pi$

Exercise 8.7B Reasoning and problem-solving

1 The diagram shows the graphs of $r = \cos\theta + \sin\theta$ and $r = 2 + \sin\theta$

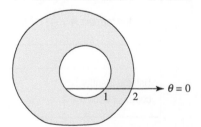

    a Prove that the curves do not intersect.

    b Calculate the shaded area.

2 The diagram shows the graphs of $r = \sin\theta\sqrt{2\cos\theta}$ and $r = \sin\theta$

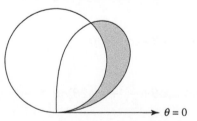

    a Find the polar coordinates of the points of intersection of the two curves.

    b Calculate the shaded area.

3 The curve shown has polar equation

$r = 2 + \cos(2\theta)$, $0 \le \theta \le \dfrac{\pi}{2}$

At the point A on the curve, $r = 1.5$

Calculate the shaded area.

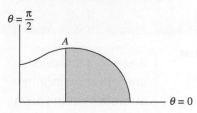

4 Calculate the area bounded by these pairs of curves.

a $r = \cos\theta$, $r = \sqrt{3}\sin\theta$

b $r = 2 - \cos\theta$, $r = 3\cos\theta$

c $r = \sqrt{2} + \sin\theta$, $r = 3\sin\theta$

d $r^2 = 1 - \sin\theta$, $r = \sqrt{2}\sin\theta$

e $r = 5 + 2\cos\theta$, $r = 4$

5 The circle with equation $x^2 + (y-2)^2 = 4$ intersects the curve with polar equation $r = 1 + \sin\theta$ at the points A and B

a Calculate the length of AB

b Calculate the area enclosed by $x^2 + (y-2)^2 = 4$ but not the curve $r = 1 + \sin\theta$

Find the shaded area in each of the curves given in questions 6–11.

6 $r = 3\sin2\theta$

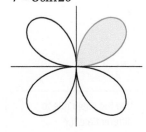

7 $r = 1 - \sin2\theta$

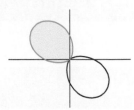

8 $r = 1 - \cos\dfrac{\theta}{2}$, $-\pi < \theta \le \pi$

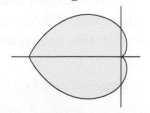

9 $r = 1 - 2\cos2\theta$

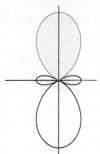

10 $r = \sqrt{\theta\cos\theta}$, $0 \le \theta \le \dfrac{\pi}{2}$

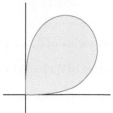

11 $r = \sqrt{2} - 2\sin\theta$

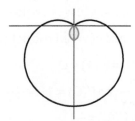

Fluency and skills

A function $f(x)$ can be expanded using the Maclaurin series given that

- $f(x)$ can be expanded as a **convergent** infinite series of terms
- each of the terms in $f(x)$ can be differentiated
- each of the differentiated terms has a finite value when $x = 0$

> A convergent series is one where an infinite number of terms has a finite sum.

Key point

The Maclaurin series, or expansion, for $f(x)$ is

$$f(x) \equiv f(0) + xf'(0) + \frac{x^2}{2!}f''(0) + \frac{x^3}{3!}f'''(0) + \frac{x^4}{4!}f''''(0) + \ldots + \frac{x^r}{r!}f^{(r)}(0) + \ldots$$

You should know the range of values of x for which the Maclaurin series is valid for different functions.

Function	Range of x where series is valid
e^x	all values of x
$\sin x$	all values of x
$\cos x$	all values of x
$(1+x)^n$	$-1 < x < 1$ for $n \in \mathbb{R}$
$\ln(1+x)$	$-1 < x \leq 1$

Example 1

a Explain why a Maclaurin series of $f(x) = \ln(x)$ is not possible.

b Derive the Maclaurin series of $f(x) = \ln(1+x)$

c The first three terms of the series for $\tan x$ are $x + \frac{x^3}{3} + \frac{2x^5}{15}$. Use this series and your answer to part **b** to find the expansion of $\tan(2x)\ln(1+x)$ as far as the term in x^5

a The first constant $f(0) = \ln(0) = -\infty$, which is not finite.

Hence a Maclaurin series of $y = \ln(x)$ is not possible.

b $f(x) = \ln(1+x)$ so $f(0) = \ln(1) = 0$

$f'(x) = (1+x)^{-1} = \frac{1}{1+x}$, so $f'(0) = 1$ •——— Use the chain rule to differentiate $f(x)$

$f''(x) = -(1+x)^{-2}$ so $f''(0) = -1$

$f'''(x) = 2(1+x)^{-3}$ so $f'''(0) = 2$

$f''''(x) = -6(1+x)^{-4}$ so $f''''(0) = -6 = -3!$ •——— The pattern is now clear and can be proved, for example, by induction.

$f'''''(x) = 24(1+x)^{-5}$ so $f'''''(0) = 4!$

(Continued on the next page)

$$f(x) \equiv f(0) + xf'(0) + \frac{x^2}{2!}f''(0) + \frac{x^3}{3!}f'''(0) + \frac{x^4}{4!}f''''(0) + \ldots$$

Use the Maclaurin expansion.

$$f(x) = \ln(1+x) \equiv x - \frac{x^2}{2!} + \frac{2x^3}{3!} - \frac{6x^4}{4!} + \frac{24x^5}{5!} - \ldots$$

$$\equiv x - \frac{x^2}{2} + \frac{x^3}{3} - \frac{x^4}{4} + \frac{x^5}{5} - \ldots$$

c $\tan(2x)\ln(1+x) \equiv \left(2x + \frac{(2x)^3}{3} + \frac{2(2x)^5}{15} + \ldots\right)\left(x - \frac{x^2}{2} + \frac{x^3}{3} - \frac{x^4}{4} + \frac{x^5}{5} - \ldots\right)$

$$\equiv \left(2x + \frac{8x^3}{3} + \frac{64x^5}{15} + \ldots\right)\left(x - \frac{x^2}{2} + \frac{x^3}{3} - \frac{x^4}{4} + \frac{x^5}{5} - \ldots\right)$$

$$\equiv 2x^2 + x^3\left[2x - \frac{1}{2}\right] + x^4\left[\frac{2}{3} + \frac{8}{3}\right] + x^5\left[-\frac{1}{2} - \frac{8}{6}\right] + \ldots$$

You only need to multiply the terms as far as x^5

$$\equiv 2x^2 - x^3 + \frac{10}{3}x^4 - \frac{11}{6}x^5 + \ldots$$

Example 2

Derive the Maclaurin series of $f(x) = \ln(2 + x)$ by adapting the series for $\ln(1 + x)$ you found in Example 1

$$\ln(1+x) \equiv x - \frac{x^2}{2} + \frac{x^3}{3} - \frac{x^4}{4} + \frac{x^5}{5} + \ldots$$

$$\ln(2+x) \equiv \ln\left[2\left(1 + \frac{x}{2}\right)\right]$$

Take out a factor of 2

$$\equiv \ln 2 + \ln\left(1 + \frac{x}{2}\right)$$

Use laws of logarithms.

$$\equiv \ln 2 + \left(\frac{x}{2}\right) - \frac{\left(\frac{x}{2}\right)^2}{2} + \frac{\left(\frac{x}{2}\right)^3}{3} - \frac{\left(\frac{x}{2}\right)^4}{4} + \frac{\left(\frac{x}{2}\right)^5}{5} + \ldots$$

Expand $\ln\left(1 + \frac{x}{2}\right)$ by replacing x with $\frac{x}{2}$ in the expansion of $\ln(1 + x)$

$$\equiv \ln 2 + \frac{x}{2} - \frac{x^2}{8} + \frac{x^3}{24} - \frac{x^4}{64} + \frac{x^5}{160} - \ldots$$

General term

Sequences and series can be represented by a formula for the general term, sometimes called the rth term or nth term. For example, the formula $3r - 2$ represents the sequence of terms 1, 4, 7, 10, …

Sometimes you will only know the first few terms of a sequence, and need to find the rth term. You do this by looking for patterns which you can write as a formula.

Example 3

Use your knowledge of these Maclaurin series to write down their general terms.

a e^{2x} **b** $\ln(1+3x)$

a $e^x = 1 + x + \dfrac{x^2}{2!} + \ldots + \dfrac{x^r}{r!} + \ldots$

So the general term of the sequence for e^{2x} is $\dfrac{(2x)^r}{r!}$ or $\dfrac{2^r x^r}{r!}$

b $\ln(1+x) = x - \dfrac{x^2}{2} + \dfrac{x^3}{3} - \ldots + (-1)^{r+1}\dfrac{x^r}{r} + \ldots$

So the rth term for $\ln(1+3x)$ is $(-1)^{r+1}\dfrac{(3x)^r}{r}$ or $(-1)^{r+1}\dfrac{3^r x^r}{r}$

Example 4

a $f(x) \equiv e^x \cos x$

  i Show that $f''(x) = -2e^x \sin x$

  ii Find $f'''(x)$ and $f''''(x)$

b Use the Maclaurin series to find the first four non-zero terms in the expansion of $f(x)$

a i $f(x) \equiv e^x \cos x$

  $f'(x) = e^x(\cos x - \sin x)$ •———————— Use the product rule to differentiate.

  $f''(x) = e^x(\cos x - \sin x - \sin x - \cos x) = -2e^x \sin x$

  ii $f'''(x) = -2e^x(\sin x + \cos x)$

  $f''''(x) = -2e^x(\sin x + \cos x + \cos x - \sin x) = -4e^x \cos x$

b $f(x) \equiv f(0) + xf'(0) + \dfrac{x^2}{2!}f''(0) + \dfrac{x^3}{3!}f'''(0) + \dfrac{x^4}{4!}f''''(0) + \ldots$

  $\equiv e^0 \cos 0 + xe^0(\cos 0 - \sin 0) + \dfrac{x^2}{2!}(-2\,e^0 \sin 0)$ •———— Substitute values for f(0), f'(0), etc.

  $+ \dfrac{x^3}{3!}[-2e^0(\sin 0 + \cos 0)] + \dfrac{x^4}{4!}(-4e^0 \cos 0) + \ldots$

  $f(x) \equiv 1 + x(1) + \dfrac{x^2}{2!}(0) + \dfrac{x^3}{3!}(-2) + \dfrac{x^4}{4!}(-4) + \ldots$

  $\equiv 1 + x - \dfrac{x^3}{3} - \dfrac{x^4}{6} + \ldots$

Limits

A limiting value, or limit, is a specific value that a function approaches or tends towards as the variable approaches a particular value. You came across this idea when differentiating from first principles.

The derivative of a function $f(x)$ is defined as $f'(x) = \lim\limits_{h \to 0} \dfrac{f(x+h) - f(x)}{h}$

It gives the gradient of the curve at any value of x

In this case, $f'(x)$ is the limit as h tends to zero, but you can also find a limit approaching other numbers, for example, 0, 1 or $\dfrac{\pi}{2}$

You can sometimes work out the limit of a function easily:

$$\lim_{n \to 0}(2 + e^n) = 2 + e^0 = 2 + 1 = 3 \qquad \lim_{n \to \frac{\pi}{2}}(1 - \cos n) = 1 - \cos\frac{\pi}{2} = 1 - 0 = 1$$

In other cases, you may need to manipulate the function to find the limit. For example, when finding $\lim\limits_{n \to \infty} \dfrac{n+10}{n}$ you must manipulate the expression before you find the limit:

$$\lim_{n \to \infty}\frac{n+10}{n} = \lim_{n \to \infty}\frac{n}{n} + \frac{10}{n} = \lim_{n \to \infty}\frac{n}{n} + \frac{10}{n} = 1 + 0 = 1$$

Example 5

Find these limits.

a $\lim\limits_{n \to \infty} \dfrac{2n+5}{n}$ **b** $\lim\limits_{n \to \infty} \dfrac{2n^2+5}{n}$ **c** $\lim\limits_{n \to \infty} \dfrac{2n+5}{n^2}$ **d** $\lim\limits_{n \to \infty}\left(\dfrac{2n^2+n-35}{5n^2-3n+7}\right)$

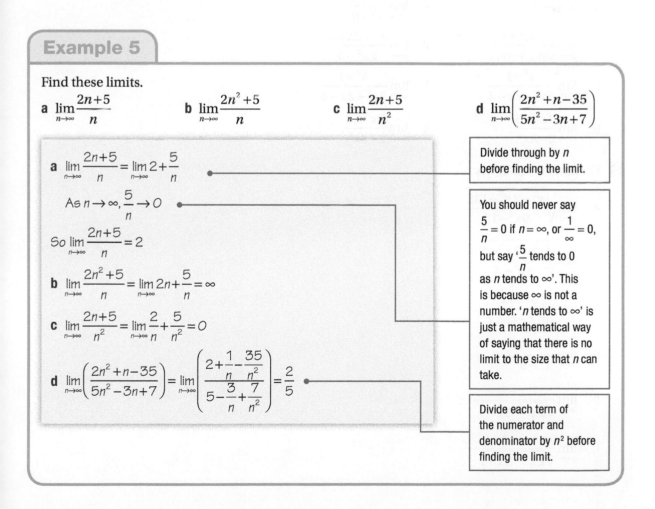

a $\lim\limits_{n \to \infty} \dfrac{2n+5}{n} = \lim\limits_{n \to \infty} 2 + \dfrac{5}{n}$

As $n \to \infty$, $\dfrac{5}{n} \to 0$

So $\lim\limits_{n \to \infty} \dfrac{2n+5}{n} = 2$

b $\lim\limits_{n \to \infty} \dfrac{2n^2+5}{n} = \lim\limits_{n \to \infty} 2n + \dfrac{5}{n} = \infty$

c $\lim\limits_{n \to \infty} \dfrac{2n+5}{n^2} = \lim\limits_{n \to \infty} \dfrac{2}{n} + \dfrac{5}{n^2} = 0$

d $\lim\limits_{n \to \infty}\left(\dfrac{2n^2+n-35}{5n^2-3n+7}\right) = \lim\limits_{n \to \infty}\left(\dfrac{2+\dfrac{1}{n}-\dfrac{35}{n^2}}{5-\dfrac{3}{n}+\dfrac{7}{n^2}}\right) = \dfrac{2}{5}$

Divide through by n before finding the limit.

You should never say $\dfrac{5}{n} = 0$ if $n = \infty$, or $\dfrac{1}{\infty} = 0$, but say '$\dfrac{5}{n}$ tends to 0 as n tends to ∞'. This is because ∞ is not a number. 'n tends to ∞' is just a mathematical way of saying that there is no limit to the size that n can take.

Divide each term of the numerator and denominator by n^2 before finding the limit.

1 Use your knowledge of these Maclaurin series to write down their general terms.

  a e^{-x} **b** $\ln(1-2x)$

  c $\sin\left(\dfrac{x}{5}\right)$ **d** $(1+3x)^n$

2 Use the Maclaurin expansion to derive a series for $\sin x$. Include the general term.

3 Use the Maclaurin expansion to derive a series for $\cos x$. Include the general term.

4 Find the first three non-zero terms in the Maclaurin expansion of $f(x) = \ln\left(\dfrac{1-2x}{1+2x}\right)$

5 Derive the first three non-zero terms in the Maclaurin expansion of $f(x) = \cos^2 x$

6 Derive the first three non-zero terms in the Maclaurin expansion of $f(x) = \sqrt{1+3x^2}$

7 Find the first three non-zero terms in the Maclaurin expansion of $f(x) = e^{2x}\sin x$

8 Find the first three non-zero terms in the Maclaurin expansion of $f(x) = e^x \ln(1+x)$

9 Evaluate these limits.

  a $\displaystyle\lim_{x\to\infty} 1 + e^{-x}$ **b** $\displaystyle\lim_{x\to1} \dfrac{3x+5}{x^2-9}$

  c $\displaystyle\lim_{x\to0} \dfrac{3\cos x}{\sin x + \cos x}$ **d** $\displaystyle\lim_{x\to\infty} \dfrac{x^2+2x+3}{2x^2-5}$

  e $\displaystyle\lim_{x\to0} \dfrac{x^2+1}{3x^2+2x-4}$ **f** $\displaystyle\lim_{x\to\infty} \dfrac{x^2+1}{3x^2+2x-4}$

  g $\displaystyle\lim_{x\to\infty} \dfrac{3x^2+8}{2-3x^2}$ **h** $\displaystyle\lim_{x\to-1} \dfrac{3x^2+8}{2-3x^2}$

10 Evaluate these limits.

  a $\displaystyle\lim_{x\to2} \sqrt{\dfrac{x^3-2x^2+135}{9x^3-12}}$

  b $\displaystyle\lim_{x\to\infty} \sqrt{\dfrac{x^3-2x^2+1}{9x^3+4}}$

  c $\displaystyle\lim_{x\to0} \dfrac{3\cos x}{\sin x - \cos x}$

  d $\displaystyle\lim_{x\to1} \dfrac{3x+2}{x^2-9}$

  e $\displaystyle\lim_{x\to0} \dfrac{x+2}{\sqrt{2x^2-3x+1}}$

  f $\displaystyle\lim_{x\to\infty} \dfrac{x+2}{\sqrt{2x^2-3x+1}}$

Reasoning and problem-solving

When you investigate a limit and end up with an indeterminate value such as $\dfrac{0}{0}$ or $\dfrac{\infty}{\infty}$ you can use Maclaurin expansions to evaluate the limit.

Strategy

To solve problems involving Maclaurin series and limits

(1) Derive the Maclaurin series from first principles or adapt a standard Maclaurin series.

(2) Substitute the series into the expression.

(3) Use the Maclaurin series to work out the limit of the function.

Example 6

$$f(x) \equiv \frac{4x^4}{3(\ln(1+x^2)-x^2)}$$

Adapt the known Maclaurin series for $\ln(1+x)$ in $f(x)$ to calculate $\lim\limits_{x \to 0} f(x)$

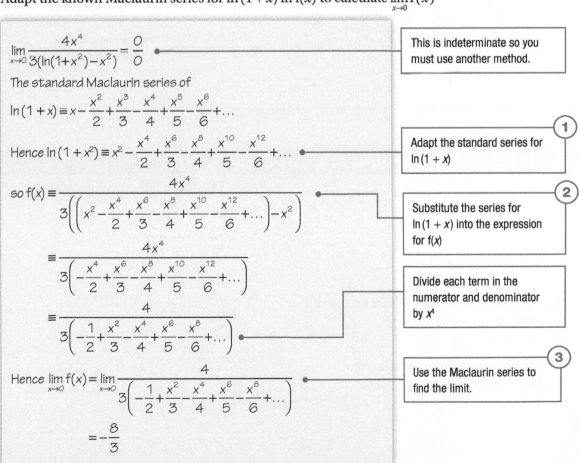

$$\lim_{x \to 0} \frac{4x^4}{3(\ln(1+x^2)-x^2)} = \frac{0}{0}$$

This is indeterminate so you must use another method.

The standard Maclaurin series of

$$\ln(1+x) \equiv x - \frac{x^2}{2} + \frac{x^3}{3} - \frac{x^4}{4} + \frac{x^5}{5} - \frac{x^6}{6} + \dots$$

Hence $\ln(1+x^2) \equiv x^2 - \dfrac{x^4}{2} + \dfrac{x^6}{3} - \dfrac{x^8}{4} + \dfrac{x^{10}}{5} - \dfrac{x^{12}}{6} + \dots$

(1) Adapt the standard series for $\ln(1+x)$

so $f(x) \equiv \dfrac{4x^4}{3\left(\left(x^2 - \dfrac{x^4}{2} + \dfrac{x^6}{3} - \dfrac{x^8}{4} + \dfrac{x^{10}}{5} - \dfrac{x^{12}}{6} + \dots\right) - x^2\right)}$

(2) Substitute the series for $\ln(1+x)$ into the expression for $f(x)$

$$\equiv \frac{4x^4}{3\left(-\dfrac{x^4}{2} + \dfrac{x^6}{3} - \dfrac{x^8}{4} + \dfrac{x^{10}}{5} - \dfrac{x^{12}}{6} + \dots\right)}$$

$$\equiv \frac{4}{3\left(-\dfrac{1}{2} + \dfrac{x^2}{3} - \dfrac{x^4}{4} + \dfrac{x^6}{5} - \dfrac{x^8}{6} + \dots\right)}$$

Divide each term in the numerator and denominator by x^4

Hence $\lim\limits_{x \to 0} f(x) = \lim\limits_{x \to 0} \dfrac{4}{3\left(-\dfrac{1}{2} + \dfrac{x^2}{3} - \dfrac{x^4}{4} + \dfrac{x^6}{5} - \dfrac{x^8}{6} + \dots\right)}$

(3) Use the Maclaurin series to find the limit.

$$= -\frac{8}{3}$$

Example 7

$f(x) = \dfrac{e^{2x} - e^x}{x}$

a Find the first four terms and the general term of the expansion of $f(x)$

b Use your expansion to find $\lim\limits_{x \to 0} f(x)$

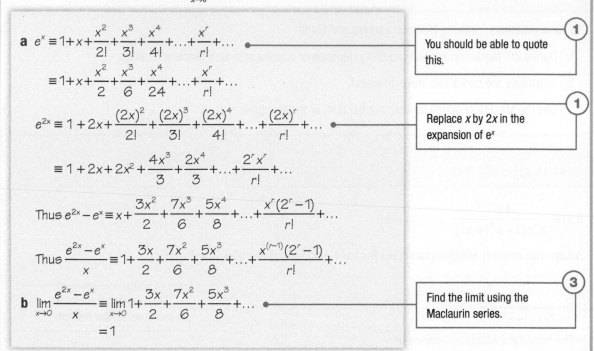

a $e^x \equiv 1 + x + \dfrac{x^2}{2!} + \dfrac{x^3}{3!} + \dfrac{x^4}{4!} + \ldots + \dfrac{x^r}{r!} + \ldots$

You should be able to quote this.

$\equiv 1 + x + \dfrac{x^2}{2} + \dfrac{x^3}{6} + \dfrac{x^4}{24} + \ldots + \dfrac{x^r}{r!} + \ldots$

$e^{2x} \equiv 1 + 2x + \dfrac{(2x)^2}{2!} + \dfrac{(2x)^3}{3!} + \dfrac{(2x)^4}{4!} + \ldots + \dfrac{(2x)^r}{r!} + \ldots$

Replace x by $2x$ in the expansion of e^x

$\equiv 1 + 2x + 2x^2 + \dfrac{4x^3}{3} + \dfrac{2x^4}{3} + \ldots + \dfrac{2^r x^r}{r!} + \ldots$

Thus $e^{2x} - e^x \equiv x + \dfrac{3x^2}{2} + \dfrac{7x^3}{6} + \dfrac{5x^4}{8} + \ldots + \dfrac{x^r(2^r - 1)}{r!} + \ldots$

Thus $\dfrac{e^{2x} - e^x}{x} \equiv 1 + \dfrac{3x}{2} + \dfrac{7x^2}{6} + \dfrac{5x^3}{8} + \ldots + \dfrac{x^{(r-1)}(2^r - 1)}{r!} + \ldots$

b $\lim\limits_{x \to 0} \dfrac{e^{2x} - e^x}{x} \equiv \lim\limits_{x \to 0} 1 + \dfrac{3x}{2} + \dfrac{7x^2}{6} + \dfrac{5x^3}{8} + \ldots$

Find the limit using the Maclaurin series.

$= 1$

Exercise 8.8B Reasoning and problem-solving

1 a Use the binomial theorem to expand

$\dfrac{4}{(1 + 4x)}$

b $\displaystyle\int_0^x \dfrac{4}{(1 + 4y)}\,dy = \ln(1 + 4x)$

Use this and your answer to part **a** to obtain the first four terms in the series of $\ln(1 + 4x)$

c Check your solution by using the Maclaurin series for $\ln(1 + 4x)$

2 Use series expansions to determine these limits.

a $\lim\limits_{x \to 0} \dfrac{(1 + 2x)^{-3} - 1}{x}$

b $\lim\limits_{x \to 0} \dfrac{x}{\sin 2x}$

c $\lim\limits_{x \to 0} \dfrac{1 - \cos 4x}{x^2}$

d $\lim\limits_{x \to 0} \dfrac{x \ln(1 - x)}{e^{x^2} - 1}$

3 Use series expansions to determine these limits.

a $\lim\limits_{x\to\infty} x - \sqrt{x^2-4x}$ **b** $\lim\limits_{x\to\infty}(\sqrt[3]{(x^3-2)} - x)$

c $2\lim\limits_{x\to\infty}\ln\left(\dfrac{x^4+3}{x^4+2}\right)$ **d** $\lim\limits_{x\to 0}\dfrac{e^{2x}-3}{4e^{2x}+6}$

e $\lim\limits_{x\to\infty}\dfrac{e^{2x}-3}{4e^{2x}+6}$

4 a Find the first four non-zero terms in the expansion of $\dfrac{\ln(1+x)}{1-x}$

   b Find the first four non-zero terms in the expansion of $\dfrac{\sqrt{(2-x)}}{1+2x}$

5 a Find the partial fractions of $\dfrac{3(1-2x)}{(x+2)(1+x)}$

   b Hence expand $\dfrac{3(1-2x)}{(x+2)(1+x)}$ as a series as far as the term in x^3

   c Write down $\lim\limits_{x\to 0}\left(\dfrac{3(1-2x)}{(x+2)(1+x)}\right)$ and confirm the result using your answer from part **b**.

6 Use standard Maclaurin expansions to find $\lim\limits_{x\to 0}\dfrac{e^x-e^{2x}}{x}$

7 Use standard Maclaurin expansions to find the first three non-zero terms in the expansion of $(2-x)e^{(2-x)}$ and hence find $\lim\limits_{x\to 0}(2-x)e^{(2-x)}$

8 Find the first three non-zero terms in the expansion of $\dfrac{x}{e^x-1}$. Hence find $\lim\limits_{x\to 0}\dfrac{x}{e^x-1}$

9 a Write down the first three non-zero terms of the expansion, in ascending powers of x, of $1+e^{-x}$

b Find the first **two** non-zero terms in the expansion, in ascending powers of x, of $\ln\left(\dfrac{1+e^{-x}}{2-3x}\right)$

c Find $\lim\limits_{x\to 0}\dfrac{\ln\left(\dfrac{1+e^{-x}}{2-3x}\right)}{4x}$

10 a Write down the first three non-zero terms in the expansions of e^{x^2} and $\sin 2x$

b Find the expansion of $\ln\left(\dfrac{\sin 2x}{2x}\right)$ as far as the term in x^4

c Evaluate $\lim\limits_{x\to 0}\dfrac{\ln\left(\dfrac{\sin 2x}{2x}\right)}{(e^{x^2}-1)}$

11 Make use of known series expansions to obtain the expansion of $\dfrac{1}{2}(e^x - e^{-x})$ up to the term in x^5

Hence evaluate $\lim\limits_{x\to 0}\dfrac{6x}{(e^x-e^{-x})}$

12 Use Maclaurin series to expand $\ln\left(\dfrac{2+x}{1-x}\right)$ up to the term in x^3

13 a Show that
$$\ln(\cos x) \equiv -\frac{x^2}{2} - \frac{x^4}{12} - \frac{x^6}{45} + \dots$$

b Hence find $\lim\limits_{x\to 0}\dfrac{\ln(\cos x)+x^2}{x^2}$

14 Evaluate $\lim\limits_{x\to\infty} x - \sqrt{x^2-3}$

Chapter summary

- An improper integral is a definite integral where either:
  - one or both of the limits is $\pm\infty$
  - the integrand (expression to be integrated) is undefined at one of the limits of the integral
  - the integrand is undefined at some point between the limits of the integral.
- To evaluate an improper integral, replace the limit where the integrand is undefined with a variable and then consider what happens to the integral as the variable tends to the original value of the limit.
- If the integral is undefined at more than one point, then split it into two integrals.
- Learn and use these important limits:
  - For any real number k, $x^k e^{-x} \to 0$ when $x \to \infty$
  - For any real number k, $x^k \ln x \to 0$ when $x \to 0+$
- The mean value of a function $f(x)$ in the range $a \le x \le b$ is given by $\dfrac{1}{b-a}\int_a^b f(x)\,dx$
- The volume of the solid formed by rotating the curve $y = f(x)$ between $x = a$ and $x = b$ a full turn around the x-axis is given by $V = \int_a^b \pi y^2\,dx$
- The volume of the solid formed by rotating the curve $x = f(y)$ between $y = a$ and $y = b$ a full turn around the y-axis is given by $V = \int_a^b \pi x^2\,dy$
- The volume of revolution for a curve defined by parametric equations $y = f(t)$ and $x = g(t)$ is
  - $V = \int_{t_1}^{t_2} \pi y^2 \dfrac{dx}{dt}\,dt$ for rotation around the x-axis
  - $V = \int_{t_1}^{t_2} \pi x^2 \dfrac{dy}{dt}\,dt$ for rotation around the y-axis.
- Inverse trigonometric functions can be differentiated using the chain rule:
  - $\dfrac{d(\arcsin x)}{dx} = \dfrac{1}{\sqrt{1-x^2}}$, $\dfrac{d(\arccos x)}{dx} = -\dfrac{1}{\sqrt{1-x^2}}$, $\dfrac{d(\arctan x)}{dx} = \dfrac{1}{1+x^2}$
- Trigonometric functions can be used as substitutions in integration:
  - For an integral involving $\sqrt{a^2 - x^2}$, try the substitution $x = a\sin u$
  - For an integral involving $a^2 + x^2$, try the substitution $x = a\tan u$
- The hyperbolic functions are defined using exponentials:
  - $\sinh x = \dfrac{1}{2}(e^x - e^{-x})$, $\cosh x = \dfrac{1}{2}(e^x + e^{-x})$, $\tanh x = \dfrac{e^x - e^{-x}}{e^x + e^{-x}}$
- Particularly useful hyperbolic identities are
  - $\cosh^2 x - \sinh^2 x \equiv 1$
  - $\sinh 2x = 2\sinh x \cosh x$ and $\cosh 2x = \cosh^2 x + \sinh^2 x$
- The reciprocal hyperbolic functions are
  - $\operatorname{cosech} x = \dfrac{1}{\sinh x}$, $\operatorname{sech} x = \dfrac{1}{\cosh x}$, $\coth x = \dfrac{1}{\tanh x}$

- Hyperbolic functions can be differentiated and integrated:
  - $\dfrac{d(\sinh x)}{dx} = \cosh x, \; \dfrac{d(\cosh x)}{dx} = \sinh x, \; \dfrac{d(\tanh x)}{dx} = \operatorname{sech}^2 x$
  - $\dfrac{d(\operatorname{sech} x)}{dx} = -\operatorname{sech} x \tanh x, \; \dfrac{d(\operatorname{cosech} x)}{dx} = -\operatorname{cosech} x \coth x, \; \dfrac{d(\coth x)}{dx} = -\operatorname{cosech}^2 x$
  - $\int \sinh x = \cosh x + c, \int \cosh x = \sinh x + c$
- Inverse hyperbolic functions can be differentiated using the chain rule:
  - $\dfrac{d(\operatorname{arcosh} x)}{dx} = \dfrac{1}{\sqrt{x^2-1}}, \dfrac{d(\operatorname{arsinh} x)}{dx} = \dfrac{1}{\sqrt{x^2+1}}$
- Hyperbolic functions can be used as substitutions in integration:
  - For an integral involving $\sqrt{x^2+a^2}$, try the substitution $x = a \sinh u$
  - For an integral involving $\sqrt{x^2-a^2}$, try the substitution $x = a \cosh u$
- A fraction $\dfrac{f(x)}{g(x)}$ where degree of $f(x) \geq$ degree of $g(x)$ is called an **improper** fraction and can be written in the form $P(x) + \dfrac{Q(x)}{g(x)}$
- $\dfrac{f(x)}{(\alpha x^2 + \beta)(\gamma x + \delta)}$ can be split into partial fractions of the form $\dfrac{Ax+B}{\alpha x^2 + \beta} + \dfrac{C}{\gamma x + \delta}$
- If a point P has polar coordinates (r, θ), then
  - r is the length of OP
  - θ is the angle between OP and the initial line.
- A half-line is a straight line that extends infinitely from a point.
- The area of a sector of a polar curve $r = f(\theta)$ between the half-lines $\theta = a$ and $\theta = b$ is given by $A = \dfrac{1}{2}\displaystyle\int_a^b r^2 d\theta$
- Maclaurin series are valid for specific ranges of x

Function	Range of x where series is valid
e^x	all values of x
$\sin x$	all values of x
$\cos x$	all values of x
$(1+x)^n$	$-1 < x < 1$ for $n \in \mathbb{R}$
$\ln(1+x)$	$-1 < x \leq 1$

- A limiting value, or limit, is a specific value that a function approaches or tends towards as the variable approaches a particular value.
- You can also use Maclaurin expansions to evaluate limits.

Check and review

You should now be able to...	Try Questions
✔ Recognise and evaluate an improper integral.	1, 2
✔ Decide whether an improper integral exists.	3, 4
✔ Calculate the mean value of a function.	5–9
✔ Calculate the volume of revolution when rotated around the x-axis.	10
✔ Calculate the volume of revolution when rotated around the y-axis.	11
✔ Calculate more complicated volumes of revolution by adding or subtracting volumes.	12
✔ Calculate the volume of revolution for curves defined parametrically.	13, 14
✔ Derive and use the derivatives of inverse trigonometric functions.	15, 16
✔ Use trigonometric functions as substitutions in integration.	17
✔ Derive and use hyperbolic identities.	18, 19
✔ Differentiate hyperbolic functions.	20
✔ Differentiate inverse hyperbolic functions.	21, 22
✔ Use hyperbolic functions as substitutions in integration.	23
✔ Write an improper fraction in the form $P(x) + \dfrac{Q(x)}{g(x)}$	24
✔ Integrate a rational function by first splitting into partial fractions.	25, 26
✔ Sketch polar curves.	27, 29
✔ Use integration to calculate the area of a sector of a polar curve.	30
✔ Find the point of intersection between two polar curves.	31
✔ Calculate the area between two polar curves.	32
✔ Derive the Maclaurin series for a function and find its range of validity.	28–32

1   Which of these are improper integrals? Explain your answers.

    A: $\displaystyle\int_0^3 \frac{1}{1-x}\,dx$ **B:** $\displaystyle\int_0^4 \frac{1}{x+2}$ **C:** $\displaystyle\int_1^\infty \frac{1}{x}\,dx$

2   Evaluate these improper integrals.

    a $\displaystyle\int_2^\infty \frac{2}{x^2}\,dx$ **b** $\displaystyle\int_{-6}^3 \frac{2}{\sqrt{3-x}}\,dx$

3   Show that the integral $\displaystyle\int_0^3 \ln x\,dx$ exists and find its value.

4   Show that the integral $\displaystyle\int_1^\infty \frac{2}{x}\,dx$ does not exist.

5   Find the mean value of the function $f(x)=8x^3+x^2-4$ for $1\le x\le 3$

6   Show that the mean value of the function $g(x)=\dfrac{1+x}{\sqrt{x}}$ in the interval $[2, 8]$ is $A\sqrt{2}$, where A is a constant to be found.

7   Work out the mean value of $\dfrac{16}{x^3}-4x^3$ in the interval $\dfrac{1}{2}\le x\le 2$

8   Show that the mean value of the function $\dfrac{2-x}{2x^3}$ for x in the interval $[a, 1]$ is $\dfrac{1}{2a^2}$.

9   The velocity of a particle after t seconds is given by $v=\dfrac{1}{20}(5t^4-3t^2)\,\mathrm{ms^{-1}}$.

    a Calculate the mean velocity over the first 4 seconds.

    b Calculate the mean acceleration over the first 4 seconds.

10 Find the volume of the solid formed when the section of the curve $y=\dfrac{3}{\sqrt{2x+1}}$, $x\ne -\dfrac{1}{2}$ between $x=0$ and $x=1$ is rotated $360°$ around the x-axis.

11 The shaded region is bounded by the curve with equation $y=\arctan(2x)$, the y-axis and the line $y=\dfrac{\pi}{4}$

    Calculate the volume of revolution when the shaded region is rotated 2π radians around the y-axis.

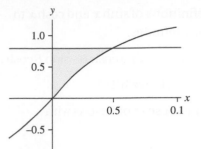

12 The region R is bounded by the curve with equation $y=\cos(x)$, the line $y=1-x$ and the x-axis.

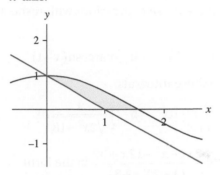

    Find the volume of the solid formed when R is rotated $360°$ around the x-axis.

13 The curve C is defined by the parametric equations $x=7t^3$, $y=t^2$. Calculate the volume of revolution when the section of the curve between $t=0$ and $t=1$ is rotated $360°$ around

    a The x-axis,   **b** The y-axis.

14 A curve is defined by the parametric equations $x=1+t^2$, $y=3t-t^2$

    Calculate the volume of revolution when a section of the curve between $(1, 0)$ and $(2, 2)$ is rotated 2π radians around

    a The x-axis,   **b** The y-axis.

15 Prove that $\dfrac{d(\arccos x)}{dx}=-\dfrac{1}{\sqrt{1-x^2}}$

16 Differentiate these expressions with respect to x

    a $x^2\arctan x$ **b** $\arcsin\left(\dfrac{x^2}{2}\right)$

17 Use a substitution in each of these integrations.

    a $\displaystyle\int \frac{1}{\sqrt{4-x^2}}\,dx$ **b** $\displaystyle\int \frac{1}{x^2+16}\,dx$

18 Use the definitions of $\sinh x$ and $\cosh x$ to prove these identities.

a $\sinh(A-B) \equiv \sinh A \cosh B - \sinh B \cosh A$

b $\cosh^2 x \equiv \frac{1}{2}(1+\cosh 2x)$

19 Differentiate these expressions with respect to x

a $\cosh 2x$ **b** $x^2 \sinh x$

20 Show that $\dfrac{d(\operatorname{artanh} x)}{dx} = \dfrac{1}{1-x^2}$

21 Differentiate these expressions with respect to x

a $\operatorname{arsinh}(2x^2)$ **b** $x\operatorname{arcosh}(x-1)$

22 Evaluate these integrals.

a $\displaystyle\int_0^5 \frac{1}{\sqrt{25+x^2}}\,dx$ **b** $\displaystyle\int_3^9 \frac{1}{\sqrt{2x^2-18}}\,dx$

23 a Write $\dfrac{2x^3-x^2-12x+32}{(x-2)(x+3)}$ in the form

$P(x) + \dfrac{Q(x)}{g(x)} + \dfrac{R(x)}{h(x)}$

b Hence evaluate

$\displaystyle\int_3^4 \frac{2x^3-x^2-12x+32}{(x-2)(x+3)}\,dx$

24 Calculate the exact value of

$\displaystyle\int_0^3 \frac{x^2+5x+15}{(4-x)(x^2+1)}\,dx$

25 For each polar equation

i Sketch the curves for $r\geq 0$ and $0\leq\theta<2\pi$

ii State the maximum and the minimum value of r for $r\geq 0$ and $0\leq\theta<2\pi$

a $r=8\cos 2\theta$ **b** $r=-3\sin\theta$

c $r=\dfrac{\theta}{4}$ **d** $r=7+5\sin\theta$

26 Calculate the total area enclosed within the polar curves for $r\geq 0$ and $0\leq\theta<2\pi$

a $r=3+\cos\theta$ **b** $r=2\sin 4\theta$

27 The graphs of $r=\sqrt{2}+\sin\theta$ and $r=3\sin\theta$ are shown. Calculate the shaded area.

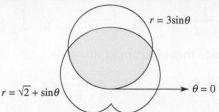

28 Use known series expansions to find the first three non-zero terms of the Maclaurin series for $e^x\ln(1+2x)$

29 Use your knowledge of standard Maclaurin series to write down the general terms in the expansion of these series.

a $e^{\frac{x}{3}}$ **b** $\ln(1-x^2)$ **c** $\cos\left(\dfrac{x}{3}\right)$

d $\sin(4x+5)$ **e** $\left(1-\dfrac{x}{6}\right)^n$ **f** xe^x

30 Write down the range of values of x for which these series are valid.

a $(2-x^2)^{-1}$ **b** $\ln\left(1+\left(\dfrac{x^2}{4}\right)\right)$

31 Write down the range of values of x for which these series are valid.

a $\left(2+\dfrac{x}{3}\right)^{-4}$ **b** $\ln(1-3x)$

c $\left(2+\dfrac{x}{3}\right)^{-4}(\ln(1-3x))$

32 Use differentiation and the Maclaurin expansion to find the first three non-zero terms in the expansions of these functions.

a $e^{2x}\sin x$ **b** $\dfrac{4}{(1+2x)}$

c $\dfrac{4}{\ln(2+x)}$ **d** $\cos 2x - \sin 2x$

e 3^x **f** $\dfrac{2\cos x}{5+\sin x}$

Investigation

A geometric solid that has finite volume but infinite surface area is called Gabriel's horn. The solid is formed by rotating the function $f(x) = \frac{1}{x}$ about the x-axis between $x = 1$ and infinity.

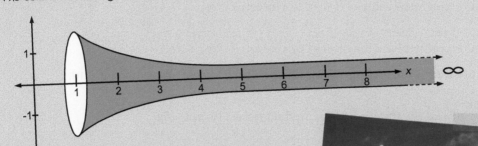

Try this to calculate the finite volume of the solid of revolution.

Research:
- How the formula for the surface area of revolution is derived,
- How the formula for the surface area of revolution leads to the answer of Gabriel's horn having infinite surface area.

Consider the lower bound of your integrals to be a where $0 \leq a \leq 1$

The results imply that, although the horn could contain a finite amount of paint, this would not be sufficient to paint its surface. This is sometimes known as the painter's paradox.

Investigation

The mean value of a trigonometric wave will be zero over one complete cycle because exactly half of the area enclosed by the function and the horizontal axis is positive and the other half is negative. One way to overcome this problem is to find the root mean square value, that is, the square root of the mean value of the function squared.

Investigate the root mean square of various trigonometric functions such as: $f(x) = A \sin x$ in the interval $0 \leq x \leq 2\pi$

That is find the value of $\sqrt{\frac{1}{2\pi} \int_0^{2\pi} A^2 \sin^2(\omega t)\, dt}$

Root mean square values have many applications in electrical/electronic engineering where engineers work with the flow of alternating and direct current. Other waveforms such as those illustrated may also be used in these fields of engineering.

Investigate their root mean square values.

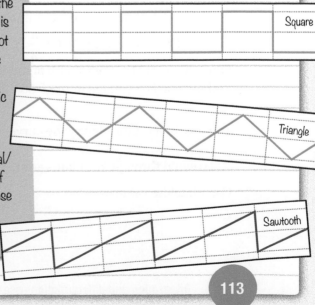

Square

Triangle

Sawtooth

1 You are given that $f(x)=\dfrac{x^3-3\sqrt{x}}{2x^2}$

  a Write $f(x)$ in the form Ax^m+Bx^n, where A, B, m and n are constants to be found. **[2]**

  b Calculate the mean value of $f(x)$ in the interval $[1,4]$. **[5]**

2 Find the mean value of the function $g(x)=x\sqrt{x}+\dfrac{3}{x^2}$ for $1\le x\le 3$

  Give your answer in the form $a+b\sqrt{3}$, where a and b are constants to be found. **[4]**

3 The mean value of the function $1+2x^3$ in the interval $[0,a]$ is 109

  Evaluate a **[5]**

4 The region R is bounded by the curve with equation $y=xe^{\frac{x}{2}}$, the x-axis and the line $x=2$

  a Calculate the area of R **[4]**

  b Calculate the volume of revolution when R is rotated $360°$ around the x-axis. **[6]**

5 The curve C is defined by the parametric equations

  $y=3t^2+1$, $x=5t$

  The region bounded by C, the x-axis and the lines $x=5$ and $x=10$ is rotated through π radians around the x-axis. Calculate the volume of revolution of the solid formed. **[6]**

6 **a** Sketch the graph of $y=\cosh 2x$ **[2]**

  The region R is bounded by the curve of $y=\cosh 2x$ and lines $x=\pm 1$

  b Calculate the volume of the solid formed when R is rotated through $90°$ about the x-axis. **[6]**

7 The curve C is defined by the parametric equations

  $x=2\sin\theta,\ y=\sin 2\theta,\ 0\le\theta\le\dfrac{\pi}{2}$

  Find, in terms of π, the volume of the solid formed when the region bounded by C and the y-axis is rotated through $360°$ around the y-axis. **[8]**

8 The region R is bounded by the curve with equation $y=\tan x$, the x-axis and the line $x=\dfrac{\pi}{4}$

  R is rotated through $k\pi$ radians around the x-axis and the volume of the solid formed is $\dfrac{\pi}{3}-\dfrac{\pi^2}{12}$

  Calculate the value of k **[6]**

9 Find $\displaystyle\int \frac{3}{5\sinh x - 4\cosh x}\,\mathrm{d}x$ **[10]**

10 The diagram shows the polar curve with
equations $r = 1 + \cos\theta$ for $0 \leq \theta \leq \pi$, along with the line
$\theta = \dfrac{\pi}{3}$. The curve and the line intersect at the origin, O,
and at the point P

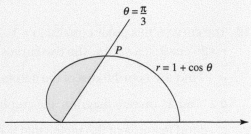

a Find the coordinates of the point P

b Find the area of the shaded region that is bounded
by the line and the curve. **[8]**

11 a Find $\displaystyle\int_0^1 \frac{1}{\sqrt{1-x^2}}\,\mathrm{d}x$ **b** Find $\displaystyle\int_1^2 \frac{1}{\sqrt{x^2-1}}\,\mathrm{d}x$ **[12]**

12 a Sketch the curve with polar equation $r = 4\sin 3\theta,\ 0 \leq \theta \leq \pi$

b Find the area enclosed by one loop of this curve. **[9]**

13 a Given $y = \sinh^{-1} x$, prove that $\dfrac{\mathrm{d}y}{\mathrm{d}x} = \dfrac{1}{\sqrt{1+x^2}}$

b Show that $\displaystyle\int_0^2 \sinh^{-1}x\,\mathrm{d}x = 2\ln(2+\sqrt5)+1-\sqrt5$ **[9]**

14 Show that $\displaystyle\int_1^4 \frac{2x+1}{\sqrt{x^2-2x+10}}\,\mathrm{d}x = 6(\sqrt2-1)+3\ln(1+\sqrt2)$ **[11]**

15 a Find $\displaystyle\int \frac{1}{x^2-4x+13}\,\mathrm{d}x$ **b** Evaluate $\displaystyle\int_0^\infty x^2\mathrm{e}^{-x}\,\mathrm{d}x$ **[15]**

16 The diagram shows the polar curves with equations $r = \sin 2\theta$ and $r = \cos\theta$ for $0 \leq \theta \leq \dfrac{\pi}{2}$
The curves intersect at the origin, O, and at the point P

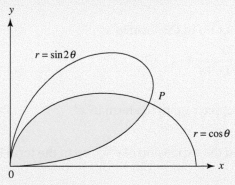

a Find the coordinates of the point P

b Show that the area of the shaded region that is bounded by the two curves is given
by $\dfrac{\pi}{8} - \dfrac{3\sqrt3}{32}$ **[13]**

17 a Find $\displaystyle\int_{\frac{\pi}{3}}^{\frac{\pi}{2}} \frac{\sin\theta}{1+\cos\theta}\,d\theta$ **b** Show that $\displaystyle\int_{1}^{3} \frac{3x-x^2}{(x+1)(x^2+3)}\,dx = \frac{\pi}{2\sqrt{3}} - \ln 2$ **[16]**

18 The curve C has polar equation $r=1+\cos\theta$, $0\le\theta\le 2\pi$. The curve D has polar equation $r=2-\cos\theta$, $0\le\theta\le 2\pi$. The two curves intersect at the points P and Q

   a Find the coordinates of the points P and Q

   b Sketch, on one diagram, the graphs of C and D

   c Find the area of the region that is both outside D and inside C **[13]**

19 The diagram shows part of the graph of $y=\dfrac{1}{\sqrt{4-x^2}}$

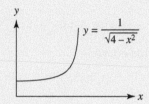

   a Find the area of the region bounded by the curve, the x-axis, the y-axis and the line $x=\sqrt{2}$

   b Find the volume of the solid generated when this region is rotated through 360° about the x-axis.

   c Find also the volume generated when this same region is rotated through 360° about the y-axis.   **[19]**

20 a Use Maclaurin's theorem to show that $\cos(3x)=1-\dfrac{9}{2}x^2+\dfrac{27}{8}x^4-\ldots$ **[4]**

   b Give the rule for the general term in the expansion of $\cos(3x)$ **[2]**

21 a Find the expansion of $\sqrt{1-3x}$ up to the term in x^3 **[3]**

   b Hence find the limit of $\dfrac{2x}{1-\sqrt{1-3x}}$ as $x\to 0$ **[3]**

22 Find the Maclaurin expansion of $\sec x$ up to the term in x^3 **[5]**

23 Use Maclaurin's theorem to find the expansion of $e^{\sin 2x}$ up to the term in x^2 **[5]**

9

Differential equations

Mathematics, and particularly calculus, plays a large part in the science and engineering that underpins space exploration. For centuries, the idea of space travel was only a remote possibility, and the mathematics needed to support the engineering required was only just being developed. The physical principles that these mathematicians worked with are relatively straight forward. For example, rocket launches rely on the application of Newton's third law to the jet engines that are used.

One of the differential equations that space engineers work with is the Tsiolkovsky rocket equation. This is named after the man who developed it, Konstantin Tsiolkovsky, a Russian physicist who was born in 1857 and died in 1935. The equation allows space engineers to calculate how much propellant is necessary to lift a rocket off the ground.

Orientation

What you need to know	What you will learn	What this leads to
Maths Ch16 • Differential equations.	• How to solve first order differential equations. • How to solve second order differential equations. • How differential equations are used to model simple harmonic motion. • How differential equations can be used to model situations involving rates of change. • How to work with differential equations that involve more than two variables.	**Careers** • Engineering. • Economics. • Ecology.

First order equations

Fluency and skills

An equation that involves only a first order derivative, such as $\dfrac{dy}{dx}$, is called a **first order differential equation**. You have already met first order differential equations where the terms involving x can be factorised out from the terms involving y. This method is called **separating the variables**. There are a number of methods for solving first order differential equations. One method involves using an **integrating factor**.

Example 1

Solve the differential equation $x^2 \dfrac{dy}{dx} + 2xy = 4x^3$, giving your answer in the form $y = f(x)$

$$\frac{d}{dx}\left[x^2 y\right] = 4x^3$$ — The LHS can be rewritten as the derivative of a product.

$$x^2 y = x^4 + c$$ — Integrate each side.

$$y = \frac{x^4 + c}{x^2}$$ — Rearrange to get an expression for y in terms of x

Key point

An expression that can be integrated by spotting that it is a perfect differential of a product is called an **exact equation**.

Suppose instead you had been asked to solve the differential equation $\dfrac{dy}{dx} + \dfrac{2}{x}y = 4x$. This is not as easy as you can no longer spot the product rule on the left-hand side. However, by multiplying through by x^2, you can transform the equation into an exact equation. The multiplier, x^2 is known as the **integrating factor**. You need to know how to find the integrating factor.

First order differential equations that can be solved using an integrating factor can often be written in the form

$$\frac{dy}{dx} + P(x)y = f(x)$$

where $P(x)$ and $f(x)$ are functions involving only the variable x

To find the integrating factor, first notice that if you differentiate a product in the form $e^{\int P(x)dx} y$

then

$$\frac{d}{dx}\left[e^{\int P(x)dx} y\right] = e^{\int P(x)dx}\frac{dy}{dx} + P(x)e^{\int P(x)dx} y$$

so if you take the equation $\frac{dy}{dx} + P(x)y = f(x)$ and multiply throughout by $e^{\int P(x)dx}$, the equation

becomes $e^{\int P(x)dx}\frac{dy}{dx} + P(x)e^{\int P(x)dx} y = e^{\int P(x)dx} f(x)$

The left-hand side is now an exact equation, and you have

$$\frac{d}{dx}\left[e^{\int P(x)dx} y\right] = e^{\int P(x)dx} f(x)$$

which you can solve by integrating each side.

Key point

The expression $e^{\int P(x)dx}$ is called the integrating factor.

To solve the equation $\frac{dy}{dx} + \frac{2}{x}y = 4x$, $P(x) = \frac{2}{x}$, and the integrating factor is $e^{\int \frac{2}{x}dx} = e^{2\ln x} = e^{\ln x^2} = x^2$

Multiplying the equation by x^2 gives you $x^2\frac{dy}{dx} + x^2\frac{2}{x}y = x^2 4x$, ie $x^2\frac{dy}{dx} + 2xy = 4x^3$, and you are

back to Example 1

Example 2

Solve the differential equation $\frac{dy}{dx} - y\tan x = 3\sin^2 x$, for $0 \le x < \frac{\pi}{2}$

This equation is in the form $\frac{dy}{dx} + P(x)y = f(x)$, where $P(x) = -\tan x$, and $f(x) = 3\sin^2 x$

$e^{\int P(x)dx} = e^{\int -\tan x\,dx} = e^{\ln(\cos x)} = \cos x$ — Find the integrating factor.

$\cos x\frac{dy}{dx} - y\cos x\tan x = 3\cos x\sin^2 x$ — Multiply through by $\cos x$

Since $\cos x\tan x = \cos x\frac{\sin x}{\cos x} = \sin x$

$\cos x\frac{dy}{dx} - y\sin x = 3\cos x\sin^2 x$ — Use the fact it is now an exact equation.

$\frac{d}{dx}[y\cos x] = 3\cos x\sin^2 x$

$y\cos x = \sin^3 x + c$ — Integrate each side. You can leave the solution in implicit form.

Key point

A solution which includes a constant of integration is called a **general solution**. If you can find a value for the constant you get a **particular solution**.

Example 3

a Find the general solution to the equation $x\dfrac{dy}{dx} - y = 2x^3$

b Given that $y = 4$ when $x = 1$, find the particular solution.

a $\dfrac{dy}{dx} - \dfrac{1}{x}y = 2x^2$ • ———————————— Rearrange the equation into the form $\dfrac{dy}{dx} + P(x)y = f(x)$

$e^{\int P(x)dx} = e^{\int -\frac{1}{x}dx} = e^{-\ln x} = \dfrac{1}{x}$ • ———————— Find the integrating factor.

$\dfrac{1}{x}\dfrac{dy}{dx} - \dfrac{1}{x^2}y = 2x$ • ———————————— Multiply through by $\dfrac{1}{x}$

$\dfrac{d}{dx}\left[\dfrac{1}{x}y\right] = 2x$ • ———————————— Use the fact it is now an exact equation.

$\dfrac{y}{x} = x^2 + c$ • ———————————————————— Integrate both sides.

b $\dfrac{4}{1} = 1^2 + c$ • ———————————————— Substitute $y = 4$, $x = 1$

$c = 3$ • ———————————————————————— Solve for c

$\dfrac{y}{x} = x^2 + 3 \ or \ y = x^3 + 3x$ • ——————— You can leave the answer in implicit or explicit form.

Exercise 9.1A Fluency and skills

1 Find the general solution to each of the following differential equations.

a $\dfrac{dy}{dx} + 3y = e^{-x}$

b $\dfrac{dy}{dx} + 2xy = 8x$

c $\dfrac{dy}{dx} + y\tan x = \sin^2 x \cos^2 x$

d $\dfrac{dy}{dx} - \dfrac{y}{x} = \dfrac{x}{x-5}$

e $3x\dfrac{dy}{dx} + y = \sqrt{x}$

f $x\dfrac{dy}{dx} - y = x^2 \ln x$

g $\cos x\dfrac{dy}{dx} - y\sin x = 4\sec^2 x$

h $\dfrac{1}{x}\dfrac{dy}{dx} + 2\dfrac{(x+1)}{x}y = 2e^{(x-1)^2}$

2 Find the particular solution to each of the following differential equations, giving each of your answers in the form $y = f(x)$

a $\dfrac{dy}{dx} - 2y = 2xe^{3x}$, given that $y = -1$ when $x = 0$

b $\dfrac{dy}{dx} + \dfrac{3}{x}y = \dfrac{4}{x^2}$, given that $y = 5$ when $x = 1$

c $\dfrac{dy}{dx} + y\cot x = 4\cos^3 x$, given that $y = \dfrac{1}{4}$ when $x = \dfrac{\pi}{6}$

d $\dfrac{dy}{dx} - 2y\cos 2x = 2e^{\sin 2x}$, given that $y = 1$ when $x = 0$

e $(x+1)\dfrac{dy}{dx} + 2y = \dfrac{14}{(x+1)}$, given that $y = 3$ when $x = -2$

f $\sin x \dfrac{dy}{dx} + y\cos x = -3\cos x \sin 2x$, given that $y = 2$ when $x = \dfrac{\pi}{4}$

g $x\dfrac{dy}{dx} - y = x^3 \ln x$, given that $y = 2$ when $x = 1$

h $\coth x \dfrac{dy}{dx} + y = 5e^{5x}\operatorname{cosech} x$, given that $y = 4$ when $x = 0$

3 a Find the general solution to the equation $\cos x \dfrac{dy}{dx} + 2y\sin x = \sin^2 x \cos x$

b Given that $y = \dfrac{1}{2}$ when $x = \dfrac{\pi}{4}$, find the particular solution.

4 Find the solution to the differential equation $\dfrac{dy}{dx} - y\tan x = 3x^2 \sec x$, given that $y = 2$ when $x = 0$.

5 Find the solution to the differential equation $x\dfrac{dy}{dx} - 2y = x^3 \ln x$, given that $y = 5$ when $x = 1$

6 a Find the general solution to the equation $\dfrac{dy}{dx} + 2y = 5\sin x$

b Given that $y = 1$ when $x = 0$, find the particular solution.

7 a Find the general solution of the differential equation $\cos x \dfrac{dy}{dx} + 2y\sin x = e^x \cos^3 x$

b Find the particular solution for which $y = 5$ at $x = 0$, giving your solution in the form $y = f(x)$

8 a Find the general solution of the differential equation $\tan x \dfrac{dy}{dx} + y = 12\sin 2x \tan x$

b Hence show that the particular solution to this equation for which $y = 2$ at $x = \dfrac{\pi}{4}$ is given by $y\sin x = 8\sin^3 x - \sqrt{2}$

9 a Find $\displaystyle\int x^3 e^x \, dx$

b Hence show that the general solution to the differential equation $x\dfrac{dy}{dx} + (x+2)y = 2x^2$ is given by $x^2 y = 2(x^3 - 3x^2 + 6x - 6) + ce^{-x}$

10 a Show that $\displaystyle\int \sec^3 x \, dx = \dfrac{1}{2}\left[\sec x \tan x + \ln(\sec x + \tan x)\right] + c$

b Hence show that the general solution to the differential equation $x\dfrac{dy}{dx} + (1 - x\tan x)y = 2\sec^4 x$ is $y = \dfrac{\sec x \tan x + \ln(\sec x + \tan x) + c}{x\cos x}$

11 Find the general solution of the differential equation $x + (t\ln t)\dfrac{dx}{dt} = 2te^{2t},\ t > 0$

12 a Find the general solution of the differential equation $(1+t)\dfrac{dx}{dt} + 3x = \ln(1+t),\ t > -1$

b Show that the particular solution for which $x = \dfrac{8}{9}$ at $t = 0$ is given by $x = \dfrac{3\ln(1+t) - 1}{9} + \dfrac{1}{(1+t)^3}$

13 a Given that $u = \dfrac{1}{y^2}$ and $\dfrac{dy}{dx} = y + 2xy^3$, show that $\dfrac{du}{dx} + 2u = -4x$

b Hence find the solution to the differential equation $\dfrac{dy}{dx} = y + 2xy^3$, given that $y = \dfrac{1}{2}$ when $x = 0$

To answer a question that involves first order differential equations

1. Decide whether to separate the variables or use an integrating factor.
2. Find the general solution of the differential equation.
3. Substitute the given values to find the constant, and hence obtain the particular solution.
4. Answer the specific question in context.

Example 4

A population of bacteria grows from an initial size of 1000. After t hours the size of the population is P. The connection between P and t can be modelled by the equation $\dfrac{dP}{dt} = \dfrac{P(4000-P)}{4000}$

a Solve this equation to show that $P = \dfrac{4000e^t}{3+e^t}$

b Find the size of the population after 15 minutes.

c Prove that the number of bacteria can never exceed 4000

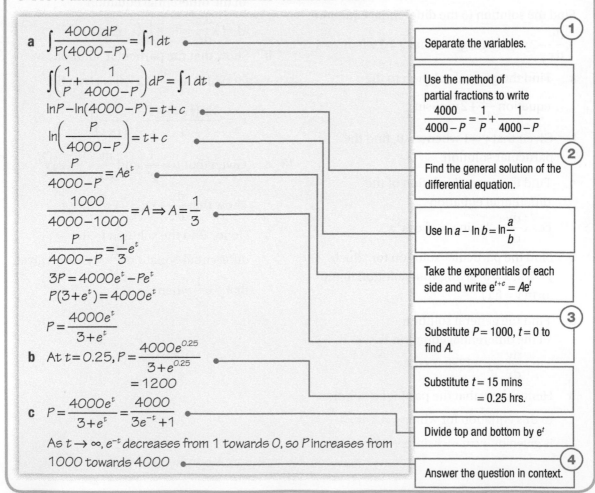

a $\displaystyle\int \frac{4000\,dP}{P(4000-P)} = \int 1\,dt$

Separate the variables. ①

$\displaystyle\int\left(\frac{1}{P}+\frac{1}{4000-P}\right)dP = \int 1\,dt$

Use the method of partial fractions to write $\dfrac{4000}{4000-P} = \dfrac{1}{P}+\dfrac{1}{4000-P}$

$\ln P - \ln(4000-P) = t + c$

$\ln\left(\dfrac{P}{4000-P}\right) = t + c$

Find the general solution of the differential equation. ②

$\dfrac{P}{4000-P} = Ae^t$

$\dfrac{1000}{4000-1000} = A \Rightarrow A = \dfrac{1}{3}$

Use $\ln a - \ln b = \ln\dfrac{a}{b}$

$\dfrac{P}{4000-P} = \dfrac{1}{3}e^t$

$3P = 4000e^t - Pe^t$

$P(3+e^t) = 4000e^t$

Take the exponentials of each side and write $e^{t+c} = Ae^t$

$P = \dfrac{4000e^t}{3+e^t}$

Substitute $P = 1000$, $t = 0$ to find A. ③

b At $t = 0.25$, $P = \dfrac{4000e^{0.25}}{3+e^{0.25}}$

$= 1200$

Substitute $t = 15$ mins $= 0.25$ hrs.

c $P = \dfrac{4000e^t}{3+e^t} = \dfrac{4000}{3e^{-t}+1}$

Divide top and bottom by e^t

As $t \to \infty$, e^{-t} decreases from 1 towards 0, so P increases from 1000 towards 4000

Answer the question in context. ④

Example 5

A raindrop falls vertically through a cloud. Initially the raindrop is at rest. At time t seconds, the velocity, v m s^{-1} of the raindrop, satisfies the equation $(2+t)\dfrac{dv}{dt}+v=10(2+t)$

a Solve this equation to find an expression for v in terms of t

b Find the time at which the velocity of the raindrop is 21 m s^{-1}.

c Criticise the model.

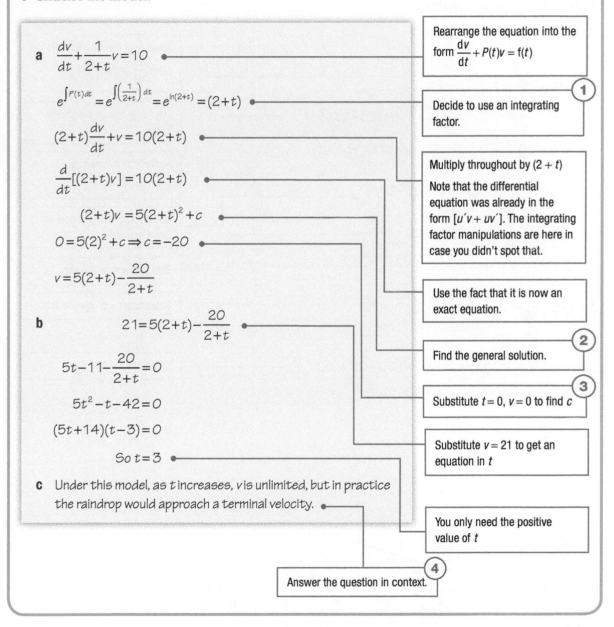

a $\dfrac{dv}{dt}+\dfrac{1}{2+t}v=10$

Rearrange the equation into the form $\dfrac{dv}{dt}+P(t)v=f(t)$

$e^{\int P(t)dt}=e^{\int\left(\frac{1}{2+t}\right)dt}=e^{\ln(2+t)}=(2+t)$

1 Decide to use an integrating factor.

$(2+t)\dfrac{dv}{dt}+v=10(2+t)$

$\dfrac{d}{dt}[(2+t)v]=10(2+t)$

Multiply throughout by $(2+t)$

Note that the differential equation was already in the form $[u'v+uv']$. The integrating factor manipulations are here in case you didn't spot that.

$(2+t)v=5(2+t)^2+c$

$0=5(2)^2+c \Rightarrow c=-20$

$v=5(2+t)-\dfrac{20}{2+t}$

Use the fact that it is now an exact equation.

b $21=5(2+t)-\dfrac{20}{2+t}$

2 Find the general solution.

$5t-11-\dfrac{20}{2+t}=0$

3 Substitute $t=0$, $v=0$ to find c

$5t^2-t-42=0$

$(5t+14)(t-3)=0$

Substitute $v=21$ to get an equation in t

So $t=3$

c Under this model, as t increases, v is unlimited, but in practice the raindrop would approach a terminal velocity.

You only need the positive value of t

4 Answer the question in context.

Example 6

A fuel filled rocket of mass 25 kg burns fuel at a rate of 2 kg s⁻¹. The rocket is initially at rest. After t seconds, the velocity v m s⁻¹ of the rocket satisfies the equation $(25-2t)\dfrac{\mathrm{d}v}{\mathrm{d}t}+v=100$ for $t \le 12$

a Show that $v = 100 - 20\sqrt{25-2t}$

b Find the speed of the rocket when $t = 12$

c Sketch a graph of v against t

d What happens for $t > 12.5$?

a $\dfrac{\mathrm{d}v}{\mathrm{d}t}+\dfrac{1}{25-2t}v=\dfrac{100}{25-2t}$
Rearrange the equation into the form $\dfrac{\mathrm{d}v}{\mathrm{d}t} + P(t)v = f(t)$

$e^{\int P(t)\,\mathrm{d}t} = e^{\int \frac{1}{25-2t}\mathrm{d}t} = e^{-\frac{1}{2}\ln(25-2t)} = (25-2t)^{-\frac{1}{2}}$
Decide to use an integrating factor.

$(25-2t)^{-\frac{1}{2}}\dfrac{\mathrm{d}v}{\mathrm{d}t}+(25-2t)^{-\frac{3}{2}}v=100(25-2t)^{-\frac{3}{2}}$
Multiply throughout by $(25-2t)^{-\frac{1}{2}}$

$\dfrac{\mathrm{d}}{\mathrm{d}t}\left[(25-2t)^{-\frac{1}{2}}v\right]=100(25-2t)^{-\frac{3}{2}}$
Use the fact that it is now an exact equation.

$(25-2t)^{-\frac{1}{2}}v =100(25-2t)^{-\frac{1}{2}}+c$
Find the general solution.

$0=\dfrac{100}{\sqrt{25}}+c \Rightarrow c=-20$
Substitute $t=0$, $v=0$ to find c

$(25-2t)^{-\frac{1}{2}}v =100(25-2t)^{-\frac{1}{2}}-20$

$v=100-20\sqrt{25-2t}$

b $v=100-20\sqrt{25-2(12)} = 100-20=80$
Substitute $t=12$

So the speed of the rocket is 80 m s⁻¹.

c

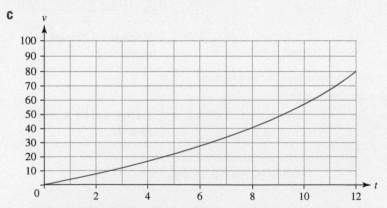

d For $t > 12.5$ the square root becomes negative, so the solution does not make sense. The rocket has burned all of its fuel by $t = 12.5$, so the equation no longer applies.

1 A pan of boiling water is placed in a room where the temperature is 20 °C. A model for the subsequent temperature, T°C of the water, after t minutes is given by Newton's law of cooling as $\dfrac{dT}{dt} = -\dfrac{(T-20)}{8}$

  a Solve this differential equation to show that $T = 20\left(1 + 4e^{-\frac{1}{8}t}\right)$

  b Calculate, to 1 decimal place, the temperature after two minutes.

  c Calculate, to 1 decimal place, the time at which the temperature reaches 25 °C.

2 The number of fish in a lake grows from an initial size of 1000. After t years the size of the population is N. The connection between N and t can be modelled by the equation
$$\frac{dN}{dt} = \frac{N(6000-N)}{12000}$$

  a Solve this equation to show that
$$N = \frac{6000e^{\frac{1}{2}t}}{5 + e^{\frac{1}{2}t}}$$

  b Find the size of the population after 3 years.

  c Prove that the number of fish can never exceed 6000

3 A population grows from an initial size of 10 000. After t years the size of the population is P. The connection between P and t can be modelled by the equation
$$\frac{dP}{dt} = 0.02P(t+1)$$

  a Solve this equation to show that
$$P = 10\,000e^{0.01(t^2+2t)}$$

  b Find the size of the population after six years.

4 In an electric circuit the current, I amps at time t seconds is modelled by the equation
$$\frac{dI}{dt} + 2I = 6$$
Initially $I = 8$

  a Solve this equation to show that $I = 3 + 5e^{-2t}$

  b Show that the current never drops below 3 amps.

  c Sketch a graph of I against t

5 A hailstone falls vertically through a cloud. Initially the hailstone is at rest. At time t seconds, the velocity, v m s^{-1} of the hailstone satisfies the equation $\dfrac{dv}{dt} = 10 - \dfrac{2v}{1+2t}$

  a Solve this equation to find an expression for v in terms of t

  b Find, correct to one decimal place, the time at which the velocity of the hailstone is 24 m s^{-1}.

  c Criticise the model.

6 As part of an industrial process a chemical is dissolved in a solution. t minutes after the start of the process, the mass, C kg, of the chemical that has dissolved can be modelled by the equation $\dfrac{dC}{dt} + \dfrac{C}{20+t} = 4$
Initially $C = 10$ kg.

  a Solve this equation to show that
$$C = 2(20+t) - \frac{600}{20+t}$$

  b Find the time at which $C = 40$

7 Water is draining from a tank. The depth of water in the tank is initially 2 metres, and after t minutes, the depth is x metres. The depth can be modelled by the equation
$$\frac{dx}{dt} = -\frac{1}{2}\left(x + e^{-\frac{1}{2}t}\cos t\right)$$

  a Solve this equation to find an expression for x in terms of t

  b Find the depth of the water in the tank after five minutes.

8 A particle is projected vertically upwards with a velocity of 13 m s^{-1}. During the motion it absorbs moisture so that when the particle is a distance x metres above the point of projection, its velocity, v m s^{-1} is given by the equation $2v\dfrac{dv}{dx} + \dfrac{2}{1+2x}v^2 = -4$

  a Solve this equation to show that
$$v^2(1+2x) = 170 - (1+2x)^2$$

  b Calculate the greatest height reached by the particle.

Second order equations

Fluency and skills

An equation that involves a second order derivative, such as $\dfrac{d^2 y}{dx^2}$, but nothing of higher order, is called a **second order differential equation**. Some second order differential equations can be written in the form

$$a\dfrac{d^2 y}{dx^2} + b\dfrac{dy}{dx} + cy = f(x), \text{ where } a, b \text{ and } c \text{ are constants.}$$

You can solve these types of equation in two stages. The first stage is to solve the related **homogeneous equation**, that is, the equation

$$a\dfrac{d^2 y}{dx^2} + b\dfrac{dy}{dx} + cy = 0$$

> Homogeneous means '= 0'

Example 1

> The full formal method to solve this is as follows. You will see later that there is also a much quicker method.

Solve the differential equation $\dfrac{d^2 y}{dx^2} + \dfrac{dy}{dx} - 6y = 0$

Rewrite the equation as $\dfrac{d^2 y}{dx^2} + 3\dfrac{dy}{dx} - 2\dfrac{dy}{dx} - 6y = 0$	This just uses the fact that $\dfrac{dy}{dx} = 3\dfrac{dy}{dx} - 2\dfrac{dy}{dx}$
	This is very much like solving a quadratic equation.
So $\dfrac{d}{dx}\left[\dfrac{dy}{dx} + 3y\right] - 2\left[\dfrac{dy}{dx} + 3y\right] = 0$	
	Since $\dfrac{d^2 y}{dx^2} + 3\dfrac{dy}{dx} = \dfrac{d}{dx}\left[\dfrac{dy}{dx} + 3y\right]$
Substitute $u = \dfrac{dy}{dx} + 3y$ to get $\dfrac{du}{dx} - 2u = 0$	
So $e^{-2x}\dfrac{du}{dx} - 2e^{-2x}u = 0$	Multiplying throughout by the integrating factor, e^{-2x}
$ue^{-2x} = C$	Integrate each side.
So $u = Ce^{2x} \Rightarrow \dfrac{dy}{dx} + 3y = Ce^{2x}$	Substitute $u = \dfrac{dy}{dx} + 3y$
$e^{3x}\dfrac{dy}{dx} + 3e^{3x}y = Ce^{5x}$	Multiply throughout by the integrating factor, e^{3x}
$e^{3x}y = \dfrac{1}{5}Ce^{5x} + A$	Integrate both sides.
$y = \dfrac{1}{5}Ce^{2x} + Ae^{-3x}$	Rearrange.
The solution is $y = Ae^{-3x} + Be^{2x}$	Relabel the constant $\dfrac{1}{5}C$ as B

You don't need to go through all this working every time. By assuming that solutions exist of the form $y = Ce^{mx}$, where C and m are constants, you can cut out lots of steps of the working.

If $y = Ce^{mx}$, then $\dfrac{dy}{dx} = Cme^{mx}$ and $\dfrac{d^2y}{dx^2} = Cm^2e^{mx}$

If you substitute each of these into the differential equation, then $\dfrac{d^2y}{dx^2} + \dfrac{dy}{dx} - 6y = 0$ becomes $Cm^2e^{mx} + Cme^{mx} - 6Ce^{mx} = 0$, which simplifies to $m^2 + m - 6 = 0$

This is called the auxiliary equation, and when solving these problems, you can use the solution to this equation to complete the first stage of the solution of the differential equation.

So $(m+3)(m-2) = 0 \Rightarrow m = -3$ or $m = 2$, giving solutions $y = Ce^{-3x}$ and $y = Ce^{2x}$

Since either of these solutions satisfies the homogeneous equation, any linear combination of these solutions will also satisfy the homogeneous equation, and the general solution is, therefore, $y = Ae^{-3x} + Be^{2x}$

> As there are two solutions you need two different constants.

This method only works if the auxiliary equation has real distinct roots. There are two other cases for the roots of the auxiliary equation. The equation could have one repeated root, or the roots could be a complex conjugate pair. These three cases need three different complementary functions.

Table 1	
Roots of auxiliary equation $ax^2 + bx + c = 0$	**Complementary function**
Real distinct roots, m_1 and m_2	$y = Ae^{m_1 x} + Be^{m_2 x}$
Repeated root, m	$y = (Ax + B)e^{mx}$
Complex roots $m \pm in$	$y = e^{mx}(A\cos nx + B\sin nx)$

You will be able to prove why these work in the next exercise.

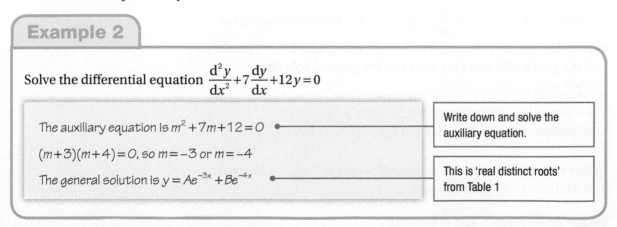

Example 2

Solve the differential equation $\dfrac{d^2y}{dx^2} + 7\dfrac{dy}{dx} + 12y = 0$

The auxiliary equation is $m^2 + 7m + 12 = 0$ — Write down and solve the auxiliary equation.

$(m+3)(m+4) = 0$, so $m = -3$ or $m = -4$

The general solution is $y = Ae^{-3x} + Be^{-4x}$ — This is 'real distinct roots' from Table 1

Example 3

Solve the differential equation $\dfrac{d^2y}{dx^2} - 8\dfrac{dy}{dx} + 16y = 0$

> The auxiliary equation is $m^2 - 8m + 16 = 0$
>
> $(m-4)^2 = 0$, so $m = 4$
>
> The general solution is $y = (Ax + B)e^{4x}$ •————— This is 'repeated root' from Table 1

Example 4

Solve the differential equation $\dfrac{d^2y}{dx^2} - 4\dfrac{dy}{dx} + 13y = 0$

> The auxiliary equation is $m^2 - 4m + 13 = 0$
>
> $(m-2)^2 = -9$, so $m = 2 \pm 3i$
>
> The general solution $y = e^{2x}(A\cos 3x + B\sin 3x)$ •————— This is 'complex roots' from Table 1

The second stage is used to solve equations that are not homogeneous.

To solve an equation like $\dfrac{d^2y}{dx^2} + \dfrac{dy}{dx} - 6y = e^{4x}$, the first step is to solve

the related homogeneous equation $\dfrac{d^2y}{dx^2} + \dfrac{dy}{dx} - 6y = 0$

This solution, $y = Ae^{-3x} + Be^{2x}$, is now called the **complementary function**.

> You have already solved this equation in Example 1

You know that this gives the answer 'zero' when substituted into

$\dfrac{d^2y}{dx^2} + \dfrac{dy}{dx} - 6y$

For the second stage you need to look for a function that will give

e^{4x} when substituted into $\dfrac{d^2y}{dx^2} + \dfrac{dy}{dx} - 6y$. This function is called the

particular integral. Adding together the **complementary function** and the **particular integral** gives you the **general solution**.

Key point

The general solution is obtained by adding together the complementary function and the particular integral.

In this example, the particular integral is easy to find. It is likely to be a multiple of e^{4x}, so try $y = ae^{4x}$

$y = ae^{4x} \Rightarrow \dfrac{dy}{dx} = 4ae^{4x}$ and $\dfrac{d^2y}{dx^2} = 16ae^{4x}$

Substituting gives

$$\frac{d^2y}{dx^2} + \frac{dy}{dx} - 6y = 16ae^{4x} + 4ae^{4x} - 6ae^{4x}$$

$$= 14ae^{4x}$$

This gives

$$14ae^{4x} = e^{4x} \text{ and } a = \frac{1}{14}$$

The **complementary function** is $y = Ae^{-3x} + Be^{2x}$ and the

particular integral is $y = \frac{1}{14}e^{4x}$

Adding the two together, gives the **general solution**

$$y = Ae^{-3x} + Be^{2x} + \frac{1}{14}e^{4x}$$

Not all particular integrals are that easy to find. The following table offers suggestions as to what to try.

Table 2	
Form of f(x)	**Form of particular integral**
C – constant	c – constant
$Mx + C$	$mx + c$
$P_n(x)$ – a polynomial of degree n	$p_n(x)$ – a polynomial of degree n
Ce^{mx}	ce^{mx}
$A\cos kx + B\sin kx$	$a\cos kx + b\sin kx$
$A\cosh kx + B\sinh kx$	$a\cosh kx + b\sinh kx$

Example 5

Solve the differential equation $\dfrac{d^2y}{dx^2} - 4\dfrac{dy}{dx} + 4y = 12x - 8$

The auxiliary equation is $m^2 - 4m + 4 = 0$

$(m-2)^2 = 0$, so $m = 2$

So the complementary function is $y = (Ax+B)e^{2x}$ — This is the 'repeated root' type of equation from Table 1

For the particular integral try $y = ax + b$ — This is from Table 2

$\dfrac{dy}{dx} = a$ and $\dfrac{d^2y}{dx^2} = 0$

So $\dfrac{d^2y}{dx^2} - 4\dfrac{dy}{dx} + 4y = -4a + 4(ax+b) = 4ax + (-4a+4b) = 12x - 8$

$4a = 12$, and $-4a + 4b = -8$ — Equate coefficients.

$a = 3$ and $b = 1$

So the particular integral is $y = 3x + 1$

and the general solution is $y = (Ax+B)e^{2x} + 3x + 1$ — Add together the complementary function and the particular integral to get the general solution.

1 Find the general solution to each of the following differential equations.

a $\dfrac{d^2y}{dx^2}-6\dfrac{dy}{dx}+8y=0$

b $\dfrac{d^2y}{dx^2}+8\dfrac{dy}{dx}+16y=0$

c $\dfrac{d^2y}{dx^2}-4\dfrac{dy}{dx}+5y=0$

d $\dfrac{d^2y}{dx^2}+3\dfrac{dy}{dx}=0$

e $\dfrac{d^2y}{dx^2}+\dfrac{dy}{dx}-12y=0$

f $\dfrac{d^2y}{dx^2}+y=0$

g $2\dfrac{d^2y}{dx^2}+5\dfrac{dy}{dx}-3y=0$

h $4\dfrac{d^2y}{dx^2}+4\dfrac{dy}{dx}+5y=0$

2 Find the general solution to each of the following differential equations.

a $\dfrac{d^2y}{dx^2}+2\dfrac{dy}{dx}-15y=19-30x$

b $\dfrac{d^2y}{dx^2}-6\dfrac{dy}{dx}+9y=3e^{2x}$

c $\dfrac{d^2y}{dx^2}+2\dfrac{dy}{dx}+17y=8\cos3x-6\sin3x$

d $\dfrac{d^2y}{dx^2}-16y=32x-48$

e $\dfrac{d^2y}{dx^2}+3\dfrac{dy}{dx}-4y=4e^{-3x}$

f $\dfrac{d^2y}{dx^2}+2\dfrac{dy}{dx}+y=x^2+4x+7$

g $3\dfrac{d^2y}{dx^2}-8\dfrac{dy}{dx}-3y=5e^{2x}$

h $4\dfrac{d^2y}{dx^2}+4\dfrac{dy}{dx}+y=5x+18$

3 a Show that the differential equation

$\dfrac{d^2y}{dx^2}-2a\dfrac{dy}{dx}+a^2y=0$ can be written as

$\dfrac{du}{dx}-a\times u=0$, where a is a constant, and

$u=\dfrac{dy}{dx}-a\times y$

b Solve the differential equation
$\dfrac{du}{dx}-a\times u=0$

c Hence show that $y=(Ax+B)e^{ax}$, where A and B are constants.

4 If the auxiliary equation has complex conjugate roots $m\pm in$, use Euler's formula to deduce that the general solution
$y=Ae^{(m+in)x}+Be^{(m-in)x}$ can be expressed as
$y=e^{mx}(\alpha\cos nx+\beta\sin nx)$ for constants α and β

Reasoning and problem-solving

Strategy

To answer a question that involves a second order differential equation

1. Write down the auxiliary equation.
2. Find the complementary function.
3. Decide on the form of the particular integral.
4. Find the particular integral.
5. Write down the general solution.
6. Use the information given in the question to find the values of the constants.
7. Answer the specific question(s) in context.

Example 6

Solve the differential equation $\dfrac{d^2 y}{dx^2} + 2\dfrac{dy}{dx} + 10y = 13\cos x + 16\sin x$ given that $y(0) = 1$ and $y'(0) = 8$

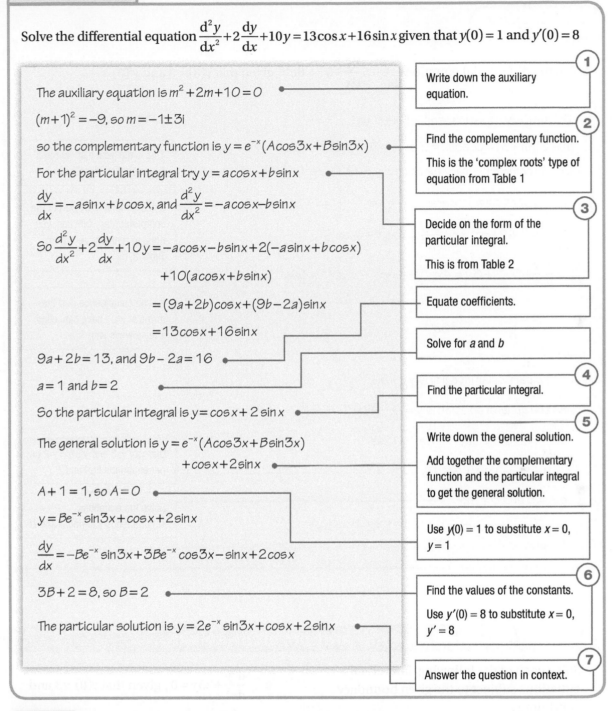

The auxiliary equation is $m^2 + 2m + 10 = 0$

1 Write down the auxiliary equation.

$(m+1)^2 = -9$, so $m = -1 \pm 3i$

so the complementary function is $y = e^{-x}(A\cos 3x + B\sin 3x)$

2 Find the complementary function.

This is the 'complex roots' type of equation from Table 1

For the particular integral try $y = a\cos x + b\sin x$

$\dfrac{dy}{dx} = -a\sin x + b\cos x$, and $\dfrac{d^2 y}{dx^2} = -a\cos x - b\sin x$

3 Decide on the form of the particular integral.

This is from Table 2

So $\dfrac{d^2 y}{dx^2} + 2\dfrac{dy}{dx} + 10y = -a\cos x - b\sin x + 2(-a\sin x + b\cos x)$

$\qquad\qquad\qquad + 10(a\cos x + b\sin x)$

$\qquad\qquad = (9a + 2b)\cos x + (9b - 2a)\sin x$

Equate coefficients.

$\qquad\qquad = 13\cos x + 16\sin x$

Solve for a and b

$9a + 2b = 13$, and $9b - 2a = 16$

$a = 1$ and $b = 2$

4 Find the particular integral.

So the particular integral is $y = \cos x + 2\sin x$

The general solution is $y = e^{-x}(A\cos 3x + B\sin 3x)$

$\qquad\qquad\qquad + \cos x + 2\sin x$

5 Write down the general solution.

Add together the complementary function and the particular integral to get the general solution.

$A + 1 = 1$, so $A = 0$

$y = Be^{-x}\sin 3x + \cos x + 2\sin x$

Use $y(0) = 1$ to substitute $x = 0$, $y = 1$

$\dfrac{dy}{dx} = -Be^{-x}\sin 3x + 3Be^{-x}\cos 3x - \sin x + 2\cos x$

$3B + 2 = 8$, so $B = 2$

6 Find the values of the constants.

Use $y'(0) = 8$ to substitute $x = 0$, $y' = 8$

The particular solution is $y = 2e^{-x}\sin 3x + \cos x + 2\sin x$

7 Answer the question in context.

Key point

When you have found the values of the constants in the general solution, the final answer is called the **particular solution**.

If the form of the particular integral you try matches the complementary function, it will not work. You simply end up with '$0 = 0$'. As long as this is not the 'repeated root' case in Table 1 you should simply multiply the function you try for the particular integral by x. In the case of the 'repeated root' multiplying by x will also give '$0 = 0$'. In this case, instead of multiplying by x, you multiply by x^2

These cases are illustrated in Example 7, and in question 6 of Exercise 9.2B.

Example 7

Solve the differential equation $\dfrac{d^2y}{dx^2} - 5\dfrac{dy}{dx} + 6y = 6e^{2x}$, given that $y(0) = 3$ and $y'(0) = -2$

The auxiliary equation is $m^2 - 5m + 6 = 0$

$(m-2)(m-3) = 0$, so $m = 2$ or $m = 3$

So the complementary function is $y = Ae^{2x} + Be^{3x}$ ●————

> This is the 'real distinct roots' type equation from Table 1

For the particular integral try $y = axe^{2x}$ ●————

$\dfrac{dy}{dx} = ae^{2x} + 2axe^{2x}$, and $\dfrac{d^2y}{dx^2} = 4ae^{2x} + 4axe^{2x}$

So $\dfrac{d^2y}{dx^2} - 5\dfrac{dy}{dx} + 6y = 4ae^{2x} + 4axe^{2x} - 5(ae^{2x} + 2axe^{2x}) + 6axe^{2x}$

> Since e^{2x} is already part of the complementary function, you need to multiply ae^{2x}, from Table 2, by x

$= -ae^{2x}$ ●————

$= 6e^{2x}$

> It is no coincidence that the terms in xe^{2x} have cancelled. They always will.

$a = -6$

So the particular integral is $y = -6xe^{2x}$

and the general solution is $y = Ae^{2x} + Be^{3x} - 6xe^{2x}$

$\Rightarrow \dfrac{dy}{dx} = 2Ae^{2x} + 3Be^{3x} - 6e^{2x} - 12xe^{2x}$

$A + B = 3$, and $2A + 3B - 6 = -2$ ●————

> Use $y(0) = 3$ and $y'(0) = -2$ to get equations in A and B

so $A = 5$ and $B = -2$ ●————

The particular solution is $y = 5e^{2x} - 2e^{3x} - 6xe^{2x}$

> Solve the equations simultaneously to find A and B

Exercise 9.2B Reasoning and problem-solving

1 Solve each of the following differential equations subject to the given boundary conditions.

a $\dfrac{d^2y}{dx^2} - 7\dfrac{dy}{dx} + 6y = 0$, given that $y(0) = 3$ and $y'(0) = 8$

b $\dfrac{d^2y}{dx^2} + 4\dfrac{dy}{dx} + 4y = 0$, given that $y(0) = 1$ and $y'(0) = 2$

c $\dfrac{d^2y}{dx^2} + 25y = 0$, given that $y(0) = 3$ and $y\left(\dfrac{\pi}{10}\right) = 6$

d $\dfrac{d^2y}{dx^2} + \dfrac{dy}{dx} - 6y = 0$, given that $y(0) = 1$ and $y'(0) = 12$

e $\dfrac{d^2y}{dx^2}-6\dfrac{dy}{dx}+9y=0$, given that $y(0)=5$ and $y'(0)=3$

f $\dfrac{d^2y}{dx^2}-\dfrac{dy}{dx}-12y=0$, given that $y(0)=12$ and $y'(0)=6$

g $9\dfrac{d^2y}{dx^2}-12\dfrac{dy}{dx}+4y=0$, given that $y(0)=12$ and $y'(0)=8$

h $4\dfrac{d^2y}{dx^2}-4\dfrac{dy}{dx}+5y=0$, given that $y(0)=6$ and $y'(0)=-1$

2 Solve each of the following differential equations subject to the given boundary conditions.

a $\dfrac{d^2y}{dx^2}-7\dfrac{dy}{dx}+10y=12e^x$, given that $y(0)=13$ and $y'(0)=8$

b $\dfrac{d^2y}{dx^2}-12\dfrac{dy}{dx}+36y=72$, given that $y(0)=3$ and $y'(0)=-2$

c $\dfrac{d^2y}{dx^2}+4\dfrac{dy}{dx}+5y=16\cos x$, given that $y(0)=5$ and $y'(0)=-6$

d $\dfrac{d^2y}{dx^2}+2\dfrac{dy}{dx}-8y=30-24x$, given that $y(0)=1$ and $y'(0)=0$

e $9\dfrac{d^2y}{dx^2}-4y=4x^2+2$, given that $y(0)=-8$ and $y'(0)=10$

f $\dfrac{d^2y}{dx^2}-4\dfrac{dy}{dx}+5y=\cos 2x+8\sin 2x$, given that $y(0)=9$ and $y\left(\dfrac{\pi}{2}\right)=4e^\pi-1$

g $4\dfrac{d^2y}{dx^2}-9y=26+27x-9x^2$, given that $y(0)=8$ and $y'(0)=-9$

h $16\dfrac{d^2y}{dx^2}-8\dfrac{dy}{dx}+y=12e^{-\frac{1}{4}x}$, given that $y(0)=10$ and $y'(0)=2$

3 Given that the differential equation $\dfrac{d^2y}{dx^2}-2\dfrac{dy}{dx}-8y=12e^{-2x}$ has a particular integral of the form $y=axe^{-2x}$, determine the value of the constant a and find the general solution of the differential equation.

4 Given that the differential equation $\dfrac{d^2y}{dx^2}-2\dfrac{dy}{dx}+y=6e^x$ has a particular integral of the form $y=ax^2e^x$, determine the value of the constant a and find the general solution of the differential equation.

5 Find the solution of the differential equation $\dfrac{d^2y}{dx^2}+5\dfrac{dy}{dx}-6y=21e^x+12$ for which $y=-2$ and $\dfrac{dy}{dx}=17$ at $x=0$

6 Find the solution of the differential equation $\dfrac{d^2y}{dx^2}-6\dfrac{dy}{dx}+9y=34e^{3x}$ for which $y=3$ and $\dfrac{dy}{dx}=11$ at $x=0$

7 Solve the differential equation $\dfrac{d^3y}{dx^3}-6\dfrac{d^2y}{dx^2}+11\dfrac{dy}{dx}-6y=12e^{4x}$, given that $y(0)=4$, $y'(0)=11$ and $y''(0)=35$

8 a Given that $x=e^u$ and $x^2\dfrac{d^2y}{dx^2}-4x\dfrac{dy}{dx}+6y=12$, show that $\dfrac{d^2y}{du^2}-5\dfrac{dy}{du}+6y=12$

b Hence solve the equation $x^2\dfrac{d^2y}{dx^2}-4x\dfrac{dy}{dx}+6y=12$, given that $y(1)=7$ and $y(2)=14$

Simple harmonic motion

Fluency and skills

Differential equations can be used to model many different situations. In particular, they can be used to describe things that oscillate. This includes the motion of a weight hanging on the end of a spring, moving up and down, or a pendulum swinging from side to side, or the depth of the water in a harbour as the tide moves in and out. Oscillatory motions like these follow a regular, cyclic pattern. If you plot a displacement–time graph of the motion of almost any oscillatory system, you will find that it will produce something like a sine graph.

Example 1

Solve the differential equation $\dfrac{d^2 x}{dt^2} + 4x = 0$, given that $x(0) = 4$ and $x\left(\dfrac{\pi}{4}\right) = -4$, and hence sketch the solution.

The auxiliary equation is $m^2 + 4 = 0 \Rightarrow m^2 = -4 \Rightarrow m = \pm 2i$

So the general solution is $x = A\cos 2t + B\sin 2t$

$A = 4$ •

$B = -4$ •

> Use $x(0) = 4$ to substitute $t = 0$, $x = 4$

The particular solution is $x = 4\cos 2t - 4\sin 2t$

$$= 4\sqrt{2}\left(\frac{1}{\sqrt{2}}\cos 2t - \frac{1}{\sqrt{2}}\sin 2t\right)$$

$$= 4\sqrt{2}\cos\left(2t + \frac{\pi}{4}\right)$$

> Use $x\left(\dfrac{\pi}{4}\right) = -4$ to substitute $t = \dfrac{\pi}{4}$, $x = -4$

> Use harmonic form.

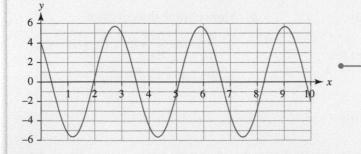

> Try plotting this on your graphical calculator.

Motion that satisfies a differential equation of the form
$\frac{d^2x}{dt^2} + k^2x = 0$ is called **simple harmonic motion**.

Example 2

Solve the differential equation $\frac{d^2x}{dt^2} + 2\frac{dx}{dt} + 17x = 34$, given that $x(0) = 2$ and $x'(0) = 20$, and hence sketch the solution.

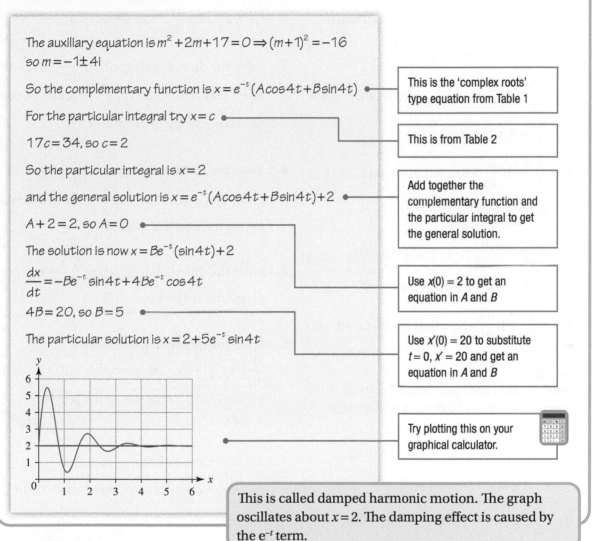

The auxiliary equation is $m^2 + 2m + 17 = 0 \Rightarrow (m+1)^2 = -16$
so $m = -1 \pm 4i$

So the complementary function is $x = e^{-t}(A\cos 4t + B\sin 4t)$ •——— This is the 'complex roots' type equation from Table 1

For the particular integral try $x = c$ •

$17c = 34$, so $c = 2$ ——— This is from Table 2

So the particular integral is $x = 2$

and the general solution is $x = e^{-t}(A\cos 4t + B\sin 4t) + 2$ •——— Add together the complementary function and the particular integral to get the general solution.

$A + 2 = 2$, so $A = 0$ •

The solution is now $x = Be^{-t}(\sin 4t) + 2$

$\frac{dx}{dt} = -Be^{-t}\sin 4t + 4Be^{-t}\cos 4t$ ——— Use $x(0) = 2$ to get an equation in A and B

$4B = 20$, so $B = 5$ •

The particular solution is $x = 2 + 5e^{-t}\sin 4t$ ——— Use $x'(0) = 20$ to substitute $t = 0$, $x' = 20$ and get an equation in A and B

Try plotting this on your graphical calculator.

This is called damped harmonic motion. The graph oscillates about $x = 2$. The damping effect is caused by the e^{-t} term.

Damping can be further classified into three different cases that correspond to the three different possibilities for the auxiliary equation.

- If the auxiliary equation has complex roots, this is called light damping, and the body oscillates as it approaches zero amplitude.
- If the auxiliary equation has distinct real roots, this is called heavy damping, and there is no oscillation as the body approaches zero amplitude.
- If the auxiliary equation has a repeated root, this is called critical damping, and the body approaches zero amplitude as quickly as possible, without oscillating.

1 Solve each of the following differential equations subject to the given boundary conditions.

   a $\dfrac{d^2x}{dt^2} + 25x = 0$, given that $x = 7$ when $t = 0$ and $x = 5$ when $t = \dfrac{\pi}{10}$

   b $\dfrac{d^2x}{dt^2} + 4x = 0$, given that $x = 2\sqrt{2}$ when $t = \dfrac{\pi}{8}$ and $x = 8$ when $t = \dfrac{\pi}{4}$

   c $\dfrac{d^2x}{dt^2} - 6\dfrac{dx}{dt} + 13x = 0$, given that $x = -5$ and $\dfrac{dx}{dt} = 3$ when $t = 0$

   d $\dfrac{d^2x}{dt^2} + 10\dfrac{dx}{dt} + 41x = 0$, given that $x = -1$ and $\dfrac{dx}{dt} = 9$ when $t = 0$

   e $2\dfrac{d^2x}{dt^2} + 10\dfrac{dx}{dt} + 13x = 0$, given that $x = 6$ and $\dfrac{dx}{dt} = -8$ when $t = 0$

   f $9\dfrac{d^2x}{dt^2} - 6\dfrac{dx}{dt} + 5x = 0$, given that $x = 8$ and $\dfrac{dx}{dt} = 4$ when $t = 0$

2 Solve each of the following differential equations subject to the given boundary conditions.

   a $\dfrac{d^2x}{dt^2} + 16x = 48$, given that $x = 0$ when $t = 0$ and $x = 5$ when $t = \dfrac{\pi}{8}$

   b $\dfrac{d^2x}{dt^2} + 9x = 9t^2 + 36t + 20$, given that $x = 12$ and $\dfrac{dx}{dt} = 22$ when $t = 0$

   c $\dfrac{d^2x}{dt^2} + 4x = 12(\sin 4t - \cos 4t)$, given that $x = 3$ when $t = \dfrac{\pi}{4}$, and $x = -1$ when $t = \dfrac{\pi}{2}$

   d $\dfrac{d^2x}{dt^2} + 10\dfrac{dx}{dt} + 29x = 29t - 48$, given that $x = 6$ and $\dfrac{dx}{dt} = -45$ when $t = 0$

   e $\dfrac{d^2x}{dt^2} + 4\dfrac{dx}{dt} + 5x = 20$, given that $x = 16$ when $t = 0$ and $x = 4 - 5e^{-\pi}$ when $t = \dfrac{\pi}{2}$

   f $\dfrac{d^2x}{dt^2} + 2\dfrac{dx}{dt} + 2x = 2\cos t + \sin t$, given that $x = 7$ and $\dfrac{dx}{dt} = -12$ when $t = 0$

3 Solve the differential equation $\dfrac{d^2x}{dt^2} + 25x = 0$, given that $x(0) = 4$ and $x\left(\dfrac{\pi}{10}\right) = 3$, and hence sketch the graph of x against t

4 Solve the differential equation $\dfrac{d^2x}{dt^2} + 9x = 18$, given that $x(0) = 7$ and $x\left(\dfrac{\pi}{2}\right) = 14$, and hence sketch the graph of x against t

5 Solve the differential equation $\dfrac{d^2x}{dt^2} + 4x = 8t$, given that $x(0) = 4$ and $x'(0) = 8$

6 Solve the differential equation $\dfrac{d^2x}{dt^2} + 2\dfrac{dx}{dt} + 5x = 20$, given that $x(0) = 4$ and $x'(0) = 2$, and hence sketch the graph of x against t

7 a Solve the differential equation $\dfrac{d^2x}{dt^2} + 2\dfrac{dx}{dt} + 2x = 2e^{-t}$, given that $x(0) = 1$ and $x'(0) = -2$

   b Show that $x > 0$ and $\dfrac{dx}{dt} \le 0$, for all values of t

   c Sketch a graph of x against t

To answer a question that involves oscillations

(1) Find the general solution to the differential equation.

(2) Use the information given in the question to find the particular solution.

(3) Answer the specific questions in context.

Example 3

Note that this example requires a knowledge of Hooke's law which you will meet if you study the Further Mechanics options.

One end of a light elastic spring of natural length $4\,\text{m}$ and modulus of elasticity $72\,\text{N}$ is fixed to a point A on a smooth horizontal table. A body of mass $2\,\text{kg}$ is attached to the other end of the spring and is released from rest at the point B, where $AB = 5\,\text{m}$. After t seconds the body is at the point P, where $AP = x\,\text{m}$.

a Show that $\dfrac{d^2x}{dt^2} + 9x = 36$

b Solve the differential equation to get an expression for x in terms of t

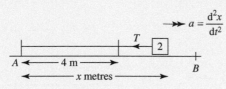

a From Hooke's law the tension in the string is given by $T = \dfrac{72(x-4)}{4}$

> Hooke's law states that $T = \dfrac{\lambda x}{l}$ where λ is the modulus of the spring, x is its extension, and l is its natural length.

Applying $F = ma$ gives $\quad -\dfrac{72(x-4)}{4} = 2\dfrac{d^2x}{dt^2}$

$\Rightarrow -18x + 72 = 2\dfrac{d^2x}{dt^2}$

So $\dfrac{d^2x}{dt^2} + 9x = 36$

b The auxiliary equation is $m^2 + 9 = 0$

> This is the complex roots type equation from Table 1

$m^2 = -9 \Rightarrow m = \pm 3i$

So the complementary function is $x = A\cos 3t + B\sin 3t$

> This is from Table 2

For the particular integral try $x = c$

$9c = 36$, so $c = 4$

So the particular integral is $x = 4$

and the general solution is $x = A\cos 3t + B\sin 3t + 4$

> Add together the complementary function and the particular integral to get the general solution.

$A + 4 = 5$, so $A = 1$

> Use $x = 5$ when $t = 0$ to get A

The solution is now $x = \cos 3t + B\sin 3t + 4$

$\dfrac{dx}{dt} = -3\sin 3t + 3B\cos 3t$

> Use $t = 0$, the body is at rest, so $\dfrac{dx}{dt} = 0$

$0 = 3B$, so $B = 0$

The particular solution is $x = \cos 3t + 4$

Example 4

A particle P of mass $2\,\text{kg}$ moves along the positive x-axis under the action of a force directed towards the origin. At time t seconds, the displacement of P from O is x metres, and P is moving away from O with a speed of $v\,\text{m s}^{-1}$. The force has magnitude $40x\,\text{N}$. The particle, P, is also subject to a resistive force of magnitude $16v\,\text{N}$.

a Show that the equation of motion of P is $\dfrac{d^2x}{dt^2}+8\dfrac{dx}{dt}+20x=0$

Given also that $x(0)=0$ and $x\left(\dfrac{\pi}{4}\right)=3e^{-\pi}$

b Solve the differential equation, and find an expression for x in terms of t

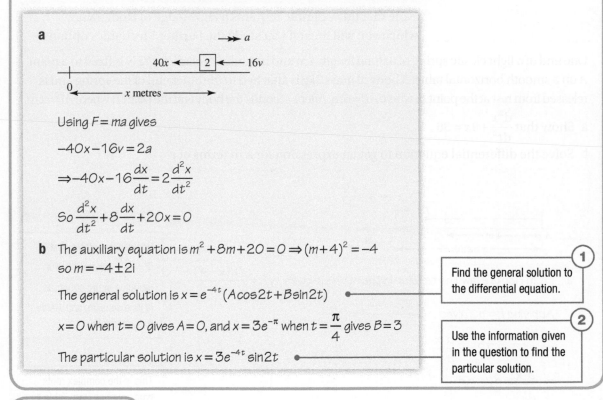

a

Using $F=ma$ gives

$-40x-16v=2a$

$\Rightarrow -40x-16\dfrac{dx}{dt}=2\dfrac{d^2x}{dt^2}$

So $\dfrac{d^2x}{dt^2}+8\dfrac{dx}{dt}+20x=0$

b The auxiliary equation is $m^2+8m+20=0 \Rightarrow (m+4)^2=-4$

so $m=-4\pm2i$

The general solution is $x=e^{-4t}(A\cos 2t+B\sin 2t)$

$x=0$ when $t=0$ gives $A=0$, and $x=3e^{-\pi}$ when $t=\dfrac{\pi}{4}$ gives $B=3$

The particular solution is $x=3e^{-4t}\sin 2t$

(1) Find the general solution to the differential equation.

(2) Use the information given in the question to find the particular solution.

Example 5

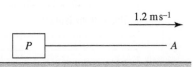

$1.2\,\text{ms}^{-1}$

A box, P, is attached to one end of a light elastic string. The box is initially at rest on a smooth horizontal table, when the other end of the spring, A, starts to move away from P with constant velocity $1.2\,\text{m s}^{-1}$. The displacement, y metres, of P from its initial position, at a time t seconds from the start, satisfies the differential equation $\dfrac{d^2y}{dt^2}+25y=10t$

a Find an expression for y in terms of t

b Calculate the time when P is next at rest.

c Sketch the graph of y against t

(Continued on the next page)

a The auxiliary equation is $m^2 + 25 = 0$

$m^2 = -25$, so $m = \pm 5i$

So the complementary function is $y = A\cos5t + B\sin5t$

For the particular integral try $y = at + b$

Then $y' = a$ and $y'' = 0$

$25(at + b) = 10t$

giving $a = \dfrac{2}{5}$ and $b = 0$

So the particular integral is $y = \dfrac{2}{5}t$

The general solution is $y = A\cos5t + B\sin5t + \dfrac{2}{5}t$

$A = 0$

The solution is now $y = B\sin5t + \dfrac{2}{5}t$

$\dfrac{dy}{dt} = 5B\cos5t + \dfrac{2}{5}$

$5B + \dfrac{2}{5} = 0$, so $B = -\dfrac{2}{25}$

The particular solution is $y = -\dfrac{2}{25}\sin5t + \dfrac{2}{5}t$

b $\dfrac{dy}{dt} = -\dfrac{2}{5}\cos5t + \dfrac{2}{5}$

$-\dfrac{2}{5}\cos5t + \dfrac{2}{5} = 0$

$\cos 5t = 1$

$t = \dfrac{2\pi}{5}\,\text{s}$

c

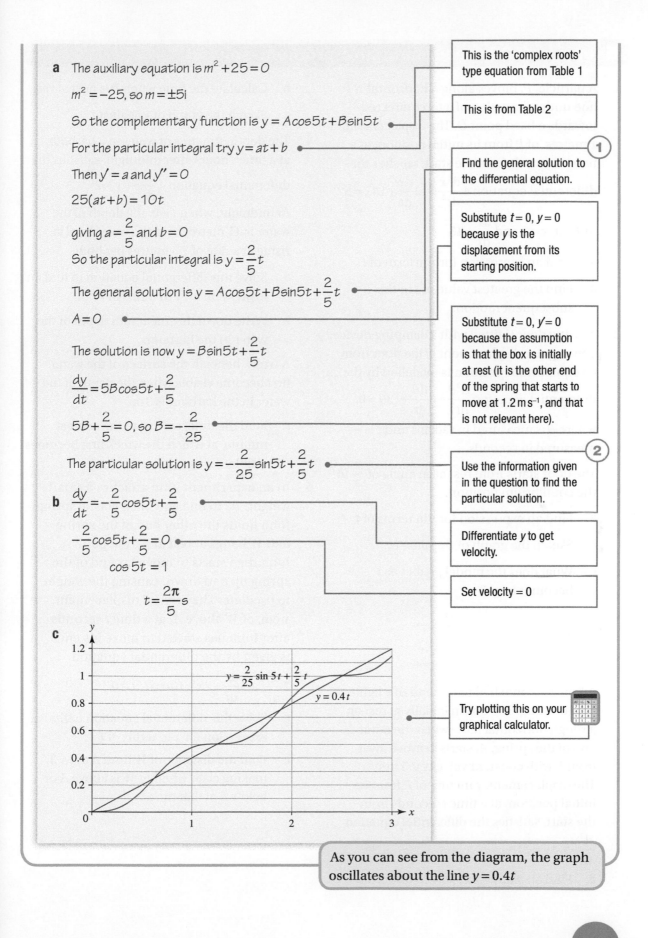

$y = \dfrac{2}{25}\sin 5t + \dfrac{2}{5}t$

$y = 0.4t$

Annotation boxes:

- This is the 'complex roots' type equation from Table 1
- This is from Table 2

1 Find the general solution to the differential equation.

- Substitute $t = 0$, $y = 0$ because y is the displacement from its starting position.
- Substitute $t = 0$, $y' = 0$ because the assumption is that the box is initially at rest (it is the other end of the spring that starts to move at $1.2\,\text{m s}^{-1}$, and that is not relevant here).

2 Use the information given in the question to find the particular solution.

- Differentiate y to get velocity.
- Set velocity = 0
- Try plotting this on your graphical calculator.

As you can see from the diagram, the graph oscillates about the line $y = 0.4t$

1   A particle, P, moves along a horizontal line under the action of a force directed towards a fixed point O. The displacement, x metres, of P from its initial position, at a time t seconds from the start, satisfies the differential equation $4\dfrac{d^2x}{dt^2}+12\dfrac{dx}{dt}+13x=52$

At $t=0$, $x=2$ and $\dfrac{dx}{dt}=13$

   a   Find an expression for x in term of t

   b   Find the greatest value of x in the subsequent motion.

2   A swing door is fitted with a damping device. The angular displacement of the door from its equilibrium position is modelled by the differential equation $\dfrac{d^2\theta}{dt^2}+6\dfrac{d\theta}{dt}+13\theta=0$, where θ is measured in radians and t is measured in seconds.

   The door starts from rest at an angle of $\dfrac{\pi}{4}$ to the equilibrium position.

   a   Find an expression for θ in terms of t

   b   Sketch the graph of θ against t

   c   What does the model predict as t becomes large?

3

   A box, P, is attached to one end of a light elastic string. The box in initially at rest on a rough horizontal table when the other end of the spring, A, starts to move away from P with constant velocity $0.5\ \text{m s}^{-1}$. The displacement, y metres, of P from its initial position, at a time t seconds from the start, satisfies the differential equation $\dfrac{d^2y}{dt^2}+4y=2t+1$

   a   Find an expression for y in terms of t

   b   Calculate the time when P is next at rest.

   c   Sketch the graph of y against t

4   The depth, y metres of water in a harbour, at a time t hours after midnight, satisfies the differential equation $4\dfrac{d^2y}{dt^2}+y=7$

   At midnight, when $t=0$, the depth of the water is 11 metres, and the water level is rising at a rate of 1.5 metres per hour.

   a   Solve this differential equation to find an expression for y in terms of t

   b   Write down the maximum depth of the water in the harbour.

   A wreck beneath the surface of the water first become visible when the depth of the water in the harbour is $3\,\text{m}$.

   c   Find the time, correct to the nearest minute, at which the wreck first becomes visible.

5   In an experiment John attaches a small weight, W, to one end of an elastic spring. John holds the other end of the spring, and W hangs at rest at a fixed point A. John then starts to move his end of the spring up and down, causing the weight to oscillate. The vertical displacement, x cm, of W above A, at a time t seconds after John has started to move his end is given by the differential equation $2\dfrac{d^2x}{dt^2}+5\dfrac{dx}{dt}+2x=10\cos t,\ t\geq 0$

   a   Solve this differential equation to find an expression for x in terms of t

   b   Find the distance of W from A at $t=3$, making clear whether W is above A or below A at that time.

6 When an alternating electromotive force of $2\cos 2t$ is applied across an electrical circuit, the current, I amps at a time t seconds satisfies the differential equation

$$\frac{\mathrm{d}^2 I}{\mathrm{d}t^2} + 16I = 36\sin 2t$$

At $t = 0$, $I = 0$ and $\dfrac{\mathrm{d}I}{\mathrm{d}t} = 14$

a Find an expression for I in term of t

b Find the time when the current first returns to zero.

7 The differential equation

$$\frac{\mathrm{d}^2 x}{\mathrm{d}t^2} + \frac{\mathrm{d}x}{\mathrm{d}t} + \frac{x}{2} = \frac{3}{4}\mathrm{e}^{-\frac{1}{2}t}$$ describes the motion of a particle along the x-axis, where x, measured in metres, is the displacement of the particle from the origin at time t seconds. At time $t = 0$, $x = 5$ and $\dfrac{\mathrm{d}x}{\mathrm{d}t} = -\dfrac{3}{2}$

a Solve this differential equation to get an expression for x in terms of t

b Prove that the particle never reaches the origin.

c Calculate the value of t when the particle first comes to rest.

8 The differential equation $\dfrac{\mathrm{d}^2 V}{\mathrm{d}t^2} + 36V = \sin 6t$ models the voltage, V, in an electrical circuit at a time t seconds. Initially $V = 0$ and $\dfrac{\mathrm{d}V}{\mathrm{d}t} = 0$

a Given that the particular integral is of the form $\lambda t \cos 6t$, find the value of the constant λ

b Find the particular solution of the differential equation.

c What happens to V as t becomes large?

9 A particle moves along a horizontal line under the action of a force directed towards a fixed point O. The displacement, x metres, of P from its initial position, at a time t seconds from the start, satisfies the differential equation $\dfrac{\mathrm{d}^2 x}{\mathrm{d}t^2} + 2\dfrac{\mathrm{d}x}{\mathrm{d}t} + 2x = 0$

At $t = 0$, $x = 1$ and $\dfrac{\mathrm{d}x}{\mathrm{d}t} = 0$

a Find an expression for x in term of t

b Find the times at which P is at rest.

c Show that the total distance moved by P will not exceed $\coth\left(\dfrac{\pi}{2}\right)$

10

A box, B, of mass 5 kg is attached to one end of a light elastic spring of natural length 2 metres and modulus 90 N. The box is initially at rest on a smooth horizontal table, with the spring in equilibrium, when the other end of the spring, A, starts to move away from B with constant velocity $3\,\mathrm{m\,s^{-1}}$. The air resistance acting on B has magnitude $30v$ N, where $v\,\mathrm{ms^{-1}}$ is the speed of B. At time t seconds, the extension of the string is x metres and the displacement of B from its initial position is y metres.

a Show that

i $x + y = 3t$

ii $\dfrac{\mathrm{d}^2 x}{\mathrm{d}t^2} + 6\dfrac{\mathrm{d}x}{\mathrm{d}t} + 9x = 18$

iii $x = 2 - (2 + 3t)\mathrm{e}^{-3t}$

b Describe the motion of the box for large values of t

Modelling systems

Fluency and skills

Oscillations are just one area that can be modelled by differential equations. There are many others. For example, if you leave a hot cup of tea to cool, the temperature at any time can be modelled using calculus. Radioactive decay, or the absorption of a drug into the bloodstream, can be modelled with a differential equation. Throw a ball into the air and bring in air resistance or the change in the value of g due to the inverse square law, or even the rotation of the Earth, which is called the Coriolis effect, and again you need a differential equation to model the motion.

Example 1

A group of scientists is studying the population of rabbits on an island. The population is modelled by the differential equation $\dfrac{dN}{dt} = \dfrac{1}{4}N(4-N)$, where N, in thousands, is the number of rabbits and t is the time, measured in years, since the study began. At the start of the study there are 1000 rabbits on the island.

Solve the differential equation, and hence show that there can never be more than 4000 rabbits on the island. Comment on your answer.

$$\frac{dN}{dt} = \frac{1}{4}N(4-N)$$

$$\int\left(\frac{4}{N(4-N)}\right)dN = \int 1\,dt$$

Separate the variables.

$$\int\left(\frac{1}{N}+\frac{1}{4-N}\right)dN = \int 1\,dt$$

Use the method of partial fractions to write
$$\frac{4}{N(4-N)} = \frac{1}{N}+\frac{1}{4-N}$$

$$\ln c\left(\frac{N}{4-N}\right) = t$$

$$\ln c\left(\frac{1}{4-1}\right) = 0$$

Substitute $t=0$, $N=1$

$c=3$, and $\dfrac{3N}{4-N} = e^t$

Rearrange to get $N = \dfrac{4e^t}{3+e^t} \Rightarrow N = \dfrac{4}{3e^{-t}+1}$

Divide top and bottom by e^t

Now $e^{-t} > 0$, so $(3e^{-t}+1) > 1$, so $\dfrac{4}{3e^{-t}+1} < 4$, and N never exceeds 4

Being an island, there is a limited amount of food. If the population grows too large, there will not be enough food to go round, and the population will decline.

1 Water evaporates from a conical tank of depth 50 cm. Initially the tank is full. After t days, the depth, h cm, of water in the tank satisfies the differential equation $\dfrac{dh}{dt} = -\dfrac{h^2}{100}$

a Solve this differential equation to show that $h = \dfrac{100}{t+2}$

b Sketch a graph of h against t

c Criticise the model.

2 A battery is being charged. After t hours of charging, Q, the number of ampere-hours stored in the battery is modelled by the differential equation $\dfrac{dQ}{dt} = \dfrac{20-Q}{4}$
Initially $Q = 5$

a Solve this differential equation to find an expression for Q in terms of t

b Explain why, using this model, the battery can never run for more than 20 hours without being recharged.

In practice, the battery is taken off the charge when $Q = 19$

c Calculate the time, to the nearest minute, at which this occurs.

3 As part of a scientific study, biologists are studying the population of fish in a lake. The population is modelled by the differential equation $\dfrac{dN}{dt} = \dfrac{1}{6}N(3-N)$, where N, in hundreds, is the number of fish and t is the time, measured in years, since the study began. At the start of the study there are 100 fish in the lake. Solve the differential equation, and hence show that there can never be more than 300 fish in the lake. Why do you think that the population cannot exceed a certain value?

4 A population changes from an initial size of 10 000. After t months the size of the population is P. The connection between P and t can be modelled by the equation $\dfrac{dP}{dt} = -2P\cos 2t$

a Find an expression for P in terms of t

b Find the value of P when $t = 3$

c Find the maximum and minimum values of P

5 A saucepan of milk is placed on a warm hotplate. The hotplate is switched off. Initially the temperature of the milk in the saucepan is 10 °C. After t minutes the temperature, T °C, of the milk satisfies the differential equation $\dfrac{dT}{dt} = \dfrac{150}{T(t+1)^2}$

a Find an expression for T in terms of t

b Calculate the temperature when $t = 2$

c Explain why the temperature of the milk in the saucepan cannot exceed 20 °C.

6 A population of 8 million bacteria is injected into a body. After t days the size of the population in the body is x million, where x and t satisfy the differential equation $\dfrac{dx}{dt} = t - x + 4$

a Show that the size of the population initially starts to decline.

b Find an expression for x in terms of t

c Find the minimum size of the population.

d What happens to the population in the long term?

Strategy

To answer a question that involves modelling systems

(1) Identify the variables and write down a differential equation.

(2) Solve the differential equation.

(3) Answer any questions in context.

You are not always given the differential equation explicitly, and sometimes you might need to form it from the information you are given. For example, laws of growth and decay can be expressed in terms of a first order differential equation.

- If the rate of growth of x is proportional to x, then $\dfrac{dx}{dt} = kx$, where k is a positive constant.

- If the rate of decay of x is proportional to x, then $\dfrac{dx}{dt} = -kx$, where k is a positive constant.

Example 2

A glass of water is placed in ice. At time t minutes the rate of change of temperature of a glass of water is proportional to the difference between the temperature, $T\,°C$, of that glass of water at that time and the temperature of the ice. Initially $T = 100\,°C$.

a Show that $T = 100e^{-kt}$, where k is a constant.

b Given that $T = 50\,°C$ when $t = 10$, find the value of T when $t = 30$

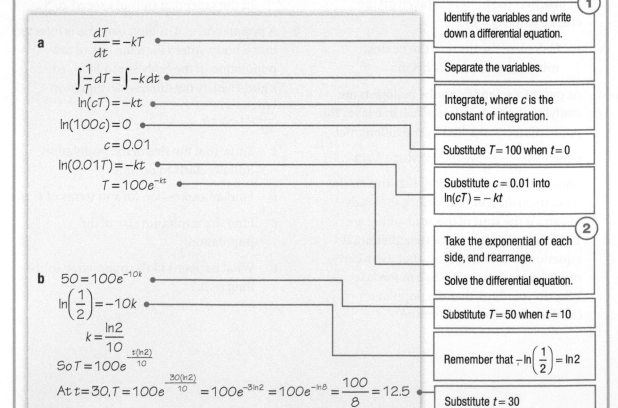

a $\quad \dfrac{dT}{dt} = -kT$ ● —— Identify the variables and write down a differential equation. ①

$\displaystyle\int \dfrac{1}{T}\,dT = \int -k\,dt$ ● —— Separate the variables.

$\ln(cT) = -kt$ ● —— Integrate, where c is the constant of integration.

$\ln(100c) = 0$ ●

$c = 0.01$

$\ln(0.01T) = -kt$ ● —— Substitute $T = 100$ when $t = 0$

$T = 100e^{-kt}$ ● —— Substitute $c = 0.01$ into $\ln(cT) = -kt$

Take the exponential of each side, and rearrange. ②

Solve the differential equation.

b $\quad 50 = 100e^{-10k}$ ● —— Substitute $T = 50$ when $t = 10$

$\ln\left(\dfrac{1}{2}\right) = -10k$ ●

$k = \dfrac{\ln 2}{10}$

$So\; T = 100e^{\frac{t(\ln 2)}{10}}$ ● —— Remember that $-\ln\left(\dfrac{1}{2}\right) = \ln 2$

At $t = 30$, $T = 100e^{\frac{30(\ln 2)}{10}} = 100e^{-3\ln 2} = 100e^{-\ln 8} = \dfrac{100}{8} = 12.5$ ● —— Substitute $t = 30$

Example 3

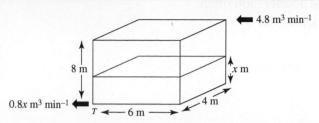

4.8 m³ min⁻¹

8 m

x m

0.8x m³ min⁻¹

T ← 6 m

4 m

The diagram shows a water tank in the shape of a cuboid of base 6 m by 4 m and depth 8 m. Water flows into the tank at a constant rate of 4.8 m³ min⁻¹. At time t minutes, the depth of the water in the tank is x metres. There is a tap at the point T at the base of the tank. When the tap is open, water leaves the tank at a rate of 0.8x m³ min⁻¹. Initially the tank is empty.

a Show that $30\dfrac{\mathrm{d}x}{\mathrm{d}t} = 6 - x$

b Solve this equation to find an expression for x in terms of t

c Sketch a graph of x against t

a Let Vm³ be the volume of water in the tank at a time t minutes.

$$\frac{\mathrm{d}V}{\mathrm{d}t} = 4.8 - 0.8x \qquad (1)$$

> Identify the variables and write down a differential equation.

Also $V = 24x$

> Use volume of a cuboid = length × width × height.

So $\dfrac{\mathrm{d}V}{\mathrm{d}t} = 24\dfrac{\mathrm{d}x}{\mathrm{d}t} \qquad (2)$

> Differentiate V with respect to t

$$24\frac{\mathrm{d}x}{\mathrm{d}t} = 4.8 - 0.8x$$

$$30\frac{\mathrm{d}x}{\mathrm{d}t} = 6 - x$$

> Combine (1) and (2)

b $\displaystyle\int\left(\frac{1}{6-x}\right)\mathrm{d}x = \int\frac{1}{30}\,\mathrm{d}t$

> Separate the variables.

$$-\ln(6-x) = \frac{t}{30} + c$$

$$6 - x = Ae^{-\frac{t}{30}}$$

$$A = 6$$

> Substitute $t = 0$, $x = 0$ to find A

$$x = 6\left(1 - e^{-\frac{t}{30}}\right)$$

> Solve the differential equation.

c

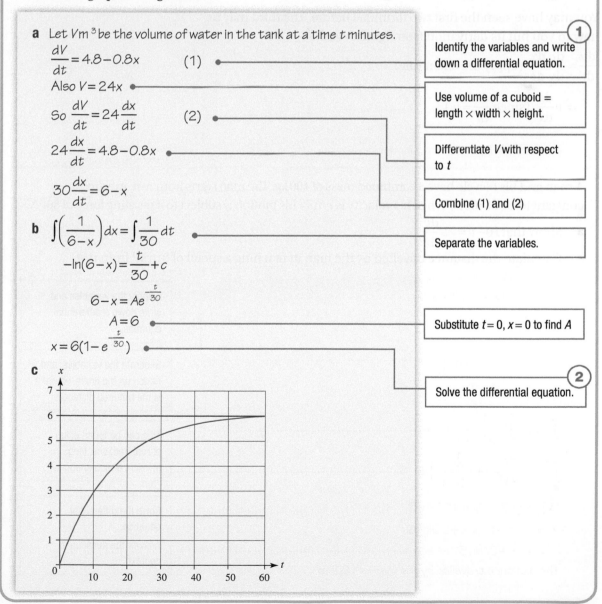

Questions involving displacement, velocity and acceleration often give rise to differential equations. You already know that

Displacement $\xrightarrow{\text{Differentiate}}$ Velocity $\xrightarrow{\text{Differentiate}}$ Acceleration

And, conversely,

Displacement $\xleftarrow{\text{Integrate}}$ Velocity $\xleftarrow{\text{Integrate}}$ Acceleration

Newton's 2nd law, which states that

force = mass × acceleration

can also be expressed as

$$F = m\frac{d^2x}{dt^2} \text{ or } F = m\frac{dv}{dt} \text{ or } F = mv\frac{dv}{dx}$$

You may have seen the first two formulae before. The third may be new to you but its derivation is straightforward.

$$\frac{dv}{dt} = \frac{dv}{dx}\frac{dx}{dt} \text{ (apply the chain rule)}$$

$$= v\frac{dv}{dx} \text{ (since } \frac{dx}{dt} = v \text{)}$$

Example 4

A man and his bicycle have a combined mass of 100 kg. The man starts from rest, exerting a constant force of 500 N. When his velocity is $v\,\text{m}\,\text{s}^{-1}$ his motion is subject to a resisting force of $50v\,\text{N}$.

a Show that $10 - v = 2v\dfrac{dv}{dx}$

b Calculate the distance travelled by the man in reaching a speed of $8\,\text{m}\,\text{s}^{-1}$ from rest.

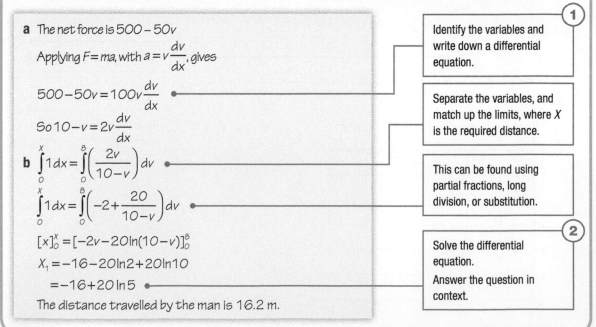

a The net force is $500 - 50v$

Applying $F = ma$, with $a = v\dfrac{dv}{dx}$, gives

$500 - 50v = 100v\dfrac{dv}{dx}$

So $10 - v = 2v\dfrac{dv}{dx}$

b $\displaystyle\int_0^X 1\,dx = \int_0^8 \left(\frac{2v}{10-v}\right)dv$

$\displaystyle\int_0^X 1\,dx = \int_0^8 \left(-2 + \frac{20}{10-v}\right)dv$

$[x]_0^X = [-2v - 20\ln(10-v)]_0^8$

$X_1 = -16 - 20\ln 2 + 20\ln 10$

$\quad = -16 + 20\ln 5$

The distance travelled by the man is 16.2 m.

① Identify the variables and write down a differential equation.

Separate the variables, and match up the limits, where X is the required distance.

This can be found using partial fractions, long division, or substitution.

② Solve the differential equation.

Answer the question in context.

1 The rate, in $cm^3 s^{-1}$, at which air is escaping from a balloon at time t seconds is proportional to the volume, $V cm^3$, of air in the balloon at that time. Initially $V = 2000$

a Show that $V = 2000e^{-kt}$, where k is a positive constant.

Given that $V = 1000$ when $t = 4$

b Show that $k = \dfrac{1}{4}\ln 2$

c Calculate the value of V when $t = 8$

2 A glass of boiling water is placed in a room. The temperature of the room is $20\,^{\circ}C$. At time t minutes the rate of change of temperature of a glass of water is proportional to the difference between the temperature, $T\,^{\circ}C$, of that glass of water at that time, and the temperature of the room.

a Explain why $\dfrac{dT}{dt} = -k(T - 20)$

After 10 minutes the temperature of the water is $36\,^{\circ}C$.

b Solve this differential equation to find an expression for T in terms of t

c Sketch a graph of T against t

d Calculate how many minutes it takes until the temperature of the water in the glass is $21\,^{\circ}C$.

3 A population is growing in such a way that, at a time t years, the rate at which the population is increasing is proportional to the size, x, of the population at that time. At $t = 0$, $x = 100$

a Show that $x = 100e^{kt}$, where k is a positive constant.

Given that after 10 years the size of the population is 500,

b Show that $k = \dfrac{1}{10}\ln 5$

c Calculate the size of the population after 30 years.

4

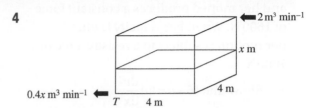

The diagram shows a water tank in the shape of a cuboid of base 4 m by 4 m, and height 6 m. Water flows into the tank at a constant rate of $2\,m^3 min^{-1}$. At time t minutes the depth of the water in the tank is x metres. There is a tap at the point T at the base of the tank. When the tap is open water leaves the tank at a rate of $0.4x\,m^3 min^{-1}$. Initially the tank is empty.

a Show that $\dfrac{dx}{dt} = \dfrac{5 - x}{40}$

b Solve this equation to find an expression for x in terms of t

c Find the time, to the nearest second, at which the depth of the water in the tank is 3 m.

d Explain why the tank will never fill to the top.

5 Water starts pouring into a large vertical tank of uniform cross-section at a constant rate of $1000\,cm^3 s^{-1}$ and is leaking out of a hole in the base of the tank at a rate proportional to the square root of the depth of the water already in the tank. The cross-sectional area of the tank is $500\,cm^2$. Initially the tank is empty. At time t seconds the depth of the water in the tank is $x\,cm$.

a Show that $\dfrac{dx}{dt} = 2 - k\sqrt{x}$, where k is a constant.

When $x = 25$, water is leaking out of the tank at the rate of $500\,cm^3 s^{-1}$.

b Show that $\dfrac{dx}{dt} = \dfrac{10 - \sqrt{x}}{5}$

c Solve the differential equation to find an expression for t in terms of x

d Find the time at which the depth of the water in the tank is 25 cm.

6 A woman and her moped have a combined mass of 200 kg. The woman starts from rest, and her moped produces a constant force of 1600 N. When her velocity is $v\,\mathrm{m\,s^{-1}}$ her motion is subject to a resistiv force of $100v\,\mathrm{N}$.

  a Show that $16 - v = 2v\dfrac{\mathrm{d}v}{\mathrm{d}x}$

  b Calculate the distance travelled by the woman in reaching a speed of $12\,\mathrm{m\,s^{-1}}$ from rest.

7 A truck of mass 3000 kg travels along a horizontal road under the action of a force of magnitude 6000 N. The truck experiences a resistiv force of magnitude $15v^2\,\mathrm{N}$, where $v\,\mathrm{m\,s^{-1}}$ is the speed of the truck.

  a Find the time taken for the truck to accelerate from a velocity of $5\,\mathrm{m\,s^{-1}}$ to a velocity of $15\,\mathrm{m\,s^{-1}}$.

  b Find the distance travelled by the truck in accelerating from a velocity of $5\,\mathrm{m\,s^{-1}}$ to a velocity of $15\,\mathrm{m\,s^{-1}}$.

8 **a** Find the values of the constants A, B and C for which
$$\frac{v^2}{1600-v^2} \equiv \frac{A}{40+v} + \frac{B}{40-v} + C$$

  b *A van and driver of total mass 600 kg is travelling along a straight horizontal road at $20\,\mathrm{ms^{-1}}$. As it passes a point O the van accelerates by increasing the power, so that the driving force of the engine is now $\dfrac{40000}{v}$ newtons, where $v\,\mathrm{ms^{-1}}$ is the speed of the van. The motion of the van is subject to air resistance of magnitude $25v$ newtons. At time t seconds after passing O, the displacement of the van from O is x metres. Show that*
$$\frac{\mathrm{d}v}{\mathrm{d}x} = \frac{1600-v^2}{24v^2}$$

  c Solve this equation to get an expression for x in terms of v

  d Hence find the distance travelled by the van, past O, when it reaches a velocity of $35\,\mathrm{ms^{-1}}$.

9 A patient is required to take a course of a certain drug. The rate of decrease in the concentration of the drug in the bloodstream at a time t hours is proportional to the amount, $x\,\mathrm{mg}$, of the drug in the bloodstream at that time. The size of the dose is x_0

  a Show that $x = x_0 e^{-kt}$, where k is a positive constant.

The patient repeats the dose of x_0 at intervals of T hours.

  b Show that the amount of the drug in the bloodstream will never exceed
$$\left(\frac{x_0}{1-e^{-kT}}\right)\mathrm{mg}$$

10 A particle P of mass 4 kg moves along the positive x-axis under the action of a force directed towards the origin. At time $t\,\mathrm{s}$, the displacement of P from O is $x\,\mathrm{m}$ and P is moving with a speed of $v\,\mathrm{m\,s^{-1}}$. The force has magnitude $13x\,\mathrm{N}$. The particle, P, is also subject to a resistiv force of magnitude $12v\,\mathrm{N}$.

  a Show that the equation of motion of P is
$$4\frac{\mathrm{d}^2x}{\mathrm{d}t^2} + 12\frac{\mathrm{d}x}{\mathrm{d}t} + 13x = 0$$

Initially $x = 2$ and P is moving away from O with a velocity of $3\,\mathrm{m\,s^{-1}}$.

  b Solve the differential equation to find an equation for x in terms of t

  c Calculate, correct to 2 decimal places, the time at which P is first at rest.

Fluency and skills

So far all of the differential equations you have looked at have involved just two variables. Many differential equations involve more than two variables. Consider, for example, an island populated by foxes and rabbits. You can denote the number of foxes by F and the number of rabbits by R. Foxes eat rabbits and are known as the **predators**. Rabbits eat grass and are eaten by the foxes, so they are known as the **prey**.

- $\dfrac{dR}{dt}$ depends on R and F. The more rabbits in the population, the more breeding and deaths there are, but the more foxes in the population, the more rabbits are eaten.

- $\dfrac{dF}{dt}$ depends on R and F. The more rabbits in the population, the more food there is for the foxes, but the more foxes in the population, the more breeding.

This gives rise to equations of the type

$$\frac{dR}{dt} = \alpha R - \beta F \quad (1) \qquad \text{and} \qquad \frac{dF}{dt} = \lambda R + \mu F \quad (2)$$

Key point

Problems that involve equations of this type are called **predator–prey** problems.

To solve equations of this type, first differentiate one of the equations with respect to t

Differentiating equation (1) gives

$$\frac{d^2 R}{dt^2} = \alpha \frac{dR}{dt} - \beta \frac{dF}{dt}$$

Now substitute for $\dfrac{dF}{dt}$ from equation (2) to get

$$\frac{d^2 R}{dt^2} = \alpha \frac{dR}{dt} - \beta(\lambda R + \mu F)$$

$$\frac{d^2 R}{dt^2} = \alpha \frac{dR}{dt} - \beta \lambda R - \mu \beta F$$

Now substitute $\beta F = \alpha R - \dfrac{dR}{dt}$ from equation (1) to get

$$\frac{d^2 R}{dt^2} = \alpha \frac{dR}{dt} - \beta \lambda R - \mu\left(\alpha R - \frac{dR}{dt}\right)$$

which rearranges to

$$\frac{d^2 R}{dt^2} - (\alpha + \mu)\frac{dR}{dt} + (\beta \lambda + \alpha \mu)R = 0$$

This is a second order differential equation in R and t, which you can solve in the usual way.

Example 1

A system of differential equations is given by

$$\frac{dx}{dt} = x + 4y \qquad (1)$$

$$\frac{dy}{dt} = 2x + 3y - 10 \qquad (2)$$

where $(x, y) = (3, 2)$ when $t = 0$

Find expressions for x and y in terms of t

Differentiating equation (2) gives

$$\frac{d^2 y}{dt^2} = 2\frac{dx}{dt} + 3\frac{dy}{dt} = 2(x + 4y) + 3\frac{dy}{dt}$$

> Substitute $\dfrac{dx}{dt} = x + 4y$ from equation (1)

$$= \frac{dy}{dt} - 3y + 10 + 8y + 3\frac{dy}{dt}$$

> Substitute $2x = \dfrac{dy}{dt} - 3y + 10$ from equation (2)

So $\dfrac{d^2 y}{dt^2} - 4\dfrac{dy}{dt} - 5y = 10$

$m^2 - 4m - 5 = 0 \Rightarrow (m - 5)(m + 1) = 0$ so $m = -1$ or $m = 5$

The complementary function is

$y = Ae^{5t} + Be^{-t}$

The particular integral is $y = -2$

and the general solution is $y = Ae^{5t} + Be^{-t} - 2$ \qquad (3)

So $y' = 5Ae^{5t} - Be^{-t}$ \qquad (4)

> Differentiate the general solution.

From equation (2), $y'(0) = 2x(0) + 3y(0) - 10$

> Substitute $t = 0$ into equation (2)

$= 2(3) + 3(2) - 10 = 2$

So $5A - B = 2$

> Substitute $t = 0$, $y' = 0$ into equation (4)

and $A + B - 2 = 2$

> Substitute $t = 0$, $y = 2$ into equation (3)

giving $A = 1$, $B = 3$

$y = e^{5t} + 3e^{-t} - 2$, and $x = e^{5t} - 6e^{-t} + 8$

> Substitute for y and y- into equation (2)

> Solve equations simultaneously.

1 Solve each of the following systems of differential equations to find expressions for y in terms of t

a $\dfrac{dx}{dt} = x - 2y; \dfrac{dy}{dt} = x + 4y$

b $\dfrac{dx}{dt} = -3y; \dfrac{dy}{dt} = 3x$

c $\dfrac{dx}{dt} = 5x + 4y; 3\dfrac{dy}{dt} = x + 4y$

d $\dfrac{dx}{dt} = 4x - y; \dfrac{dy}{dt} = 6x - 3y + 2$

e $\dfrac{dx}{dt} = 2x - 10y; \dfrac{dy}{dt} = 5x - 12y$

f $\dfrac{dx}{dt} = -3x + 4y + \cos t; \dfrac{dy}{dt} = -2x + y + \sin t$

2 Solve each of the following systems of differential equations, subject to the given boundary conditions, to find expressions for x and y in terms of t

a $\dfrac{dx}{dt} = 2x + 4y; \dfrac{dy}{dt} = x - y;$ given that
$y(0) = 3$ and $y'(0) = 4$

b $\dfrac{dx}{dt} = 2x - y + 3; \dfrac{dy}{dt} = 5x - 4y;$ given that
$y(0) = -1$ and $y'(0) = -8$

c $\dfrac{dx}{dt} = x - 5y; \dfrac{dy}{dt} = x - 3y;$ given that
$y(0) = 3$ and $y'(0) = -5$

d $\dfrac{dx}{dt} = x + 2y + t + 2; \dfrac{dy}{dt} = -2x - 3y + 3t;$
given that $y(0) = 9$ and $y'(0) = -2$

e $\dfrac{dx}{dt} = -3x - y - 3; \dfrac{dy}{dt} = 2x - y + 2;$ given
that $y(0) = 5$ and $y'(0) = 2$

f $\dfrac{dx}{dt} = 7x - 9y + 3e^{-2t}; \dfrac{dy}{dt} = 4x - 5y + e^{-2t};$
given that $y(0) = 2$ and $y'(0) = 3$

3 A system of differential equations is given by
$$\dfrac{dx}{dt} = 2x + y \qquad (1)$$
$$\dfrac{dy}{dt} = x + 2y \qquad (2)$$
where $(x, y) = (3, 1)$ when $t = 0$

Find expressions for x and y in terms of t

4 A system of differential equations is given by
$$\dfrac{dx}{dt} = -3x - 2y + t \qquad (1)$$
$$\dfrac{dy}{dt} = 2x + y + 3t - 1 \qquad (2)$$
where $(x, y) = (8, -11)$ when $t = 0$

Find expressions for x and y in terms of t

5 A system of differential equations is given by
$$\dfrac{dx}{dt} = x + 2y \qquad (1)$$
$$\dfrac{dy}{dt} = y - z \qquad (2)$$
$$\dfrac{dz}{dt} = -x \qquad (3)$$

At $t = 0$, $x = 0$, $\dfrac{dx}{dt} = 4$ and $\dfrac{d^2x}{dt^2} = 5$

a Show that $\dfrac{d^3x}{dt^3} - 2\dfrac{d^2x}{dt^2} + \dfrac{dx}{dt} - 2x = 0$

b Solve this equation to find an expression for x in terms of t

Reasoning and problem-solving

Strategy

To answer a question that involves coupled equations

(1) Differentiate either one of the pair of equations.

(2) Eliminate one of the variables between the two equations to get a second order differential equation in just one variable.

(3) Solve the second order differential equation.

(4) Answer the question, which might involve substituting back to find an expression for the second variable, or substituting values to find the constants.

(5) Answer the question(s) in context.

Example 2

As part of a reaction, a substance X changes into a substance Y, which in turn decays. At time t hours, the masses, in grams, of X and Y are denoted by x and y respectively.

The differential equations modelling the changes are $\dfrac{dx}{dt} = -2y$ and $\dfrac{dy}{dt} = x - 3y$. Initially $x = 20$ and $y = 0$

a Solve the equations to find expressions for x and y in terms of t

b Prove that there can never be equal amounts of X and Y

a $\dfrac{d^2 y}{dt^2} = \dfrac{dx}{dt} - 3\dfrac{dy}{dt}$ ← Differentiate $\dfrac{dy}{dt} = x - 3y$ ①

$\dfrac{d^2 y}{dt^2} + 3\dfrac{dy}{dt} + 2y = 0$ ← Substitute $\dfrac{dx}{dt} = -2y$. Eliminate one of the variables between the two equations to get a second order differential equation in just one variable. ②

$(m+1)(m+2) = 0$

So the general solution is

$y = Ae^{-t} + Be^{-2t}$ (1)

$y' = -Ae^{-t} - 2Be^{-2t}$ (2) ← Substitute $t = 0$ into $\dfrac{dy}{dt} = x - 3y$

$y'(0) = x(0) - 3y(0)$ ← Substitute $t = 0$, $y' = 20$ into equation (2)

So $y'(0) = 20$

$-A - 2B = 20$ ← Substitute $t = 0$, $y = 0$ into equation (1)

and $A + B = 0$

giving $A = 20$, $B = -20$ ← Solve equations simultaneously.

$y = 20e^{-t} - 20e^{-2t}$, and $x = 40e^{-t} - 20e^{-2t}$ ← Substitute for y into $\dfrac{dy}{dt} = x - 3y$. Solve the second order differential equation. ③

b If there are equal amounts of X and Y, then $x = y$

$\Rightarrow 40e^{-t} - 20e^{-2t} = 20e^{-t} - 20e^{-2t}$

So $20e^{-t} = 0$, which is impossible since $e^{-t} > 0$ for all values of t

So there can never be equal amounts of X and Y ← Answer the question in context. ⑤

1 As part of a reaction, a substance X changes into a substance Y, which in turn evaporates away. At time t hours, the masses, in grams, of X and Y are denoted by x and y respectively.

The differential equations modelling the changes are $\dfrac{dx}{dt} = -3y$ and $\dfrac{dy}{dt} = x - 4y$
Initially $x = 40$ and $y = 0$

   a Solve the equations to find expressions for x and y in terms of t

   b Find the time at which there is precisely four times as much of substance X as there is of substance Y

2 Two declining mollusc populations, X and Y, are competing for supremacy. After t years, the populations of X and Y, measured in thousands, are respectively x and y. The differential equations modelling the changing states are $\dfrac{dx}{dt} = 3y - 2x$ and $\dfrac{dy}{dt} = x - 4y$. Initially $x = 7$ and $y = 5$

   a Solve the equations to find expressions for x and y in terms of t

   b Show that the population of X will always be greater than the population of Y

3 Two species of an insect, X and Y, compete for survival in the same environment. The population of the species, measured in millions, at time t months, are x and y respectively. While both species are in existence, their populations can be modelled by the simultaneous differential equations $\dfrac{dx}{dt} = 3x - 3y$ and $\dfrac{dy}{dt} = 5y - x$
Initially $x = 50$ and $y = 30$

   a Solve the equations to find expressions for x and y in terms of t

   b Which is the first population to become extinct, and when does that happen?

4 As part of a process involving changing states of two liquids, X and Y, liquid X is poured into a container at a constant rate. Liquid X undergoes a reaction that

changes it into liquid Y, and the liquid Y is then removed from the container. At time t minutes, the masses, in grams, of X and Y in the container are denoted by x and y respectively.

The differential equations modelling the changes are $\dfrac{dx}{dt} = -4x + y + 7$ and $\dfrac{dy}{dt} = 6x - 5y$. Initially $x = 3$ and $y = 14$

   a Solve the equations to find expressions for x and y in terms of t

   b Show that, over time, the ratio of $x : y$ settles down to a constant value. Find this value.

5 Gazelles and lions are fighting for survival on a closed grass plain. Initially there are 120 gazelles and 80 lions. At time t years, the number of gazelles is given by G and the number of lions is given by L. The populations are modelled by the differential equations $\dfrac{dG}{dt} = G - L$ and $\dfrac{dL}{dt} = 3G + 5L$

   a Solve the equations to find expressions for G and L in terms of t

   b Calculate the time at which the population of gazelles becomes extinct.

6 Rabbits and foxes are introduced onto an island. Initially there are 100 rabbits and 160 foxes. At time t years after the animals are introduced, the number of rabbits is given by R, and the number of foxes is given by F. The populations are modelled by the differential equations $\dfrac{dR}{dt} = 6R - F$ and $\dfrac{dF}{dt} = 3R + 2F$

   a Solve the equations to find expressions for R and F in terms of t

   b Show that, using this model, over time, there will be approximately equal numbers of foxes and rabbits.

7 In a tropical forest, pandas like to eat bamboo plants. At time t years, there are

P hundred pandas in the forest and B thousand bamboo plants. When $t=0$, $P=1$ and $B=2$. P and B are connected by the differential equations $10\dfrac{dP}{dt}=2P+3B$ and $30\dfrac{dB}{dt}=-P$

a Solve the equations to find expressions for P and B in terms of t

b Find the number of years after which there are no bamboo plants left in the forest.

8 A small colony of bears feed on the fish in a lake. A time $t=0$, there are 4 bears in the colony, and 1000 fish in the lake. At time t years, the number of fish, x, and the number of bears, y, are modelled by the differential equations

$$\frac{dx}{dt}=-0.2y+0.2x \quad (1)$$

$$\frac{dy}{dt}=0.4y+0.1x \quad (2)$$

a Show that $\dfrac{d^2y}{dt^2}-0.6\dfrac{dy}{dt}+0.1y=0$

b Find expressions for x and y in terms of t

c According to the model, after how many years will there be no fish in the lake?

d Criticise the model in terms of how it is constructed.

9 An Alaskan forest is populated by Canadian lynx and snowshoe hares. At time t years, the number of lynx is L and the number of hares is H

It is assumed that:

- The number of lynx increases at a rate proportional to the number of hares such that $\dfrac{dL}{dt}=0.05H$
- If there were no lynx present, the number of hares would increase by 20% each year.
- When both lynx and hares are present, the lynx kill hares at a yearly rate that is equal to 115% of the current number of lynx.

a Explain why $\dfrac{dH}{dt}=1.2H-1.15L$

b Deduce that
$$\frac{d^2H}{dt^2}-1.2\frac{dH}{dt}+0.0575H=0$$

At time $t=0$, the number of lynx is 60 and the number of hares is 500

c Find expressions for the number of lynx and the number of hares at time t years.

d Criticise the model in terms of the implications of the solution.

10 A nature reserve is populated by lions and zebra. At time t years the number of lions is L and the number of zebra is Z

It is assumed that:

- The number of lions increases at a rate proportional to the number of zebra such that $\dfrac{dL}{dt}=0.2Z$
- If there were no lions present, the number of zebra would increase by 30% each year.
- When both lions and zebra are present, the lions kill zebra at a yearly rate that is equal to 110% of the current number of lions.

a Show that that $\dfrac{d^2Z}{dt^2}-1.3\dfrac{dZ}{dt}+0.22Z=0$

At time $t=0$, the number of lions is 100 and the number of zebra is 1000

b Find expressions for the numbers of lions and the number of zebra at time t years.

c Show that, for large values of t, the ratio number of lions : number of zebra $=2:11$

11 A radioactive element X decays into a radioactive element Y, which in turn decays into the stable element Z. This decay can be modelled by the equations $\dfrac{dx}{dt}=-0.2x$, $\dfrac{dy}{dt}=0.2x-0.1y$ and $\dfrac{dz}{dt}=0.1y$ where x, y and z, are the masses of X, Y and Z respectively, measured in milligrams at time t seconds. Initially $x=100$, $y=0$ and $z=0$

a Find an expression for x in terms of t

b Use your solution for x to find an expression for y in terms of t

c Use your solution for y to find an expression for z in terms of t

d Verify that $x+y+z=100$

Summary and review

Chapter summary

- An equation that involves only a first order derivative, such as $\dfrac{dy}{dx}$, is called a **first order differential equation**.
- An equation of the form $\dfrac{dy}{dx} = F(x)G(y)$ is solved by **separating the variables** to get
  $$\int F(x)\,dx = \int \frac{1}{G(y)}\,dy$$
- An equation of the form $\dfrac{dy}{dx} + P(x)y = f(x)$ is solved by multiplying throughout by an **integrating factor** $e^{\int P(x)dx}$
- A solution with constant(s) is called a **general solution**. Substitute given value(s) to get a **particular solution**.
- An equation that involves a second order derivative, such as $\dfrac{d^2y}{dx^2}$, but nothing of higher order, is called a **second order differential equation**.
- To solve an equation of the form $a\dfrac{d^2y}{dx^2} + b\dfrac{dy}{dx} + cy = f(x)$
  - First solve the **auxiliary equation** $am^2 + bm + c = 0$. The three cases for the **complementary function** that come from the solution of the auxiliary equation are given in Table 1
  - Next find the **particular integral**. What to look for as a particular integral is given in Table 2
  - To get the **general solution**, add together the complementary function and the particular integral.
- Differential equations that include two or more variables, linked to a common additional variable are called **coupled equations**.
- To solve a system of coupled equations, use differentiation to change the differential equations into a single equation of higher order connecting just two variables.

Check and review

You should now be able to...	Review Questions
✔ Solve a first order differential equation using an integrating factor.	1
✔ Solve a first order differential equation and interpret the solution in context.	2
✔ Set up a first order differential equation from a model and solve it.	3, 4
✔ Solve a second order differential equation of the form $a\dfrac{d^2y}{dx^2} + b\dfrac{dy}{dx} + cy = f(x)$	5
✔ Solve a second order differential equation and interpret the solution in context.	6
✔ Set up a second order differential equation from a model and solve it.	7
✔ Solve a system of coupled equations.	8

1 Find the solution to the differential equation $\dfrac{dy}{dx}+y\cot x=\cos^3 x$ for which $y=1$ at $x=\dfrac{\pi}{2}$, giving your answer in the form $y=\mathrm{f}(x)$

2 As part of an industrial process, salt is dissolved in a liquid. The mass, $S\,\mathrm{kg}$, of the salt dissolved, t minutes after the process begins, is modelled by the equation
$$\dfrac{dS}{dt}+\dfrac{2S}{300-t}=\dfrac{1}{2},\ 0\le t<300$$
Given that $S=0$ when $t=0$,

   a Find S in terms of t

   b Calculate the maximum mass of salt that the model predicts will be dissolved in the liquid.

3 At time t minutes, the rate of change of temperature of a cooling liquid is proportional to the temperature, $T\,^\circ\mathrm{C}$, of that liquid at that time. Initially $T=100$

   a Show that $T=100\mathrm{e}^{-kt}$, where k is a positive constant.

   Given also that $T=25$ when $t=6$,

   b Show that $k=\dfrac{1}{3}\ln 2$

   c Calculate the time at which the temperature of this liquid will reach $5^\circ\mathrm{C}$.

4 A woman and her bicycle have a combined mass of $120\,\mathrm{kg}$. The woman starts from rest, exerting a constant force of magnitude $600\,\mathrm{N}$. When her velocity is $v\,\mathrm{m\,s^{-1}}$ her motion is subject to a resisting force of $6v^2\,\mathrm{N}$.

   a Show that $100-v^2=20\dfrac{dv}{dt}$

   b Calculate the time taken by the woman to reach a speed of $9\,\mathrm{m\,s^{-1}}$ from rest.

   c Calculate the distance travelled by the woman in reaching a speed of $9\,\mathrm{m\,s^{-1}}$ from rest.

5 For the differential equation
$$\dfrac{d^2y}{dx^2}+4\dfrac{dy}{dx}+13y=90\mathrm{e}^{4x}$$

   a Find the general solution,

   b Find the particular solution for which $y(0)=6$ and $y'(0)=-3$

6

A box, B, is attached to one end of a light elastic string. The box is initially at rest on a smooth horizontal table when the other end of the string, A, starts to move away from B with constant velocity $2\,\mathrm{m\,s^{-1}}$. The displacement, y metres, of B from its initial position, at a time t seconds from the start, satisfies the differential equation
$$\dfrac{d^2y}{dt^2}+16y=8t$$

   a Find an expression for y in terms of t

   b Calculate the time when B is next at rest.

   c Sketch the graph of y against t

7 A particle P of mass $4\,\mathrm{kg}$ moves along the horizontal x-axis under the action of a force directed towards the origin. At time t seconds when the displacement of P from the origin is x metres, P is moving with a speed of $v\,\mathrm{m\,s^{-1}}$ and the force has magnitude $9x\,\mathrm{N}$. The particle is also subject to a resistive force of magnitude $12v\,\mathrm{N}$.

   a Show that $4\dfrac{d^2x}{dt^2}+12\dfrac{dx}{dt}+9x=0$

   Initially $x=6$ and $\dfrac{dx}{dt}=-11$

   b Show that $x=\mathrm{e}^{-\frac{3t}{2}}(6-2t)$

   c Write down the time when P passes through the origin.

   d Find the speed of P at that time.

8 A particle moves in a plane so that at time $t\ge 0$ its Cartesian coordinates (x,y) satisfy the equations $\dfrac{dx}{dt}=x-y-3t$ and $\dfrac{dy}{dt}=y-4x-3$

   a Show that $\dfrac{d^2x}{dt^2}-2\dfrac{dx}{dt}-3x=3t$

   b Find an expression for x and y in terms of t, given that $x(0)=2$ and $y(0)=4$

Research

Bungee jumping was first attempted in 1979, by students jumping off the Clifton suspension bridge in Bristol. Their inspiration came from vine jumping, as practised by the islanders of Vanuatu in the Pacific. The bungee cords used in Northern and Southern hemispheres tend to have different construction, with the Northern hemisphere cords giving a harder, sharper bounce than those in the southern hemisphere.

Try modelling the motion of a bungee jumper using differential equations where appropriate. Research the sorts of values you might use in this motion, including different values for the elasticity of the ropes in the different hemispheres.

Investigation

The diagram shows a potential situation that can be developed as a water feature.

Water flows into the top tank and, once it reaches the holes in the tank, it flows into the middle tank. As the water in the middle tank fills and reaches the holes in this tank, it then flows into the lower tank, which starts to fill. Water continues to fill the lower tank until it overflows.

Investigate this situation using differential equations. Develop your own mathematical model. Assume that that the rate of flow of water out of a tank depends on the volume of

water in the tank. In other words, $\frac{dV}{dt} = g(V)$.

Start your investigation with a function such as $g(V) = -kV$

Try different functions to develop a design that would be attractive.

Research

Engineers use differential equations when designing a rocket to launch a satellite into space. These equations are derived from Newton's Laws of motion. Important factors that had to be considered when the differential equations were developed included the changing mass of the rocket and the change in the pull of gravity and air resistance as the rocket gets further away from the Earth. Research the use of differential equations to model the motion of a rocket.

1 Find the general solution to the differential equation $\dfrac{dy}{dx} = e^{x+y}\cos x$ **[5 marks]**

2 Solve the differential equation $\dfrac{1}{\ln x}\dfrac{dy}{dx} + e^y = 0$ given that when $x = 1$, $y = 0$ **[6]**

3 Given that $y\dfrac{dy}{dx} - x\ln x = 0$

   a Find the general solution to the equation,   **[3]**

   b Find the particular solution if when $x = 1$, $y = 4$ **[3]**

4 Pheasants are introduced into a wooded area. It has been estimated that there are enough resources in the area to sustain a population of 500 breeding pairs.

   The rate at which the population grows is proportional to both the population size, P pairs, and to the space available to grow into $(500 - P)$ pairs.

   Thus $\dfrac{dP}{dt} = kP(500 - P)$, where k is the constant of proportionality and t is the year since their introduction.

   Initially, when $t = 0$, 10 breeding pairs were introduced to the area.

   After a year, there were 40 breeding pairs.

   a Find the general solution to the differential equation, expressing P as an explicit function of t **[5]**

   b Use the initial conditions to find both the constant of proportionality and the constant of integration.   **[4]**

   c When calculating the population, answers are rounded to the nearest whole number. In which year will this first round to 500 pairs?   **[2]**

5 Find the general solution to the differential equation $x\dfrac{dy}{dx} + 2y = x^2$ **[4]**

6 Solve the differential equation $\cos x\dfrac{dy}{dx} - y\sin x = 4x^2\cos x$ given that when $x = 0$, $y = 1$ **[6]**

7 Given that $2x\dfrac{dy}{dx} + y = \ln x$

   a Find the general solution to the equation,   **[4]**

   b Find the particular solution if when $x = e^2$, $y = e^{-1}$ **[2]**

8 A family of differential equations takes the form

   $2\dfrac{d^2y}{dx^2} + 8\dfrac{dy}{dx} + ky = 0$, where k is a constant.

   Find the general solution to the equation when

   a $k = 6$ **[3]**   **b** $k = 8$ **[4]**   **c** $k = 10$ **[4]**

9 Find the particular solution of $2\dfrac{d^2y}{dx^2} + 3\dfrac{dy}{dx} - 2y = 0$ given that when $x = 0$, $y = 5$ and $\dfrac{dy}{dx} = -5$ **[7]**

10 Find the particular solution of $9\dfrac{d^2y}{dx^2} - 6\dfrac{dy}{dx} + y = 0$ given that

when $x = 0$, $y = 6$ and $\dfrac{dy}{dx} = 4$ [7]

11 Find the particular solution of $\dfrac{d^2y}{dx^2} + 2\dfrac{dy}{dx} + 5y = 0$ given that

when $x = 0$, $y = 10$ and $\dfrac{dy}{dx} = 2$ [7]

12 Find the general solution of $\dfrac{d^2y}{dx^2} + \dfrac{dy}{dx} - 12y = 12x + 1$ [7]

13 Find the particular solution of $\dfrac{d^2y}{dx^2} - 4\dfrac{dy}{dx} + 13y = 40\sin x$ given that

when $x = 0$, $y = 5$ and $\dfrac{dy}{dx} = -1$ [7]

14 Find the particular solution of $\dfrac{d^2y}{dx^2} - 7\dfrac{dy}{dx} + 10y = 12e^x$ given that

when $x = 0$, $y = 0$ and $\dfrac{dy}{dx} = 2$ [7]

15 Newton's law of cooling says that the rate of cooling of an object is directly proportional to

the difference between its temperature and that of its surroundings. so $\dfrac{dT}{dt} = k(T - r)$, where

$T\,^\circ$C is the temperature of the object, $r\,^\circ$C is room temperature and t hours is the time since measurements began.

Crime scene investigators arrived at a murder scene. At that time, the temperature of the victim was 35 °C. An hour later, the victim's temperature had dropped to 34 °C. Room temperature was 19 °C.

A living person has a normal temperature of 37.5 °C.

a Find the particular solution to the differential equation. [7]

b Estimate how long the victim had been dead. [2]

16 An enterprise scheme in a school starts to sell a particular product.

The rate at which number of students who buy the product increases is proportional to the number of students in the school who have still to buy it.

There are 1000 students in the school and P students have bought the product at time t days after the enterprise started.

Before the first day, 5 students had already bought the product. By the end of the tenth day, 100 students had bought the product.

a Form a differential equation in P [1]

b Find the general solution to the differential equation. [3]

c Use the initial conditions to find the relationship between P and t [3]

d How many students had bought the product by the end of day 20? [1]

17 Money placed in a savings account will grow in direct proportion to the amount of money in the bank.

Initially £1000 is placed in the account. At the end of year 1, there is £1005 in the account. Let £A represent the amount after a time t years.

a Form a differential equation to model the situation. [2]

b Find the particular solution to the equation. [5]

c How much will be in the bank after 5 years? [1]

d When will the amount in the bank first exceed £2000? [1]

18 A particle is released from rest at time $t = 0$ and falls freely vertically downwards. While falling, it is subject to air resistance that is proportional to its velocity, v. The particle has acceleration given by $\dfrac{\mathrm{d}v}{\mathrm{d}t} = g - kv$, where k is a constant.

    a By solving the differential equation, show that the particle approaches a terminal velocity. [7]

    b If $k = 1.25$ and $g = 10$, sketch a graph of v against t and find the terminal velocity. [3]

19 A particle moves so that at time t seconds it is x units from the origin.

    Its motion is modelled by $100\dfrac{\mathrm{d}^2 x}{\mathrm{d}t^2} = -25x$

    Initially $x = 0$ when $t = 0$. When $t = \pi$, $x = 4$

    Find the particular solution of the equation. [6]

20 A damped oscillating system is modelled by the differential equation
$\dfrac{\mathrm{d}^2 x}{\mathrm{d}t^2} + 0.3\dfrac{\mathrm{d}x}{\mathrm{d}t} + 0.15^2 x = 0$

    a Explain why the damping will be critical. [3]

    b Given that $x = 0.5$ when $t = 0$ and $\dfrac{\mathrm{d}x}{\mathrm{d}t} = 0.425$ when $t = 0$, solve the differential equation. [8]

    c Sketch a graph of x against t [2]

21 A particle of mass m attached to a spring is subject to a damping force proportional to its velocity given by $-5m\dfrac{\mathrm{d}x}{\mathrm{d}t}$ and a tension in the spring given by $-6mx$. A disturbing force $3m\sin 2t$ is applied to the particle. The equation of motion of the particle can be written as $m\dfrac{\mathrm{d}^2 x}{\mathrm{d}t^2} = -5m\dfrac{\mathrm{d}x}{\mathrm{d}t} - 6mx + 3m\sin 2t$

    a Show that $\dfrac{\mathrm{d}^2 x}{\mathrm{d}t^2} + 5\dfrac{\mathrm{d}x}{\mathrm{d}t} + 6x = 3\sin 2t$ [2]

    b Find the complementary function of this differential equation. [3]

    c Find the particular integral. [8]

    d Find the particular solution given that $x = 0$ when $t = 0$ and $\dfrac{\mathrm{d}x}{\mathrm{d}t} = 1$ when $t = 0$ [6]

22 A coupled system of differential equations is given by $\dfrac{\mathrm{d}x}{\mathrm{d}t} = x + 8y + 1$
and $\dfrac{\mathrm{d}y}{\mathrm{d}t} = -x - 5y - 4$
when $t = 0$, $x = 0$ and $y = 1$.

    a Find expressions for x and y in terms of t [17]

    b Describe what happens to the values of x and y as $t \to \infty$ [2]

Mathematics formulae
For A Level Further Maths

The following mathematical formulae will be provided for you. Pure Mathematics formulae, for this course, are provided in this section.

Pure Mathematics
Summations

$$\sum_{r=1}^{n} r^2 = \frac{1}{6}n(n+1)(2n+1) \qquad \sum_{r=1}^{n} r^3 = \frac{1}{4}n^2(n+1)^2$$

Matrix transformations

Anticlockwise rotation through θ about O: $\begin{pmatrix} \cos\theta & -\sin\theta \\ \sin\theta & \cos\theta \end{pmatrix}$

Reflection in the line $y = (\tan\theta)x$: $\begin{pmatrix} \cos2\theta & \sin2\theta \\ \sin2\theta & -\cos2\theta \end{pmatrix}$

Area of a sector

$$A = \frac{1}{2}\int r^2 \, d\theta \qquad \text{(polar coordinates)}$$

Complex numbers

$$\{r(\cos\theta + i\sin\theta)\}^n = r^n(\cos n\theta + i\sin n\theta)$$

The roots of $z^n = 1$ are given by $z = e^{\frac{2\pi k i}{n}}$ for $k = 0, 1, 2, ..., n-1$

Maclaurin's and Taylor's Series

$$f(x) = f(0) + xf'(0) + \frac{x^2}{2!}f''(0) + ... + \frac{x^r}{r!}f^{(r)}(0) + ...$$

$$e^x = \exp(x) = 1 + x + \frac{x^2}{2!} + ... + \frac{x^r}{r!} + ... \qquad \text{for all } x$$

$$\ln(1+x) = x - \frac{x^2}{2} + \frac{x^3}{3} - ... + (-1)^{r+1}\frac{x^r}{r} + ... \qquad (-1 < x \le 1)$$

$$\sin x = x - \frac{x^3}{3!} + \frac{x^5}{5!} - ... + (-1)^r\frac{x^{2r+1}}{(2r+1)!} + ... \qquad \text{for all } x$$

$$\cos x = 1 - \frac{x^2}{2!} + \frac{x^4}{4!} - ... + (-1)^r\frac{x^{2r}}{(2r)!} + ... \qquad \text{for all } x$$

$$\arctan x = x - \frac{x^3}{3} + \frac{x^5}{5} - ... + (-1)^r\frac{x^{2r+1}}{2r+1} + ... \qquad (-1 \le x \le 1)$$

Vectors

Vector products: $\mathbf{a} \times \mathbf{b} = |\mathbf{a}||\mathbf{b}| \sin\theta \, \hat{\mathbf{n}} = \begin{vmatrix} \mathbf{i} & \mathbf{j} & \mathbf{k} \\ a_1 & a_2 & a_3 \\ b_1 & b_2 & b_3 \end{vmatrix} = \begin{pmatrix} a_2 b_3 - a_3 b_2 \\ a_3 b_1 - a_1 b_3 \\ a_1 b_2 - a_2 b_1 \end{pmatrix}$

$\mathbf{a} \cdot (\mathbf{b} \times \mathbf{c}) = \begin{vmatrix} a_1 & a_2 & a_3 \\ b_1 & b_2 & b_3 \\ c_1 & c_2 & c_3 \end{vmatrix} = \mathbf{b} \cdot (\mathbf{c} \times \mathbf{a}) = \mathbf{c} \cdot (\mathbf{a} \times \mathbf{b})$

If A is the point with position vector $\mathbf{a} = a_1 \mathbf{i} + a_2 \mathbf{j} + a_3 \mathbf{k}$ and the direction vector $\mathbf{b}$ is given by $\mathbf{b} = b_1 \mathbf{i} + b_2 \mathbf{j} + b_3 \mathbf{k}$, then the straight line through A with direction vector $\mathbf{b}$ has cartesian equation

$$\frac{x - a_1}{b_1} = \frac{y - a_2}{b_2} = \frac{z - a_3}{b_3} (= \lambda)$$

The plane through A with normal vector $\mathbf{n} = n_1 \mathbf{i} + n_2 \mathbf{j} + n_3 \mathbf{k}$ has cartesian equation $n_1 x + n_2 y + n_3 z + d = 0$ where $d = -\mathbf{a} . \mathbf{n}$

The plane through non-collinear points A, B and C has vector equation

$$\mathbf{r} = \mathbf{a} + \lambda(\mathbf{b} - \mathbf{a}) + \mu(\mathbf{c} - \mathbf{a}) = (1 - \lambda - \mu)\mathbf{a} + \lambda\mathbf{b} + \mu\mathbf{c}$$

The plane through the point with position vector $\mathbf{a}$ and parallel to $\mathbf{b}$ and $\mathbf{c}$ has equation

$$\mathbf{r} = \mathbf{a} + s\mathbf{b} + t\mathbf{c}$$

The perpendicular distance of (α, β, γ) from $n_1 x + n_2 y + n_3 z + d = 0$ is $\dfrac{|n_1 \alpha + n_2 \beta + n_3 \gamma + d|}{\sqrt{n_1^2 + n_2^2 + n_3^2}}$

Hyperbolic functions

$\cosh^2 x - \sinh^2 x = 1$

$\sinh 2x = 2 \sinh x \cosh x$

$\cosh 2x = \cosh^2 x + \sinh^2 x$

$\operatorname{arcosh} x = \ln\left\{ x + \sqrt{x^2 - 1} \right\}$ $\qquad (x \geq 1)$

$\operatorname{arsinh} x = \ln\left\{ x + \sqrt{x^2 + 1} \right\}$

$\operatorname{artanh} x = \dfrac{1}{2}\ln\left(\dfrac{1+x}{1-x} \right)$ $\qquad (|x| < 1)$

Conics

	Ellipse	Parabola	Hyperbola	Rectangular Hyperbola
Standard Form	$\dfrac{x^2}{a^2}+\dfrac{y^2}{b^2}=1$	$y^2=4ax$	$\dfrac{x^2}{a^2}-\dfrac{y^2}{b^2}=1$	$xy=c^2$
Parametric Form	$(a\cos\theta,\, b\sin\theta)$	$(at^2,\, 2at)$	$(a\sec\theta,\, b\tan\theta)$ $(\pm a\cosh\theta,\, b\sinh\theta)$	$\left(ct,\dfrac{c}{t}\right)$
Eccentricity	$e<1$ $b^2=a^2(1-e^2)$	$e=1$	$e>1$ $b^2=a^2(e^2-1)$	$e=\sqrt{2}$
Foci	$(\pm ae,0)$	$(a,0)$	$(\pm ae,0)$	$\left(\pm\sqrt{2}c,\pm\sqrt{2}c\right)$
Directrices	$x=\pm\dfrac{a}{e}$	$x=-a$	$x=\pm\dfrac{a}{e}$	$x+y=\pm\sqrt{2}c$
Asymptotes	none	none	$\dfrac{x}{a}=\pm\dfrac{y}{b}$	$x=0,\, y=0$

Differentiation

$f(x)$	$f'(x)$	$f(x)$	$f'(x)$
$\arcsin x$	$\dfrac{1}{\sqrt{1-x^2}}$	$\cosh x$	$\sinh x$
$\arccos x$	$-\dfrac{1}{\sqrt{1-x^2}}$	$\tanh x$	$\operatorname{sech}^2 x$
$\arctan x$	$\dfrac{1}{1+x^2}$	$\operatorname{arsinh} x$	$\dfrac{1}{\sqrt{1+x^2}}$
$\sinh x$	$\cosh x$	$\operatorname{arcosh} x$	$\dfrac{1}{\sqrt{x^2-1}}$
		$\operatorname{artanh} x$	$\dfrac{1}{1-x^2}$

Integration (+ constant; $a>0$ where relevant)

$f(x)$	$\int f(x)\,dx$
$\sinh x$	$\cosh x$
$\cosh x$	$\sinh x$
$\tanh x$	$\ln\cosh x$
$\dfrac{1}{\sqrt{a^2-x^2}}$	$\arcsin\left(\dfrac{x}{a}\right)\quad(|x|<a)$
$\dfrac{1}{a^2+x^2}$	$\dfrac{1}{a}\arctan\left(\dfrac{x}{a}\right)$

Mathematics formulae for A Level Further Maths

$$\frac{1}{\sqrt{x^2-a^2}} \qquad \text{arcosh}\left(\frac{x}{a}\right),\ \ln\{x+\sqrt{x^2-a^2}\} \quad (x>a)$$

$$\frac{1}{\sqrt{a^2+x^2}} \qquad \text{arsinh}\left(\frac{x}{a}\right),\ \ln\{x+\sqrt{x^2+a^2}\}$$

$$\frac{1}{a^2-x^2} \qquad \frac{1}{2a}\ln\left|\frac{a+x}{a-x}\right|=\frac{1}{a}\text{artanh}\left(\frac{x}{a}\right) \quad (|x|<a)$$

$$\frac{1}{x^2-a^2} \qquad \frac{1}{2a}\ln\left|\frac{x-a}{x+a}\right|$$

Arc length

$$s=\int\sqrt{1+\left(\frac{dy}{dx}\right)^2}\,dx \qquad \text{(cartesian coordinates)}$$

$$s=\int\sqrt{\left(\frac{dx}{dt}\right)^2+\left(\frac{dy}{dt}\right)^2}\,dt \qquad \text{(parametric form)}$$

$$s=\int\sqrt{r^2+\left(\frac{dr}{d\theta}\right)^2}\,d\theta \qquad \text{(polar form)}$$

Surface area of revolution

$$s_x=2\pi\int y\sqrt{1+\left(\frac{dy}{dx}\right)^2}\,dx \qquad \text{(cartesian coordinates)}$$

$$s_x=2\pi\int y\sqrt{\left(\frac{dx}{dt}\right)^2+\left(\frac{dy}{dt}\right)^2}\,dt \qquad \text{(parametric form)}$$

$$s_x=2\pi\int r\sin\theta\sqrt{r^2+\left(\frac{dr}{d\theta}\right)^2}\,d\theta \qquad \text{(polar form)}$$

Mathematical formulae – to learn
For A Level Further Maths

Pure Mathematics

Quadratic Equations

$ax^2 + bx + c = 0$ has roots $\dfrac{-b \pm \sqrt{b^2 - 4ac}}{2a}$

Laws of indices

$a^x a^y \equiv a^{x+y}$

$a^x \div a^y \equiv a^{x-y}$

$(a^x)^y \equiv a^{xy}$

Laws of logarithms

$x = a^n \Leftrightarrow n = \log_a x$ for $a > 0$ and $x > 0$

$\log_a x + \log_a y \equiv \log_a xy$

$\log_a x - \log_a y \equiv \log_a\left(\dfrac{x}{y}\right)$

$k \log_a x \equiv \log_a (x)^k$

Coordinate geometry

A straight-line graph, gradient m passing through (x_1, y_1), has equation $y - y_1 = m(x - x_1)$

Straight lines with gradients m_1 and m_2 are perpendicular when $m_1 m_2 = -1$

Sequences

General term of an arithmetic progression:

$u_n = a + (n-1)d$

General term of a geometric progression:

$u_n = ar^{n-1}$

Trigonometry

In the triangle ABC:

Sine rule: $\dfrac{a}{\sin A} = \dfrac{b}{\sin B} = \dfrac{c}{\sin C}$

Cosine rule: $a^2 = b^2 + c^2 - 2bc\cos A$

Area $= \dfrac{1}{2}ab\sin C$

$\cos^2 A + \sin^2 A \equiv 1$

$\sec^2 A \equiv 1 + \tan^2 A$

$\operatorname{cosec}^2 A \equiv 1 + \cot^2 A$

$\sin 2A \equiv 2\sin A\cos A$

$\cos 2A \equiv \cos^2 A - \sin^2 A$

$\tan 2A \equiv \dfrac{2\tan A}{1 - \tan^2 A}$

Mensuration

Circumference and area of circle radius r and diameter d:

$C = 2\pi r = \pi d \qquad A = \pi r^2$

Pythagoras' Theorem:

In any right-angled triangle where a, b and c are the lengths of the sides and c is the hypotenuse, $c^2 = a^2 + b^2$

Area of a trapezium $= \dfrac{1}{2}(a+b)h$, where a and b are the lengths of the parallel sides and h is their perpendicular separation.

Volume of a prism = area of cross section $\times$ length

For a circle of radius r, where an angle at the centre of θ radians subtends an arc of length s and encloses an associated sector of area A:

$s = r\theta \qquad A = \dfrac{1}{2}r^2\theta$

Complex Numbers

For two complex numbers $z_1 = r_1 e^{i\theta_1}$ and $z_2 = r_2 e^{i\theta_2}$

$$z_1 z_2 = r_1 r_2\, e^{i(\theta_1 + \theta_2)}$$

$$\frac{z_1}{z_2} = \frac{r_1}{r_2} e^{i(\theta_1 - \theta_2)}$$

Loci in the Argand diagram:

$|z - a| = r$ is a circle radius r centred at a

$\arg(z - a) = \theta$ is a half line drawn from a at angle θ to a line parallel to the positive real axis.

Exponential Form: $e^{i\theta} = \cos\theta + i\sin\theta$

Matrices

For a 2 by 2 matrix $\begin{pmatrix} a & b \\ c & d \end{pmatrix}$ the determinant $\Delta = \begin{vmatrix} a & b \\ c & d \end{vmatrix} = ad - bc$

the inverse is $\dfrac{1}{\Delta} \begin{pmatrix} d & -b \\ -c & a \end{pmatrix}$

The transformation represented by matrix **AB** is the transformation represented by matrix **B** followed by the transformation represented by matrix **A**.

For matrices **A**, **B**:

$(\mathbf{AB})^{-1} = \mathbf{B}^{-1}\mathbf{A}^{-1}$

Algebra

$$\sum_{r=1}^{n} r = \frac{1}{2} n(n+1)$$

For $ax^2 + bx + c = 0$ with roots α and β:

$$\alpha + \beta = -\frac{b}{a} \qquad \alpha\beta = \frac{c}{a}$$

For $ax^3 + bx^2 + cx + d = 0$ with roots α, β and γ:

$$\sum \alpha = -\frac{b}{a} \qquad \sum \alpha\beta = \frac{c}{a} \qquad \alpha\beta\gamma = -\frac{d}{a}$$

Hyperbolic Functions

$$\cosh x \equiv \frac{1}{2}(e^x + e^{-x})$$

$$\sinh x \equiv \frac{1}{2}(e^x - e^{-x})$$

$$\tanh x \equiv \frac{\sinh x}{\cosh x}$$

Calculus and Differential Equations

Differentiation

Function	Derivative
x^n	nx^{n-1}
$\sin kx$	$k\cos kx$
$\cos kx$	$-k\sin kx$
$\sinh kx$	$k\cosh kx$
$\cosh kx$	$k\sinh kx$
e^{kx}	ke^{kx}
$\ln x$	$\dfrac{1}{x}$
$f(x) + g(x)$	$f'(x) + g'(x)$
$f(x)g(x)$	$f'(x)g(x) + f(x)g'(x)$
$f(g(x))$	$f'(g(x))g'(x)$

Integration

Function	Integral
x^n	$\dfrac{1}{n+1}x^{n+1}+c,\ n\neq -1$
$\cos kx$	$\dfrac{1}{k}\sin kx+c$
$\sin kx$	$-\dfrac{1}{k}\cos kx+c$
$\cosh kx$	$\dfrac{1}{k}\sinh kx+c$
$\sinh kx$	$\dfrac{1}{k}\cosh kx+c$
e^{kx}	$\dfrac{1}{k}e^{kx}+c$
$\dfrac{1}{x}$	$\ln|x|+c,\ x\neq 0$
$f'(x)+g'(x)$	$f(x)+g(x)+c$
$f'(g(x))g'(x)$	$f(g(x))+c$

Area under a curve $=\displaystyle\int_a^b y\,\mathrm{d}x\,(y\geq 0)$

Volumes of revolution about the x and y axes:

$$V_x=\pi\int_a^b y^2\,\mathrm{d}x \qquad V_y=\pi\int_c^d x^2\,\mathrm{d}y$$

Simple Harmonic Motion:

$$\ddot{x}=-\omega^2 x$$

Vectors

$$|x\mathbf{i} + y\mathbf{j} + z\mathbf{k}| = \sqrt{(x^2 + y^2 + z^2)}$$

Scalar product of two vectors $\mathbf{a} = \begin{pmatrix} a_1 \\ a_2 \\ a_3 \end{pmatrix}$ and $\mathbf{b} = \begin{pmatrix} b_1 \\ b_2 \\ b_3 \end{pmatrix}$ is

$$\begin{pmatrix} a_1 \\ a_2 \\ a_3 \end{pmatrix} \cdot \begin{pmatrix} b_1 \\ b_2 \\ b_3 \end{pmatrix} = a_1 b_1 + a_2 b_2 + a_3 b_3 = |\mathbf{a}| \, |\mathbf{b}| \cos\theta$$

where θ is the acute angle between the vectors $\mathbf{a}$ and $\mathbf{b}$

The equation of the line through the point with position vector $\mathbf{a}$ parallel to vector $\mathbf{b}$ is:

$$\mathbf{r} = \mathbf{a} + t\mathbf{b}$$

The equation of the plane containing the point with position vector $\mathbf{a}$ and perpendicular to vector $\mathbf{n}$ is:

$$(\mathbf{r} - \mathbf{a}) \cdot \mathbf{n} = 0$$

Mathematical notation
For A Level Further Maths

You should understand the following notation for AS and A Level Further Maths, without need for further explanation. Notation from all strands of mathematics has been included for your reference.

Set Notation

$\in$	is an element of
$\notin$	is not an element of
$\subseteq$	is a subset of
$\subset$	is a proper subset of
$\{x_1, x_2, \dots\}$	the set with elements $x_1, x_2, \dots$
$\{x : \dots\}$	the set of all x such that ...
$n(A)$	the number of elements in set A
$\varnothing$	the empty set
ε	the universal set
A'	the complement of the set A
$\mathbb{N}$	the set of natural numbers, $\{1, 2, 3, \dots\}$
$\mathbb{Z}$	the set of integers, $\{0, \pm1, \pm2, \pm3, \dots\}$
$\mathbb{Z}^+$	the set of positive integers, $\{1, 2, 3, \dots\}$
$\mathbb{Z}_0^+$	the set of non-negative integers, $\{0, 1, 2, 3, \dots\}$
$\mathbb{R}$	the set of real numbers
$\mathbb{Q}$	the set of rational numbers, $\left\{\dfrac{p}{q} : p \in \mathbb{Z},\ q \in \mathbb{Z}^+\right\}$
$\cup$	union
$\cap$	intersection
(x, y)	the ordered pair x, y
$[a, b]$	the closed interval $\{x \in \mathbb{R} : a \le x \le b\}$
$[a, b)$	the interval $\{x \in \mathbb{R} : a \le x < b\}$
$(a, b]$	the interval $\{x \in \mathbb{R} : a < x \le b\}$
(a, b)	the open interval $\{x \in \mathbb{R} : a < x < b\}$
$\mathbb{C}$	the set of complex numbers

Miscellaneous Symbols

$=$	is equal to
$\ne$	is not equal to
$\equiv$	is identical to or is congruent to
$\approx$	is approximately equal to
∞	infinity
$\propto$	is proportional to
$<$	is less than
$\le, \le$	is less than or equal to; is not greater than
$>$	is greater than
$\ge, \ge$	is greater than or equal to; is not less than
$\therefore$	therefore
$\because$	because
$p \Rightarrow q$	p implies q (if p then q)
$p \Leftarrow q$	p is implied by q (if q then p)
$p \Leftrightarrow q$	p implies and is implied by q (p is equivalent to q)
a	first term for an arithmetic or geometric sequence

l	last term for an arithmetic sequence
d	common difference for an arithmetic sequence
r	common ratio for a geometric sequence
S_n	sum to n terms of a sequence
S_∞	sum to infinity of a sequence

Operations

$a + b$	a plus b
$a - b$	a minus b
$a \times b,\ ab,\ a \cdot b$	a multiplied by b
$a \div b,\ \dfrac{a}{b}$	a divided by b
$\displaystyle\sum_{i=1}^{n} a_i$	$a_1 + a_2 + \ldots + a_n$
$\displaystyle\prod_{i=1}^{n} a_i$	$a_1 \times a_2 \times \ldots \times a_n$
$\sqrt{a}$	the non-negative square root of a
$\lvert a \rvert$	the modulus of a
$n!$	n factorial: $n! = n \times (n-1) \times \ldots \times 2 \times 1,\ n \in \mathbb{N};\ 0! = 1$
$\dbinom{n}{r},\ ^nC_r,\ _nC_r$	the binomial coefficient $\dfrac{n!}{r!(n-r)!}$ for $n, r \in \mathbb{Z}_0^+,\ r \leq n$ or $\dfrac{n(n-1)\ldots(n-r+1)}{r!}$ for $n \in \mathbb{Q}, r \in \mathbb{Z}_0^+$

Functions

$\mathrm{f}(x)$	the value of the function f at x
$\displaystyle\lim_{x \to a} \mathrm{f}(x)$	the limit of $\mathrm{f}(x)$ as x tends to a
$\mathrm{f}\colon x \mapsto y$	the function f maps the element x to the element y
f^{-1}	the inverse function of the function f
gf	the composite function of f and g which is defined by $\mathrm{gf}(x) = \mathrm{g}(\mathrm{f}(x))$
$\Delta x,\ \delta x$	an increment of x
$\dfrac{\mathrm{d}y}{\mathrm{d}x}$	the derivative of y with respect to x
$\dfrac{\mathrm{d}^n y}{\mathrm{d}x^n}$	the nth derivative of y with respect to x
$\mathrm{f}'(x) \ldots,\ \mathrm{f}^{(n)}(x)$	the first, ..., nth derivatives of $\mathrm{f}(x)$ with respect to x
$\dot{x},\ \ddot{x},\ \ldots$	the first, second, ... derivatives of x with respect to t
$\displaystyle\int y\,\mathrm{d}x$	the indefinite integral of y with respect to x
$\displaystyle\int_a^b y\,\mathrm{d}x$	the definite integral of y with respect to x between the limits $x = a$ and $x = b$

Exponential and Logarithmic Functions

e	base of natural logarithms
$\mathrm{e}^x,\ \exp x$	exponential function of x
$\log_a x$	logarithm to the base a of x
$\ln x,\ \log_e x$	natural logarithm of x

Mathematical notation for A Level Further Maths

Trigonometric Functions

$\left.\begin{array}{c}\sin, \cos, \tan, \\ \operatorname{cosec}, \sec, \cot\end{array}\right\}$ the trigonometric functions

$\left.\begin{array}{c}\sin^{-1}, \cos^{-1}, \tan^{-1} \\ \arcsin, \arccos, \arctan\end{array}\right\}$ the inverse trigonometric functions

$^{\circ}$ degrees

rad   radians

$\left.\begin{array}{c}\operatorname{cosec}^{-1}, \sec^{-1}, \cot^{-1} \\ \operatorname{arccosec}, \operatorname{arcsec}, \operatorname{arccot}\end{array}\right\}$ the inverse trigonometric functions

$\left.\begin{array}{c}\sinh, \cosh, \tanh \\ \operatorname{cosech}, \operatorname{sech}, \coth\end{array}\right\}$ the hyperbolic functions

$\left.\begin{array}{c}\sinh^{-1}, \cosh^{-1}, \tanh^{-1} \\ \operatorname{cosech}^{-1}, \operatorname{sech}^{-1}, \coth^{-1} \\ \operatorname{arsinh}, \operatorname{arcosh}, \operatorname{artanh} \\ \operatorname{arcosech}, \operatorname{arsech}, \operatorname{arcoth}\end{array}\right\}$ the inverse hyperbolic functions

Complex numbers

i, j	square root of -1
$x + iy$	complex number with real part x and imaginary part y
$r(\cos\theta + i\sin\theta)$	modulus argument form of a complex number with modulus r and argument θ
z	a complex number $z = x + iy = r(\cos\theta + i\sin\theta)$
$\operatorname{Re}(z)$	the real part of z, $\operatorname{Re}(z) = x$
$\operatorname{Im}(z)$	the imaginary part of z, $\operatorname{Im}(z) = y$
$|z|$	the modulus of z, $|z| = \sqrt{x^2 + y^2}$
$\arg(z)$	the argument of z, $\arg(z) = \theta$, $-\pi < \theta < \pi$
z^*	the complex conjugate of z, $x - iy$

Matrices

$\mathbf{M}$	a matrix $\mathbf{M}$
$\mathbf{0}$	zero matrix
$\mathbf{I}$	identity matrix
$\mathbf{M}^{-1}$	the inverse of the matrix $\mathbf{M}$
$\mathbf{M}^{\mathrm{T}}$	the transpose of the matrix $\mathbf{M}$
$\det \mathbf{M}$ or $|\mathbf{M}|$	the determinant of the square matrix $\mathbf{M}$
$\mathbf{Mr}$	Image of the column vector $\mathbf{r}$ under the transformation associated with the matrix $\mathbf{M}$

Vectors

$\mathbf{a}, \underline{a}, \underset{\sim}{a}$	the vector $\mathbf{a}, \underline{a}, \underset{\sim}{a}$
$\overrightarrow{AB}$	the vector represented in magnitude and direction by the directed line segment AB
$\hat{\mathbf{a}}$	a unit vector in the direction of $\mathbf{a}$
$\mathbf{i}, \mathbf{j}, \mathbf{k}$	unit vectors in the directions of the Cartesian coordinate axes
$|\mathbf{a}|, a$	the magnitude of $\mathbf{a}$

$\left\|\overrightarrow{AB}\right\|$, AB	the magnitude of $\overrightarrow{AB}$
$\begin{pmatrix} a \\ b \end{pmatrix}$, $a\mathbf{i} + b\mathbf{j}$	column vector and corresponding unit vector notation
$\mathbf{r}$	position vector
$\mathbf{s}$	displacement vector
$\mathbf{v}$	velocity vector
$\mathbf{a}$	acceleration vector
$\mathbf{a} \cdot \mathbf{b}$	the scalar product of $\mathbf{a}$ and $\mathbf{b}$
$\mathbf{a} \times \mathbf{b}$	the vector product of $\mathbf{a}$ and $\mathbf{b}$
$\mathbf{a} . \mathbf{b} \times \mathbf{c}$	the scalar triple product of $\mathbf{a}$, $\mathbf{b}$ and $\mathbf{c}$
ω	angular speed

Answers

Chapter 6

Exercise 6.1A

1 a $5(\cos(0.927)+i\sin(0.927))$

b $\sqrt{5}(\cos(-0.464)+i\sin(-0.464))$

c $10(\cos(0)+i\sin(0))$

d $5(\cos(\pi)+i\sin(\pi))$

e $2\left(\cos\left(\dfrac{\pi}{2}\right)+i\sin\left(\dfrac{\pi}{2}\right)\right)$

f $6\left(\cos\left(-\dfrac{\pi}{2}\right)+i\sin\left(-\dfrac{\pi}{2}\right)\right)$

g $13(\cos(1.97)+i\sin(1.97))$

h $4\sqrt{5}(\cos(-2.03)+i\sin(-2.03))$

i $2\left(\cos\left(\dfrac{\pi}{6}\right)+i\sin\left(\dfrac{\pi}{6}\right)\right)$

j $5\sqrt{2}\left(\cos\left(-\dfrac{\pi}{4}\right)+i\sin\left(-\dfrac{\pi}{4}\right)\right)$

2 a $2e^{\frac{\pi}{12}i}$ **b** $4e^{-\frac{2\pi}{3}i}$

c $3e^{-\frac{5\pi}{6}i}$ **d** $6e^{-\frac{\pi}{7}i}$

e $e^{-\frac{3\pi}{5}i}$ **f** $\sqrt{2}e^{\frac{7\pi}{8}i}$

g $\sqrt{3}e^{\frac{5\pi}{6}i}$ **h** $8e^{-\frac{5\pi}{12}i}$

3 a $2i$ **b** $\dfrac{7}{2}-\dfrac{7\sqrt{3}}{2}i$

c $1+i$ **d** $\dfrac{\sqrt{3}}{2}-\dfrac{1}{2}i$

e $-2\sqrt{2}$ **f** $-\dfrac{3}{2}+\dfrac{\sqrt{3}}{2}i$

4 a 6 **b** $\dfrac{2}{3}$

c 0 **d** $\dfrac{2\pi}{3}$

5 a $|z||\omega|=5\times\dfrac{1}{5}=1$

b $\dfrac{|z|}{|\omega|}=\dfrac{5}{\frac{1}{5}}=25$

c $\arg z+\arg\omega=\dfrac{2\pi}{7}+-\dfrac{\pi}{7}=\dfrac{\pi}{7}$

d $\arg z-\arg\omega=\dfrac{2\pi}{7}--\dfrac{\pi}{7}=\dfrac{3\pi}{7}$

6 a $3\sqrt{2}$ **b** $\sqrt{2}$

c $\dfrac{11\pi}{12}$ **d** $\dfrac{7\pi}{12}$

Exercise 6.1B

1 $e^{i\theta}=\cos\theta+i\sin\theta$

$e^{-i\theta}=\cos(-\theta)+i\sin(-\theta)$

$\quad=\cos\theta-i\sin\theta$

$e^{i\theta}-e^{-i\theta}=\cos\theta+i\sin\theta-(\cos\theta-i\sin\theta)$

$\quad=2i\sin\theta$

$\sin\theta=\dfrac{e^{i\theta}-e^{-i\theta}}{2i}$, as required

2 a $z^n+\dfrac{1}{z^n}=(e^{\theta i})^n+\dfrac{1}{(e^{\theta i})^n}$

$\quad=e^{n\theta i}+e^{-n\theta i}$

$\quad=2\cos(n\theta)$, as required

b $z^n+\dfrac{1}{z^n}=(e^{\theta i})^n-\dfrac{1}{(e^{\theta i})^n}$

$\quad=e^{n\theta i}-e^{-n\theta i}$

$\quad=2i\sin(n\theta)$, as required

3 a $z_1z_2=r_1e^{\theta_1 i}\times r_2e^{\theta_2 i}$

$\quad=r_1r_2e^{\theta_1 i}e^{\theta_2 i}$

$\quad=r_1r_2e^{(\theta_1+\theta_2)i}$

So $|z_1z_2|=r_1r_2=|z_1||z_2|$

and $\arg(z_1z_2)=\theta_1+\theta_2=\arg z_1+\arg z_2$

b $\dfrac{z_1}{z_2}=\dfrac{r_1e^{\theta_1 i}}{r_2e^{\theta_2 i}}$

$\quad=\dfrac{r_1}{r_2}e^{(\theta_1-\theta_2)i}$

So $\left|\dfrac{z_1}{z_2}\right|=\dfrac{r_1}{r_2}=\dfrac{|z_1|}{|z_2|}$

and $\arg\left(\dfrac{z_1}{z_2}\right)=\theta_1-\theta_2=\arg z_1-\arg z_2$

4 a $\text{RHS}=\left(\dfrac{e^{iA}-e^{-iA}}{2i}\right)\left(\dfrac{e^{iB}+e^{-iB}}{2}\right)+\left(\dfrac{e^{iB}-e^{-iB}}{2i}\right)\left(\dfrac{e^{iA}+e^{-iA}}{2}\right)$

$=\dfrac{(e^{i(A+B)}+e^{i(A-B)}-e^{-i(A-B)}-e^{-i(A+B)})}{4i}$

$+\dfrac{(e^{i(A+B)}+e^{-i(A-B)}-e^{i(A-B)}-e^{-i(A+B)})}{4i}$

$=\dfrac{2e^{i(A+B)}-2e^{-i(A+B)}}{4i}$

$=\dfrac{e^{i(A+B)}-e^{-i(A+B)}}{2i}$

$=\sin(A+B)$, as required

b $\text{RHS} = \left(\dfrac{e^{iA}+e^{-iA}}{2}\right)\left(\dfrac{e^{iB}+e^{-iB}}{2}\right) - \left(\dfrac{e^{iA}-e^{-iA}}{2i}\right)\left(\dfrac{e^{iB}-e^{-iB}}{2i}\right)$

$= \dfrac{e^{i(A+B)}+e^{i(A-B)}+e^{-i(A-B)}+e^{-i(A+B)}}{4}$

$\quad - \dfrac{e^{i(A+B)}-e^{i(A-B)}-e^{-i(A-B)}+e^{-i(A+B)}}{4i^2}$

$= \dfrac{e^{i(A+B)}+e^{i(A-B)}+e^{-i(A-B)}+e^{-i(A+B)}}{4}$

$\quad + \dfrac{e^{i(A+B)}-e^{i(A-B)}-e^{-i(A-B)}+e^{-i(A+B)}}{4}$

$= \dfrac{2e^{i(A+B)}+2e^{-i(A+B)}}{4}$

$= \dfrac{e^{i(A+B)}+e^{-i(A+B)}}{2}$

$= \cos(A+B),\ \text{as required}$

5 a $\text{RHS} = \left(\dfrac{e^{ix}+e^{-ix}}{2}\right)^2 - \left(\dfrac{e^{ix}-e^{-ix}}{2i}\right)^2$

$= \dfrac{e^{2ix}+2+e^{-2ix}}{4} - \dfrac{e^{2ix}-2+e^{-2ix}}{4i^2}$

$= \dfrac{e^{2ix}+2+e^{-2ix}}{4} + \dfrac{e^{2ix}-2+e^{-2ix}}{4}$

$= \dfrac{2e^{2ix}+2e^{-2ix}}{4}$

$= \dfrac{e^{2ix}+e^{-2ix}}{2}$

$= \cos 2x,\ \text{as required}$

b $\text{LHS} = \left(\dfrac{e^{ix}+e^{-ix}}{2}\right)^2 + \left(\dfrac{e^{ix}-e^{-ix}}{2i}\right)^2$

$= \dfrac{e^{2ix}+2+e^{-2ix}}{4} + \dfrac{e^{2ix}-2+e^{-2ix}}{4i^2}$

$= \dfrac{e^{2ix}+2+e^{-2ix}}{4} - \dfrac{e^{2ix}-2+e^{-2ix}}{4}$

$= \dfrac{4}{4} = 1,\ \text{as required}$

6 a $\text{LHS} = \left(\dfrac{e^{i\theta}+e^{-i\theta}}{2} + i\left(\dfrac{e^{i\theta}-e^{-i\theta}}{2i}\right)\right)^2$

$= \left(\dfrac{e^{i\theta}+e^{-i\theta}}{2} + \dfrac{e^{i\theta}-e^{-i\theta}}{2}\right)^2$

$= \left(\dfrac{2e^{i\theta}}{2}\right)^2$

$= \left(e^{i\theta}\right)^2$

$= e^{2i\theta}$

$\text{RHS} = \dfrac{e^{2i\theta}+e^{-2i\theta}}{2} + i\left(\dfrac{e^{2i\theta}-e^{-2i\theta}}{2i}\right)$

$= \dfrac{e^{2i\theta}+e^{-2i\theta}}{2} + \dfrac{e^{2i\theta}-e^{-2i\theta}}{2}$

$= \dfrac{2e^{2i\theta}}{2}$

$= e^{2i\theta} = \text{LHS so } \left(\cos\theta + i\sin\theta\right)^2 \equiv \cos 2\theta + i\sin 2\theta$

b $\text{LHS} = \left(\dfrac{e^{i\theta}+e^{-i\theta}}{2} + i\left(\dfrac{e^{i\theta}-e^{-i\theta}}{2i}\right)\right)^n$

$= \left(\dfrac{e^{i\theta}+e^{-i\theta}}{2} + \dfrac{e^{i\theta}-e^{-i\theta}}{2}\right)^n$

$= \left(\dfrac{2e^{i\theta}}{2}\right)^n$

$= \left(e^{i\theta}\right)^n$

$= e^{in\theta}$

$\text{RHS} = \dfrac{e^{in\theta}+e^{-in\theta}}{2} + i\left(\dfrac{e^{in\theta}-e^{-in\theta}}{2i}\right)$

$= \dfrac{e^{in\theta}+e^{-in\theta}}{2} + \dfrac{e^{in\theta}-e^{-in\theta}}{2}$

$= \dfrac{2e^{in\theta}}{2}$

$= e^{in\theta} = \text{LHS, so } \left(\cos\theta+i\sin\theta\right)^n \equiv \cos(n\theta)+i\sin(n\theta)$

7 $z\omega = (4\times 3)\left(\cos\left(\dfrac{\pi}{9}+\dfrac{2\pi}{9}\right)+i\sin\left(\dfrac{\pi}{9}+\dfrac{2\pi}{9}\right)\right)$

$= 12\left(\cos\left(\dfrac{\pi}{3}\right)+i\sin\left(\dfrac{\pi}{3}\right)\right)$

$= 12\left(\dfrac{1}{2}+i\left(\dfrac{\sqrt{3}}{2}\right)\right)$

$= 6+6\sqrt{3}i,\ \text{as required}$

8 a $\dfrac{z}{\omega} = \dfrac{8}{6}\left(\cos\left(\dfrac{5\pi}{12}-\dfrac{\pi}{3}\right)+i\sin\left(\dfrac{5\pi}{12}-\dfrac{\pi}{3}\right)\right)$

$= \dfrac{4}{3}\left(\cos\left(\dfrac{3\pi}{4}\right)+i\sin\left(\dfrac{3\pi}{4}\right)\right)$

$= \dfrac{4}{3}\left(-\dfrac{\sqrt{2}}{2}+i\left(\dfrac{\sqrt{2}}{2}\right)\right)$

$= \dfrac{2\sqrt{2}}{3}(i-1),\ \text{as required}$

b $z^2 = 8^2\left(\cos\left(\dfrac{5\pi}{12}+\dfrac{5\pi}{12}\right)+i\sin\left(\dfrac{5\pi}{12}+\dfrac{5\pi}{12}\right)\right)$

$= 64\left(\cos\left(\dfrac{5\pi}{6}\right)+\sin\left(\dfrac{5\pi}{6}\right)\right)$

$= 64\left(\left(-\dfrac{\sqrt{3}}{2}\right)+i\left(\dfrac{1}{2}\right)\right)$

$= -32\sqrt{3}+32i$

9 a $|z\omega| = \sqrt{2}k$

$\arg(z\omega) = \theta - \dfrac{\pi}{4}$

b $\left|\dfrac{z}{\omega}\right| = \dfrac{k}{\sqrt{2}} = \dfrac{\sqrt{2}}{2}k$

$\arg\left(\dfrac{z}{\omega}\right) = \theta + \dfrac{\pi}{4}$

Exercise 6.2A

1 a 1 **b** $-\dfrac{\sqrt{3}}{2}+\dfrac{1}{2}i$

c $\dfrac{1}{2}-\dfrac{\sqrt{3}}{2}i$ **d** $-i$

e $-\dfrac{\sqrt{2}}{2}-\dfrac{\sqrt{2}}{2}i$ **f** $-\dfrac{\sqrt{2}}{2}+\dfrac{\sqrt{2}}{2}i$

2 a $\dfrac{81}{2}\sqrt{3}+\dfrac{81}{2}\mathrm{i}$ **b** $\dfrac{\sqrt{2}}{1458}-\dfrac{\sqrt{2}}{1458}\mathrm{i}$

3 a $4\sqrt{2}+4\sqrt{2}\mathrm{i}$ **b** $\dfrac{\sqrt{3}}{8}-\dfrac{1}{8}\mathrm{i}$

   c $64\mathrm{i}$ **d** $\dfrac{1}{32}-\dfrac{\sqrt{3}}{32}\mathrm{i}$

4 a -16 **b** $64\mathrm{i}$

   c $\dfrac{1}{4}\mathrm{i}$ **d** $\dfrac{1}{16}$

5 a $-2-2\mathrm{i}$ **b** $8+8\mathrm{i}$

   c $-\dfrac{1}{8}-\dfrac{1}{8}\mathrm{i}$ **d** $-\dfrac{1}{8}\mathrm{i}$

6 a -9 **b** $-\dfrac{1}{3}\mathrm{i}$

   c $\dfrac{1}{27}\mathrm{i}$ **d** $\dfrac{1}{9}\mathrm{i}$

7 a $8-8\sqrt{3}\mathrm{i}$ **b** $-\dfrac{1}{8}\mathrm{i}$

   c $=\dfrac{1}{8}+\dfrac{\sqrt{3}}{8}\mathrm{i}$ **d** $-\dfrac{1}{8}$

8 a $\cos 6x+\mathrm{i}\sin 6x$ **b** $\cos 2x+\mathrm{i}\sin 2x$

   c $\cos(-x)+\mathrm{i}\sin(-x)$ **d** $\cos 3x+\mathrm{i}\sin 3x$

   e $\cos(2x)+\mathrm{i}\sin(2x)$ **f** $\cos 10x+\mathrm{i}\sin 10x$

9 a $\dfrac{1}{2}-\dfrac{\sqrt{3}}{2}\mathrm{i}$ **b** $\dfrac{\sqrt{2}}{2}+\dfrac{\sqrt{2}}{2}\mathrm{i}$

10 a $-1-\sqrt{3}\mathrm{i}$ **b** $-32+32\sqrt{3}\mathrm{i}$

Exercise 6.2B

1 a $\quad 2\cos^2\theta \equiv \dfrac{1}{2}(2\cos\theta)^2$

$$\equiv \dfrac{1}{2}(\mathrm{e}^{\mathrm{i}\theta}+\mathrm{e}^{-\mathrm{i}\theta})^2$$

$$\equiv \dfrac{1}{2}(\mathrm{e}^{2\mathrm{i}\theta}+2+\mathrm{e}^{-2\mathrm{i}\theta})$$

$$\equiv \dfrac{1}{2}(2\cos 2\theta+2)$$

$$\equiv \cos 2\theta+1$$

b $\quad 8\sin^3\theta \equiv \dfrac{1}{\mathrm{i}^3}(2\mathrm{i}\sin\theta)^3$

$$\equiv \dfrac{1}{-\mathrm{i}}(\mathrm{e}^{\mathrm{i}\theta}-\mathrm{e}^{-\mathrm{i}\theta})^3$$

$$\equiv \dfrac{-1}{\mathrm{i}}(\mathrm{e}^{3\mathrm{i}\theta}-\mathrm{e}^{-3\mathrm{i}\theta}-3\mathrm{e}^{\mathrm{i}\theta}+3\mathrm{e}^{-\mathrm{i}\theta})$$

$$\equiv -\dfrac{1}{\mathrm{i}}(2\mathrm{i}\sin 3\theta-6\mathrm{i}\sin\theta)$$

$$\equiv 6\sin\theta-2\sin 3\theta$$

c $\quad 4\sin^4\theta \equiv \dfrac{1}{4}(2\mathrm{i}\sin\theta)^4$

$$\equiv \dfrac{1}{4}(\mathrm{e}^{\mathrm{i}\theta}-\mathrm{e}^{-\mathrm{i}\theta})^4$$

$$\equiv \dfrac{1}{4}(\mathrm{e}^{4\mathrm{i}\theta}+\mathrm{e}^{-4\mathrm{i}\theta}-4\mathrm{e}^{2\mathrm{i}\theta}-4\mathrm{e}^{-2\mathrm{i}\theta}+6)$$

$$\equiv \dfrac{1}{4}(2\cos 4\theta-8\cos 2\theta+6)$$

$$\equiv \dfrac{1}{2}\cos 4\theta-2\cos 2\theta+\dfrac{3}{2}$$

2 a $\quad \cos^5\theta \equiv \dfrac{1}{32}(2\cos\theta)^5$

$$\equiv \dfrac{1}{32}(\mathrm{e}^{\mathrm{i}\theta}+\mathrm{e}^{-\mathrm{i}\theta})^5$$

$$\equiv \dfrac{1}{32}(\mathrm{e}^{5\mathrm{i}\theta}+\mathrm{e}^{-5\mathrm{i}\theta}+5\mathrm{e}^{3\mathrm{i}\theta}+5\mathrm{e}^{-3\mathrm{i}\theta}+10\mathrm{e}^{\mathrm{i}\theta}+10\mathrm{e}^{-\mathrm{i}\theta})$$

$$\equiv \dfrac{1}{32}(2\cos 5\theta+10\cos 3\theta+20\cos\theta)$$

$$\equiv \dfrac{1}{16}(10\cos\theta+5\cos 3\theta+\cos 5\theta)$$

So $A=\dfrac{1}{16}$

b $\quad \dfrac{1}{16}\left(10\sin\theta+\dfrac{5}{3}\sin 3\theta+\dfrac{1}{5}\sin 5\theta\right)+c$

3 a $\quad \sin^6\theta \equiv \dfrac{1}{-64}(2\mathrm{i}\sin\theta)^6$

$$\equiv -\dfrac{1}{64}(\mathrm{e}^{\mathrm{i}\theta}-\mathrm{e}^{-\mathrm{i}\theta})^6$$

$$\equiv -\dfrac{1}{64}(\mathrm{e}^{6\mathrm{i}\theta}+\mathrm{e}^{-6\mathrm{i}\theta}-6\mathrm{e}^{4\mathrm{i}\theta}-6\mathrm{e}^{-4\mathrm{i}\theta}$$
$$+15\mathrm{e}^{2\mathrm{i}\theta}+15\mathrm{e}^{-2\mathrm{i}\theta}-20)$$

$$\equiv -\dfrac{1}{64}(2\cos 6\theta-12\cos 4\theta+30\cos 2\theta-20)$$

$$\equiv -\dfrac{1}{32}(15\cos 2\theta-6\cos 4\theta+\cos 6\theta-10)$$

So $B=-\dfrac{1}{32}$

b $\quad -\dfrac{1}{32}\left(\dfrac{15}{2}\sin\theta-\dfrac{3}{2}\sin 4\theta+\dfrac{1}{6}\sin 6\theta-10\theta\right)+c$

4 a $\quad 2\sin^3\theta \equiv \dfrac{1}{-4\mathrm{i}}(2\mathrm{i}\sin\theta)^3$

$$\equiv -\dfrac{1}{4\mathrm{i}}(\mathrm{e}^{\mathrm{i}\theta}-\mathrm{e}^{-\mathrm{i}\theta})^3$$

$$\equiv -\dfrac{1}{4\mathrm{i}}\left(\mathrm{e}^{3\mathrm{i}\theta}-\mathrm{e}^{-3\mathrm{i}\theta}-3\mathrm{e}^{\mathrm{i}\theta}+3\mathrm{e}^{-\mathrm{i}\theta}\right)$$

$$\equiv -\dfrac{1}{4\mathrm{i}}(2\mathrm{i}\sin 3\theta-6\mathrm{i}\sin\theta)$$

$$\equiv \dfrac{3}{2}\sin\theta-\dfrac{1}{2}\sin 3\theta$$

b $\quad \theta=\dfrac{\pi}{6},\dfrac{5\pi}{6}$

5 a $\quad 5\cos^4\theta \equiv \dfrac{5}{16}(2\cos\theta)^4$

$$\equiv \dfrac{5}{16}(\mathrm{e}^{\mathrm{i}\theta}+\mathrm{e}^{-\mathrm{i}\theta})^4$$

$$\equiv \dfrac{5}{16}(\mathrm{e}^{4\mathrm{i}\theta}+\mathrm{e}^{-4\mathrm{i}\theta}+4\mathrm{e}^{2\mathrm{i}\theta}+4\mathrm{e}^{-2\mathrm{i}\theta}+6)$$

$$\equiv \dfrac{5}{16}(2\cos 4\theta+8\cos 2\theta+6)$$

$$\equiv \dfrac{5}{8}\cos 4\theta+\dfrac{5}{2}\cos 2\theta+\dfrac{15}{8}$$

So $A=\dfrac{5}{8},B=\dfrac{5}{2}$ and $C=\dfrac{15}{8}$

b $\quad \theta=\pm\dfrac{\pi}{4},\pm\dfrac{3\pi}{4}$

6 a $\cos 2\theta + i\sin 2\theta = (\cos\theta + i\sin\theta)^2$

$\qquad = \cos^2\theta + 2i\cos\theta\sin\theta + i^2\sin^2\theta$

$\qquad = \cos^2\theta + 2i\cos\theta\sin\theta - \sin^2\theta$

Im: $\sin 2\theta = 2\cos\theta\sin\theta$

b $\cos 3\theta + i\sin 3\theta = (\cos\theta + i\sin\theta)^3$

$\qquad = \cos^3\theta + 3\cos^2\theta(i\sin\theta)$

$\qquad\quad + 3\cos\theta(i\sin\theta)^2 + (i\sin\theta)^3$

$\qquad = \cos^3\theta + 3i\cos^2\theta\sin\theta$

$\qquad\quad - 3\cos\theta\sin^2\theta - i\sin^3\theta$

Im: $\sin 3\theta = 3\cos^2\theta\sin\theta - \sin^3\theta$

$\qquad = 3(1 - \sin^2\theta)\sin\theta - \sin^3\theta$

$\qquad = 3\sin\theta - 3\sin^3\theta - \sin^3\theta$

$\qquad = 3\sin\theta - 4\sin^3\theta$

c Re: $\cos 3\theta = \cos^3\theta - 3\cos\theta\sin^2\theta$

$\qquad = \cos^3\theta - 3\cos\theta(1 - \cos^2\theta)$

$\qquad = \cos^3\theta - 3\cos\theta + 3\cos^3\theta$

$\qquad = 4\cos^3\theta - 3\cos\theta$

d $\cos 4\theta + i\sin 4\theta = (\cos\theta + i\sin\theta)^4$

$\qquad = \cos^4\theta + 4\cos^3\theta(i\sin\theta)$

$\qquad\quad + 6\cos^2\theta(i\sin\theta)^2 + 4\cos\theta(i\sin\theta)^3 + (i\sin\theta)^4$

$\qquad = \cos^4\theta + 4i\cos^3\theta\sin\theta - 6\cos^2\theta\sin^2\theta$

$\qquad\quad - 4i\cos\theta\sin^3\theta + \sin^4\theta$

Im: $\sin 4\theta = 4\cos^3\theta\sin\theta - 4\cos\theta\sin^3\theta$

$\qquad = 4\cos\theta\sin\theta(\cos^2\theta - \sin^2\theta)$

$\qquad = 4\cos\theta\sin\theta(1 - 2\sin^2\theta)$

$\qquad = 4\cos\theta\sin\theta - 8\cos\theta\sin^3\theta$

7 a Re: $\cos 6\theta = \cos^6\theta - 15\cos^4\theta\sin^2\theta$

$\qquad + 15\cos^2\theta\sin^4\theta - \sin^6\theta$

$\qquad = \cos^6\theta - 15\cos^4\theta(1 - \cos^2\theta)$

$\qquad + 15\cos^2\theta(1 - \cos^2\theta)^2 - (1 - \cos^2\theta)^3$

$\qquad = \cos^6\theta - 15\cos^4\theta + 15\cos^6\theta$

$\qquad + 15\cos^2\theta - 30\cos^4\theta + 15\cos^6\theta$

$\qquad - 1 + 3\cos^2\theta - 3\cos^4\theta + \cos^6\theta$

$\qquad = 32\cos^6\theta - 48\cos^4\theta + 18\cos^2\theta - 1$

b Im: $\sin 6\theta = 6\cos^5\theta\sin\theta - 20\cos^3\theta\sin^3\theta + 6\cos\theta\sin^5\theta$

$\qquad = 2\sin\theta\cos\theta(3\cos^4\theta - 10\cos^2\theta\sin^2\theta + 3\sin^4\theta)$

$\qquad = 2\sin\theta\cos\theta(3(1 - \sin^2\theta)^2$

$\qquad - 10(1 - \sin^2\theta)\sin^2\theta + 3\sin^4\theta)$

$\qquad = 2\sin\theta\cos\theta(3 - 6\sin^2\theta + 3\sin^4\theta - 10\sin^2\theta$

$\qquad + 10\sin^4\theta + 3\sin^4\theta)$

$\qquad = 2\sin\theta\cos\theta(16\sin^4\theta - 16\sin^2\theta + 3)$

8 a $\cos 5\theta + i\sin 5\theta = (\cos\theta + i\sin\theta)^5$

$\qquad = \cos^5\theta + 5\cos^4\theta(i\sin\theta)$

$\qquad\quad + 10\cos^3\theta(i\sin\theta)^2$

$\qquad\quad + 10\cos^2\theta(i\sin\theta)^3$

$\qquad\quad + 5\cos\theta(i\sin\theta)^4 + (i\sin\theta)^5$

$\qquad = \cos^5\theta + 5i\cos^4\theta\sin\theta - 10\cos^3\theta\sin^2\theta$

$\qquad\quad - 10i\cos^2\theta\sin^3\theta + 5\cos\theta\sin^4\theta + i\sin^5\theta$

Re: $\cos 5\theta = \cos^5\theta - 10\cos^3\theta\sin^2\theta + 5\cos\theta\sin^4\theta$

$\qquad = \cos^5\theta - 10\cos^3\theta(1 - \cos^2\theta)$

$\qquad\quad + 5\cos\theta(1 - \cos^2\theta)^2$

$\qquad = \cos^5\theta - 10\cos^3\theta + 10\cos^5\theta$

$\qquad\quad + 5\cos\theta - 10\cos^3\theta + 5\cos^5\theta$

$\qquad = 16\cos^5\theta - 20\cos^3\theta + 5\cos\theta$

b $1, 0.309, -0.809$

9 a $\cos 4\theta + i\sin 4\theta = (\cos\theta + i\sin\theta)^4$

$\qquad = \cos^4\theta + 4\cos^3\theta(i\sin\theta) + 6\cos^2\theta(i\sin\theta)^2$

$\qquad\quad + 4\cos\theta(i\sin\theta)^3 + (i\sin\theta)^4$

$\qquad = \cos^4\theta + 4i\cos^3\theta\sin\theta - 6\cos^2\theta\sin^2\theta$

$\qquad\quad - 4i\cos\theta\sin^3\theta + \sin^4\theta$

Re: $\cos 4\theta = \cos^4\theta - 6\cos^2\theta(1 - \cos^2\theta) + (1 - \cos^2\theta)^2$

$\qquad = \cos^4\theta - 6\cos^2\theta + 6\cos^4\theta + 1 - 2\cos^2\theta + \cos^4\theta$

$\qquad = 8\cos^4\theta - 8\cos^2\theta + 1$

b $\pm 0.966, \pm 0.259$

10 $\cos 2\theta + i\sin 2\theta = (\cos\theta + i\sin\theta)^2$

$\qquad = \cos^2\theta + 2i\cos\theta\sin\theta - \sin^2\theta$

Re: $\cos 2\theta = \cos^2\theta - \sin^2\theta$

Im: $\sin 2\theta = 2\cos\theta\sin\theta$

$$\tan 2\theta = \frac{\sin 2\theta}{\cos 2\theta}$$

$$= \frac{2\cos\theta\sin\theta}{\cos^2\theta - \sin^2\theta}$$

$$= \frac{\dfrac{2\cos\theta\sin\theta}{\cos^2\theta}}{\dfrac{\cos^2\theta}{\cos^2\theta} - \dfrac{\sin^2\theta}{\cos^2\theta}}$$

$$= \frac{2\tan\theta}{1 - \tan^2\theta}$$

11 $\left[r(\cos\theta + i\sin\theta)\right]^1 = r(\cos\theta + i\sin\theta)$

and $r^1(\cos(1\theta) + i\sin(1\theta)) = r(\cos\theta + i\sin\theta)$

So true for $n = 1$

Assume true for $n = k$

$\left[r(\cos\theta + i\sin\theta)\right]^{k+1} = \left[r(\cos\theta + i\sin\theta)\right]^k \left[r(\cos\theta + i\sin\theta)\right]$

$\qquad = r^k(\cos k\theta + i\sin k\theta)\left[r(\cos\theta + i\sin\theta)\right]$

$\qquad = r^{k+1}(\cos k\theta\cos\theta + i\cos k\theta\sin\theta$

$\qquad\qquad + i\cos\theta\sin k\theta + i^2\sin k\theta\sin\theta)$

$\qquad = r^{k+1}(\cos k\theta\cos\theta - \sin k\theta\sin\theta$

$\qquad\qquad + i(\cos k\theta\sin\theta + \cos\theta\sin k\theta))$

$\qquad = r^{k+1}(\cos(k\theta + \theta) + i\sin(k\theta + \theta))$

$\qquad = r^{k+1}[\cos(k+1)\theta + i\sin(k+1)\theta]$,

$\qquad\quad$ as required

Since true for $n = 1$ and assuming true for $n = k$ implies true for $n = k + 1$ and hence true for all positive integers n

12 a $z^n = (\cos\theta + i\sin\theta)^n$

$= \cos(n\theta) + i\sin(n\theta)$

$\dfrac{1}{z^n} = z^{-n} = (\cos\theta + i\sin\theta)^{-n}$

$= \cos(-n\theta) + i\sin(-n\theta)$

$= \cos(n\theta) - i\sin(n\theta)$

Therefore $z^n + \dfrac{1}{z^n} = \cos(n\theta) + i\sin(n\theta)$

$\qquad\qquad + (\cos(n\theta) - i\sin(n\theta))$

$= 2\cos(n\theta)$, as required

b $4\cos\theta\sin^2\theta \equiv 4\cos\theta(1 - \cos^2\theta)$

$\equiv 4\cos\theta - 4\cos^3\theta$

$\equiv 4\cos\theta - 4\left(\dfrac{1}{2}\right)^3\left(z + \dfrac{1}{z}\right)^3$

$\equiv 4\cos\theta - \dfrac{1}{2}\left(z^3 + \dfrac{1}{z^3} + 3\left(z + \dfrac{1}{z}\right)\right)$

$\equiv 4\cos\theta - \dfrac{1}{2}\left(2\cos(3\theta) + 6\cos\theta\right)$

$\equiv \cos\theta - \cos(3\theta)$

13 a $z^n = (\cos\theta + i\sin\theta)^n$

$= \cos(n\theta) + i\sin(n\theta)$

$\dfrac{1}{z^n} = z^{-n} = (\cos\theta + i\sin\theta)^{-n}$

$= \cos(-n\theta) + i\sin(-n\theta)$

$= \cos(n\theta) - i\sin(n\theta)$

Therefore $z^n - \dfrac{1}{z^n} = \cos(n\theta) + i\sin(n\theta)$

$\qquad\qquad - (\cos(n\theta) - i\sin(n\theta))$

$= 2i\sin(n\theta)$, as required

b $16\sin^3\theta\cos^2\theta \equiv 16\sin^3\theta(1 - \sin^2\theta)$

$\equiv 16\left(\sin^3\theta - \sin^5\theta\right)$

$\equiv 16\left(\dfrac{1}{2i}\right)^3\left(z - \dfrac{1}{z}\right)^3 - 16\left(\dfrac{1}{2i}\right)^5\left(z - \dfrac{1}{z}\right)^5$

$\equiv 2i\left(z^3 - \dfrac{1}{z^3} - 3\left(z - \dfrac{1}{z}\right)\right)$

$\quad + \dfrac{i}{2}\left(z^5 - \dfrac{1}{z^5} - 5\left(z^3 - \dfrac{1}{z^3}\right) + 10\left(z - \dfrac{1}{z}\right)\right)$

$\equiv 2i\left(2i\sin(3\theta) - 6i\sin\theta\right)$

$\quad + \dfrac{i}{2}(2i\sin(5\theta) - 10i\sin(3\theta)$

$\quad + 20i\sin\theta)$

$\equiv -4\sin(3\theta) + 12\sin\theta - \sin(5\theta)$

$\quad + 5\sin(3\theta) - 10\sin\theta$

$\equiv 2\sin\theta + \sin(3\theta) - \sin(5\theta)$

14 a $\displaystyle\sum_{r=0}^{10} z^r = \dfrac{z^{11} - 1}{z - 1}$

$= \dfrac{\left(e^{\frac{i\pi}{5}}\right)^{11} - 1}{e^{\frac{i\pi}{5}} - 1}$

$= \dfrac{e^{\frac{11i\pi}{5}} - 1}{e^{\frac{i\pi}{5}} - 1}$

$= \dfrac{e^{2\pi i}e^{\frac{i\pi}{5}} - 1}{e^{\frac{i\pi}{5}} - 1}$

$= \dfrac{e^{\frac{i\pi}{5}} - 1}{e^{\frac{i\pi}{5}} - 1}$

$= 1$, as required

b $z = e^{\frac{i\pi}{5}} = \cos\dfrac{\pi}{5} + i\sin\dfrac{\pi}{5}$

Therefore, $\sin\dfrac{\pi}{5} + \sin\dfrac{2\pi}{5} + \ldots + \sin\dfrac{10\pi}{5} = \text{Im}\left(\displaystyle\sum_{r=0}^{10} z^r\right)$

$= \text{Im}(1)$

$= 0$

15 a $1 + e^{i\theta} + e^{2i\theta} + \ldots + e^{11i\theta} = \dfrac{(e^{i\theta})^{12} - 1}{e^{i\theta} - 1}$

$= \dfrac{e^{12i\theta} - 1}{e^{i\theta} - 1}$

$= \dfrac{e^{-\frac{i\theta}{2}}(e^{12i\theta} - 1)}{e^{\frac{i\theta}{2}} - e^{-\frac{i\theta}{2}}}$

$= \dfrac{e^{-\frac{i\theta}{2}}e^{6i\theta}(e^{6i\theta} - e^{-6i\theta})}{2i\sin\left(\dfrac{\theta}{2}\right)}$

$= \dfrac{e^{-\frac{i\theta}{2}}e^{6i\theta}(2i\sin(6\theta))}{2i\sin\left(\dfrac{\theta}{2}\right)}$

$= \dfrac{e^{\frac{11i\theta}{2}}\sin(6\theta)}{\sin\left(\dfrac{\theta}{2}\right)}$, as required

b $\displaystyle\sum_{r=0}^{11}\cos(r\theta) = \text{Re}\left(\dfrac{\left(\cos\left(\dfrac{11\theta}{2}\right) + i\sin\left(\dfrac{11\theta}{2}\right)\right)\sin(6\theta)}{\sin\left(\dfrac{\theta}{2}\right)}\right)$

$= \dfrac{\cos\left(\dfrac{11\theta}{2}\right)\sin(6\theta)}{\sin\left(\dfrac{\theta}{2}\right)}$

$$\sum_{r=0}^{11} \sin(r\theta) = \operatorname{Im}\left[\frac{\left(\cos\left(\frac{11\theta}{2}\right) + i\sin\left(\frac{11\theta}{2}\right)\right)\sin(6\theta)}{\sin\left(\frac{\theta}{2}\right)}\right]$$

$$= \frac{\sin\left(\frac{11\theta}{2}\right)\sin(6\theta)}{\sin\left(\frac{\theta}{2}\right)}$$

16 a $\displaystyle\sum_{r=1}^{3} \cos(r\theta) + i\sin(r\theta) = \sum_{r=1}^{3} z^r$, where $z = \cos\theta + i\sin\theta$

$$= \frac{z(z^3-1)}{z-1}$$

$$= \frac{e^{i\theta}(e^{3i\theta}-1)}{e^{i\theta}-1}$$

$$= \frac{e^{i\theta}e^{-\frac{i\theta}{2}}(e^{3i\theta}-1)}{e^{\frac{i\theta}{2}} - e^{-\frac{i\theta}{2}}}$$

$$= \frac{e^{\frac{i\theta}{2}}(e^{3i\theta}-1)}{2i\sin\left(\frac{\theta}{2}\right)}$$

$$= \frac{e^{\frac{i\theta}{2}}e^{\frac{3i\theta}{2}}\left(e^{\frac{3i\theta}{2}} - e^{-\frac{3i\theta}{2}}\right)}{2i\sin\left(\frac{\theta}{2}\right)}$$

$$= \frac{e^{2i\theta}\left(2i\sin\left(\frac{3\theta}{2}\right)\right)}{2i\sin\left(\frac{\theta}{2}\right)}$$

$$= \frac{e^{2i\theta}\sin\left(\frac{3\theta}{2}\right)}{\sin\left(\frac{\theta}{2}\right)}$$

$$= \frac{(\cos 2\theta + i\sin 2\theta)\sin\left(\frac{3\theta}{2}\right)}{\sin\left(\frac{\theta}{2}\right)}$$

Therefore, $\displaystyle\sum_{r=1}^{3} \cos(r\theta) \equiv \frac{\cos 2\theta \sin\left(\frac{3\theta}{2}\right)}{\sin\left(\frac{\theta}{2}\right)}$

b $\displaystyle\sum_{r=1}^{3} \sin(r\theta) = \frac{\sin 2\theta \sin\left(\frac{3\theta}{2}\right)}{\sin\left(\frac{\theta}{2}\right)}$

So $\dfrac{\cos\theta + \cos 2\theta + \cos 3\theta}{\sin\theta + \sin 2\theta + \sin 3\theta} \equiv \dfrac{\dfrac{\cos 2\theta \sin\left(\frac{3\theta}{2}\right)}{\sin\left(\frac{\theta}{2}\right)}}{\dfrac{\sin 2\theta \sin\left(\frac{3\theta}{2}\right)}{\sin\left(\frac{\theta}{2}\right)}}$

$$\equiv \frac{\cos 2\theta}{\sin 2\theta}$$

$$\equiv \cot 2\theta$$

17 $\displaystyle\sum_{r=0}^{n-1} z^r = \frac{z^n - 1}{z - 1}$

$$= \frac{\left(e^{\frac{\pi i}{n}}\right)^n - 1}{e^{\frac{\pi i}{n}} - 1}$$

$$= \frac{e^{\pi i} - 1}{e^{\frac{\pi i}{n}} - 1}$$

$$= \frac{-1 - 1}{e^{\frac{\pi i}{n}} - 1}$$

$$= \frac{-2}{e^{\frac{\pi i}{n}} - 1}$$

$$= \frac{-2e^{-\frac{\pi i}{2n}}}{e^{\frac{\pi i}{2n}} - e^{-\frac{\pi i}{2n}}}$$

$$= \frac{-2e^{-\frac{\pi i}{2n}}}{2i\sin\left(\frac{\pi}{2n}\right)}$$

$$= \frac{-e^{-\frac{\pi i}{2n}}}{i\sin\left(\frac{\pi}{2n}\right)}$$

$$= \frac{-\left(\cos\left(-\frac{\pi}{2n}\right) + i\sin\left(-\frac{\pi}{2n}\right)\right)}{i\sin\left(\frac{\pi}{2n}\right)}$$

$$= \frac{-\left(\cos\left(\frac{\pi}{2n}\right) - i\sin\left(\frac{\pi}{2n}\right)\right)}{i\sin\left(\frac{\pi}{2n}\right)}$$

$$= \frac{i\sin\left(\frac{\pi}{2n}\right) - \cos\left(\frac{\pi}{2n}\right)}{i\sin\left(\frac{\pi}{2n}\right)}$$

$$= 1 - \frac{1}{i}\cot\left(\frac{\pi}{2n}\right)$$

$$= 1 + i\cot\left(\frac{\pi}{2n}\right)$$

18 a $\displaystyle\sum_{r=1}^{\infty} \frac{\cos r\theta + i\sin r\theta}{2^r} = \sum_{r=1}^{\infty}\left(\frac{1}{2}(\cos\theta + i\sin\theta)\right)^r$

$$= \sum_{r=1}^{\infty} z^r, \text{ where } z = \frac{1}{2}(\cos\theta + i\sin\theta)$$

The series converges since $|z| = \dfrac{1}{2} < 1$

$$\sum_{r=1}^{\infty} z^r = \frac{z}{1-z}$$

$$= \frac{\frac{1}{2}e^{i\theta}}{1 - \frac{1}{2}e^{i\theta}}$$

$$= \frac{e^{i\theta}}{2 - e^{i\theta}}, \text{ as required}$$

b $\dfrac{e^{i\theta}}{2-e^{i\theta}} = \dfrac{\cos\theta + i\sin\theta}{2-(\cos\theta + i\sin\theta)}$

$= \dfrac{\cos\theta + i\sin\theta}{2 - \cos\theta - i\sin\theta}$

$= \dfrac{(\cos\theta + i\sin\theta)(2 - \cos\theta + i\sin\theta)}{(2 - \cos\theta - i\sin\theta)(2 - \cos\theta + i\sin\theta)}$

$= \dfrac{(2\cos\theta - \cos^2\theta - \sin^2\theta) + i(2\sin\theta - \sin\theta\cos\theta + \sin\theta\cos\theta)}{(2 - \cos\theta)^2 + \sin^2\theta}$

$= \dfrac{(2\cos\theta - (\cos^2\theta + \sin^2\theta)) + 2i\sin\theta}{4 + \cos^2\theta - 4\cos\theta + \sin^2\theta}$

$= \dfrac{(2\cos\theta - 1) + 2i\sin\theta}{5 - 4\cos\theta}$

So $\dfrac{1}{2}\cos\theta + \dfrac{1}{4}\cos 2\theta + \dfrac{1}{8}\cos 3\theta + \ldots = \dfrac{2\cos\theta - 1}{5 - 4\cos\theta}$

and $\displaystyle\sum_{r=1}^{\infty} \dfrac{\sin r\theta}{2^r} = \dfrac{2\sin\theta}{5 - 4\cos\theta}$

19 a $\displaystyle\sum_{r=1}^{n} \cos(r\theta) + i\sin(r\theta) = \sum_{r=1}^{n}\big(\cos(r\theta) + i\sin(r\theta)\big)^r$

$= (\cos\theta + i\sin\theta) + (\cos\theta + i\sin\theta)^2 + \ldots + (\cos\theta + i\sin\theta)^n$

$= \dfrac{(\cos\theta + i\sin\theta)(1 - (\cos(\theta) + i\sin(\theta))^r)}{1 - (\cos(\theta) + i\sin(\theta))}$

$= \dfrac{(\cos\theta + i\sin\theta)(1 - (\cos(r\theta) + i\sin(r\theta)))}{1 - (\cos(\theta) + i\sin(\theta))}$

$= \dfrac{(\cos\theta + i\sin\theta)\left(2\sin^2\left(\dfrac{n\theta}{2}\right) - 2i\sin\left(\dfrac{n\theta}{2}\right)\cos\left(\dfrac{n\theta}{2}\right)\right)}{2\sin^2\left(\dfrac{\theta}{2}\right) - 2i\sin\left(\dfrac{\theta}{2}\right)\cos\left(\dfrac{\theta}{2}\right)}$

$= \dfrac{(\cos\theta + i\sin\theta)\,2\sin\left(\dfrac{n\theta}{2}\right)\left(\sin\left(\dfrac{n\theta}{2}\right) - i\cos\left(\dfrac{n\theta}{2}\right)\right)}{2\sin\left(\dfrac{\theta}{2}\right)\left(\sin\left(\dfrac{\theta}{2}\right) - i\cos\left(\dfrac{\theta}{2}\right)\right)}$

$= \dfrac{(\cos\theta + i\sin\theta)\sin\left(\dfrac{n\theta}{2}\right)\left(\sin\left(\dfrac{n\theta}{2}\right) - i\cos\left(\dfrac{n\theta}{2}\right)\right)\left(\sin\left(\dfrac{\theta}{2}\right) + i\cos\left(\dfrac{\theta}{2}\right)\right)}{\sin\left(\dfrac{\theta}{2}\right)\left(\sin\left(\dfrac{\theta}{2}\right) - i\cos\left(\dfrac{\theta}{2}\right)\right)\left(\sin\left(\dfrac{\theta}{2}\right) + i\cos\left(\dfrac{\theta}{2}\right)\right)}$

$= \dfrac{(\cos\theta + i\sin\theta)\left(\sin\left(\dfrac{n\theta}{2}\right)\sin\left(\dfrac{\theta}{2}\right) + \cos\left(\dfrac{n\theta}{2}\right)\cos\left(\dfrac{\theta}{2}\right) + i\left(\cos\left(\dfrac{\theta}{2}\right)\sin\left(\dfrac{n\theta}{2}\right) - \cos\left(\dfrac{n\theta}{2}\right)\sin\left(\dfrac{\theta}{2}\right)\right)\right)\sin\left(\dfrac{n\theta}{2}\right)}{\sin\left(\dfrac{\theta}{2}\right)\left(\sin^2\left(\dfrac{\theta}{2}\right) + \cos^2\left(\dfrac{\theta}{2}\right)\right)}$

$= \dfrac{(\cos\theta + i\sin\theta)\left(\sin\left(\dfrac{(n-1)\theta}{2}\right) + i\cos\left(\dfrac{(n-1)\theta}{2}\right)\right)\sin\left(\dfrac{n\theta}{2}\right)}{\sin\left(\dfrac{\theta}{2}\right)}$

Considering only the real parts of this sum gives

$\displaystyle\sum_{r=1}^{n} \cos(r\theta)$

$= \dfrac{(\cos\theta + i\sin\theta)\left(\cos\theta\sin\left(\dfrac{(n-1)\theta}{2}\right) - \cos\left(\dfrac{(n-1)\theta}{2}\right)\sin\theta\right)\sin\left(\dfrac{n\theta}{2}\right)}{\sin\left(\dfrac{\theta}{2}\right)}$

$= \dfrac{\cos\left(\dfrac{(n+1)\theta}{2}\right)\sin\left(\dfrac{n\theta}{2}\right)}{\sin\left(\dfrac{\theta}{2}\right)}$

b $\dfrac{\sin\left(\dfrac{(n+1)\theta}{2}\right)\sin\left(\dfrac{n\theta}{2}\right)}{\sin\left(\dfrac{\theta}{2}\right)}$

Exercise 6.3A

1 a $k = 0: z = 1$

$k = 1: z = e^{\frac{2\pi i}{3}} = -\dfrac{1}{2} + \dfrac{\sqrt{3}}{2}i$

$k = 2: z = e^{\frac{4\pi i}{3}} = -\dfrac{1}{2} - \dfrac{\sqrt{3}}{2}i$

b

2 a $k = 0: z = 1$

$k = 1: z = e^{\frac{\pi i}{2}} = i$

$k = 2: z = e^{\frac{2\pi i}{2}} = -1$

$k = 3: z = e^{\frac{3\pi i}{2}} = -i$

b

3. a $k = 0: z = 1$

$k = 1: z = e^{\frac{2\pi i}{5}}$

$k = 2: z = e^{\frac{4\pi i}{5}}$

$k = 3: z = e^{\frac{6\pi i}{5}}$

$k = 4: z = e^{\frac{8\pi i}{5}}$

b

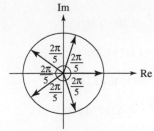

4. a $k=0: z=2$

$k=1: z=2e^{\frac{2\pi i}{3}}=-1+\sqrt{3}i$

$k=2: z=2e^{\frac{4\pi i}{3}}=-1-\sqrt{3}i$

b $k=0: z=e^{\frac{\pi}{6}i}=\frac{\sqrt{3}}{2}+\frac{1}{2}i$

$k=1: z=e^{\frac{5\pi}{6}i}=-\frac{\sqrt{3}}{2}+\frac{1}{2}i$

$k=2: z=e^{\frac{\pi}{6}i}=-i$

c $k=0: z=\sqrt{7}e^{-\frac{\pi i}{4}}=\frac{\sqrt{14}}{2}-\frac{\sqrt{14}}{2}i$

$k=1: z=\sqrt{7}e^{\frac{\pi i}{4}}=\frac{\sqrt{14}}{2}+\frac{\sqrt{14}}{2}i$

$k=2: z=\sqrt{7}e^{\frac{3\pi i}{4}}=-\frac{\sqrt{14}}{2}+\frac{\sqrt{14}}{2}i$

$k=3: z=\sqrt{7}e^{\frac{5\pi i}{4}}=-\frac{\sqrt{14}}{2}-\frac{\sqrt{14}}{2}i$

d $k=0: z=3e^{-\frac{\pi i}{6}}=\frac{3\sqrt{3}}{2}-\frac{3}{2}i$

$k=1: z=3e^{\frac{\pi i}{2}}=3i$

$k=2: z=3e^{\frac{7\pi i}{6}}=-\frac{3\sqrt{3}}{2}-\frac{3}{2}i$

e $k=0: z=e^{-\frac{\pi i}{6}}=\frac{\sqrt{6}}{2}-\frac{\sqrt{2}}{2}i$

$k=1: z=e^{\frac{\pi i}{6}}=\frac{\sqrt{6}}{2}+\frac{\sqrt{2}}{2}i$

$k=2: z=e^{\frac{5\pi i}{6}}=-\frac{\sqrt{6}}{2}+\frac{\sqrt{2}}{2}i$

$k=3: z=e^{\frac{\pi i}{2}}=\sqrt{2}i$

$k=4: z=e^{\frac{7\pi i}{6}}=-\frac{\sqrt{6}}{2}-\frac{\sqrt{2}}{2}i$

$k=5: z=e^{\frac{3\pi i}{2}}=-\sqrt{2}i$

5 a $k=0: z=2e^{-\frac{\pi}{8}i}$

$k=1: z=2e^{\frac{3\pi}{8}i}$

$k=2: z=2e^{\frac{7\pi}{8}i}$

$k=3: z=2e^{\frac{11\pi}{8}i}$ or $2e^{-\frac{5\pi}{8}i}$

b $k=0: z=2e^{\frac{\pi}{10}i}$

$k=1: z=2e^{\frac{\pi i}{2}}$

$k=2: z=2e^{\frac{9\pi}{10}i}$

$k=3: z=2e^{\frac{13\pi}{10}i}$ or $2e^{-\frac{7\pi}{10}i}$

$k=4: z=2e^{\frac{17\pi}{10}i}$ or $2e^{-\frac{3\pi}{10}i}$

6 a $k=0: z=2e^{\frac{\pi}{12}i}$

$k=1: z=2e^{\frac{3\pi}{4}i}$

$k=-1: z=2e^{-\frac{7\pi}{12}i}$

b $k=0: z=2e^{\frac{\pi}{4}i}$

$k=1: z=2e^{\frac{11\pi}{12}i}$

$k=-1: z=2e^{-\frac{5\pi}{12}i}$

c $k=0: z=2e^{-\frac{\pi}{4}i}$

$k=1: z=2e^{\frac{5\pi}{12}i}$

$k=-1: z=2e^{-\frac{11\pi}{12}i}$

d $k=0: z=2e^{-\frac{\pi}{12}i}$

$k=1: z=2e^{\frac{7\pi}{12}i}$

$k=-1: z=2e^{-\frac{3\pi}{4}i}$

7 a $k=0: z=\sqrt{3}\left(\cos(0.18)+i\sin(0.18)\right)$

$k=1: z=\sqrt{3}\left(\cos(1.75)+i\sin(1.75)\right)$

$k=-1: z=\sqrt{3}\left(\cos(-1.39)+i\sin(-1.39)\right)$

$k=2: z=\sqrt{3}\left(\cos(-2.96)+i\sin(-2.96)\right)$

b $k=0: z=\sqrt{3}\left(\cos(0.60)+i\sin(0.60)\right)$

$k=1: z=\sqrt{3}\left(\cos(2.17)+i\sin(2.17)\right)$

$k=-1: z=\sqrt{3}\left(\cos(-0.97)+i\sin(-0.97)\right)$

$k=-2: z=\sqrt{3}\left(\cos(-2.54)+i\sin(-2.54)\right)$

c $k=0: z=\sqrt{3}\left(\cos(-0.58)+i\sin(-0.58)\right)$

$k=1: z=\sqrt{3}\left(\cos(1)+i\sin(1)\right)$

$k=2: z=\sqrt{3}\left(\cos(2.57)+i\sin(2.57)\right)$

$k=-1: z=\sqrt{3}\left(\cos(-2.14)+i\sin(-2.14)\right)$

d $k=0: z=\sqrt{3}\left(\cos(-0.21)+i\sin(-0.21)\right)$

$k=1: z=\sqrt{3}\left(\cos(1.36)+i\sin(1.36)\right)$

$k=2: z=\sqrt{3}\left(\cos(2.93)+i\sin(2.93)\right)$

$k=-1: z=\sqrt{3}\left(\cos(-1.78)+i\sin(-1.78)\right)$

8 a
$$k=0: z = 3e^{-\frac{\pi}{5}i}$$
$$k=1: z = 3e^{\frac{\pi}{5}i}$$
$$k=2: z = 3e^{\frac{3\pi}{5}i}$$
$$k=-1: z = 3e^{-\frac{3\pi}{5}i}$$
$$k=3: z = 3e^{\pi i}$$

b
$$k=0: z = \frac{1}{2}e^{-\frac{\pi}{5}i}$$
$$k=1: z = \frac{1}{2}e^{\frac{\pi}{5}i}$$
$$k=2: z = \frac{1}{2}e^{\frac{3\pi}{5}i}$$
$$k=-1: z = \frac{1}{2}e^{-\frac{3\pi}{5}i}$$
$$k=3: z = \frac{1}{2}e^{\pi i}$$

c
$$k=0: z = 2e^{-\frac{\pi}{12}i}$$
$$k=1: z = 2e^{\frac{\pi}{4}i}$$
$$k=2: z = 2e^{\frac{7\pi}{12}i}$$
$$k=3: z = 2e^{\frac{11\pi}{12}i}$$
$$k=-1: z = 2e^{-\frac{5\pi}{12}i}$$
$$k=-2: z = 2e^{-\frac{3\pi}{4}i}$$

d
$$k=0: 1.057e^{-0.160875}$$
$$k=1: 1.057e^{1.4099}$$
$$k=2: 1.057e^{2.9807}$$
$$k=-1: 1.057e^{-1.7317}$$

9 a
$$k=0: z = 2e^{\left(\frac{\pi}{12}\right)i} = \frac{\sqrt{6}+\sqrt{2}}{2} + \frac{\sqrt{6}-\sqrt{2}}{2}i$$
$$k=1: z = 2e^{\left(\frac{7\pi}{12}\right)i} = \frac{\sqrt{2}-\sqrt{6}}{2} + \frac{\sqrt{6}+\sqrt{2}}{2}i$$
$$k=2: z = 2e^{\left(\frac{13\pi}{12}\right)i} = -\frac{\sqrt{6}+\sqrt{2}}{2} + \frac{\sqrt{2}-\sqrt{6}}{2}i$$
$$k=-1: z = 2e^{\left(-\frac{5\pi}{12}\right)i} = \frac{\sqrt{6}-\sqrt{2}}{2} - \frac{\sqrt{6}+\sqrt{2}}{2}i$$

b
$$k=0: z = 2e^{\left(\frac{\pi}{6}\right)i} = \sqrt{3}+i$$
$$k=1: z = 2e^{\left(\frac{2\pi}{3}\right)i} = -1+\sqrt{3}i$$
$$k=2: z = 2e^{\left(\frac{7\pi}{6}\right)i} = -\sqrt{3}-i$$
$$k=-1: z = 2e^{\left(-\frac{\pi}{3}\right)i} = 1-\sqrt{3}i$$

c
$$k=0: z = 2e^{\left(-\frac{\pi}{24}\right)i} = 1.982-0.261i$$
$$k=1: z = 2e^{\left(\frac{11\pi}{24}\right)i} = 0.261+1.982i$$
$$k=2: z = 2e^{\left(\frac{23\pi}{24}\right)i} = -1.982+0.261i$$
$$k=-1: z = 2e^{\left(-\frac{13\pi}{3}\right)i} = -0.261-1.982i$$

d
$$k=0: z = 2.10e^{\left(-\frac{3\pi}{16}\right)i} = 1.746-1.167i$$
$$k=1: z = 2.10e^{\left(\frac{5\pi}{16}\right)i} = 1.167+1.746i$$
$$k=2: z = 2.10e^{\left(\frac{13\pi}{16}\right)i} = -1.746+1.167i$$
$$k=-1: z = 2.10e^{\left(-\frac{11\pi}{16}\right)i} = -1.167-1.746i$$

10
$$k=0: z = 2e^{\frac{-\pi i}{6}} = \sqrt{3}-i$$
$$k=1: z = 2e^{\frac{7\pi i}{6}} = -\sqrt{3}-i$$
$$k=2: z = 2e^{\frac{15\pi i}{6}} = 2i$$

11
$$k=0: z = 16\sqrt{2}e^{\frac{5\pi i}{8}}$$
$$k=1: z = 16\sqrt{2}e^{\frac{17\pi i}{8}} = 16\sqrt{2}e^{\frac{\pi i}{8}}$$
$$k=2: z = 16\sqrt{2}e^{\frac{29\pi i}{8}} = 16\sqrt{2}e^{-\frac{5\pi i}{8}}$$
$$k=3: z = 16\sqrt{2}e^{\frac{41\pi i}{8}} = 16\sqrt{2}e^{-\frac{7\pi i}{8}}$$

12 a
$$k=0: z = e^{\frac{\pi i}{6}} = \sqrt{6}+\sqrt{2}i$$
$$k=1: z = e^{\frac{2\pi i}{3}} = -\sqrt{2}+\sqrt{6}i$$
$$k=2: z = e^{\frac{7\pi i}{6}} = -\sqrt{6}-\sqrt{2}i$$
$$k=3: z = e^{\frac{5\pi i}{3}} = \sqrt{2}-\sqrt{6}i$$

12 b

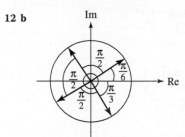

13 a
$$k=0: z = 2e^{\frac{\pi i}{18}}$$
$$k=1: z = 2e^{\frac{13\pi i}{18}}$$
$$k=2: z = 2e^{\frac{25\pi i}{18}} = 2e^{-\frac{11\pi i}{18}}$$

b
$$k=0: z = \sqrt{2}e^{-\frac{\pi i}{16}}$$
$$k=1: z = \sqrt{2}e^{\frac{7\pi i}{16}}$$
$$k=2: z = \sqrt{2}e^{\frac{15\pi i}{16}}$$
$$k=3: z = \sqrt{2}e^{\frac{23\pi i}{16}} = \sqrt{2}e^{-\frac{9\pi i}{16}}$$

14 a
$$k=0: z = \sqrt{2}\left(\cos\left(-\frac{\pi}{9}\right)+i\sin\left(-\frac{\pi}{9}\right)\right)$$
$$k=1: z = \sqrt{2}\left(\cos\frac{5\pi}{9}+i\sin\frac{5\pi}{9}\right)$$
$$k=2: z = \sqrt{2}\left(\cos\left(-\frac{7\pi}{9}\right)+i\sin\left(-\frac{7\pi}{9}\right)\right)$$

b $k=0: z=\sqrt{2}\left(\cos\dfrac{5\pi}{36}+\text{i}\sin\dfrac{5\pi}{36}\right)$

$k=1: z=\sqrt{2}\left(\cos\dfrac{17\pi}{36}+\text{i}\sin\dfrac{17\pi}{36}\right)$

$k=2: z=\sqrt{2}\left(\cos\dfrac{29\pi}{36}+\text{i}\sin\dfrac{29\pi}{36}\right)$

$k=3: z=\sqrt{2}\left(\cos\left(-\dfrac{31\pi}{36}\right)+\text{i}\sin\left(-\dfrac{31\pi}{36}\right)\right)$

$k=4: z=\sqrt{2}\left(\cos\left(-\dfrac{19\pi}{36}\right)+\text{i}\sin\left(-\dfrac{19\pi}{36}\right)\right)$

$k=5: z=\sqrt{2}\left(\cos\left(-\dfrac{7\pi}{36}\right)+\text{i}\sin\left(-\dfrac{7\pi}{36}\right)\right)$

Exercise 6.3B

1 a $1+\omega+\omega^2=\dfrac{1(1-\omega^3)}{1-\omega}$

$=\dfrac{1(1-1)}{1-\omega}$

$=0$

b i 0   **ii** 1   **iii** −1   **iv** 0

2 a $1+\omega+\omega^2+\omega^3+\omega^4=\dfrac{1(1-\omega^5)}{1-\omega}$

$=\dfrac{1(1-1)}{1-\omega}$

$=0$

b i −1   **ii** 0   **iii** 0

3 a $B(0,3)$ and $C\left(-\dfrac{3\sqrt{3}}{2},-\dfrac{3}{2}\right)$

b i $\dfrac{27\sqrt{3}}{4}$ square units

ii $9\sqrt{3}$ units

4 a $B(-5,0)$ and $C\left(\dfrac{5}{2},-\dfrac{5\sqrt{3}}{2}\right)$

b $15\sqrt{3}$ units

c $A'\left(-\dfrac{5\sqrt{3}}{2},-\dfrac{5}{2}\right),\ B'\left(\dfrac{5\sqrt{3}}{2},-\dfrac{5}{2}\right),\ C'(0,5)$

5 a $k=0: z=\sqrt{2}e^{\frac{\pi\text{i}}{6}}=\dfrac{\sqrt{6}}{2}+\dfrac{\sqrt{2}}{2}\text{i}$

$k=1: z=\sqrt{2}e^{\frac{4\pi\text{i}}{6}}=-\dfrac{\sqrt{2}}{2}+\dfrac{\sqrt{6}}{2}\text{i}$

$k=2: z=\sqrt{2}e^{\frac{7\pi\text{i}}{6}}=-\dfrac{\sqrt{6}}{2}-\dfrac{\sqrt{2}}{2}\text{i}$

$k=3: z=\sqrt{2}e^{\frac{10\pi\text{i}}{6}}=\dfrac{\sqrt{2}}{2}-\dfrac{\sqrt{6}}{2}\text{i}$

b Distance $\left(\dfrac{\sqrt{6}}{2},\dfrac{\sqrt{2}}{2}\right)$ to $\left(-\dfrac{\sqrt{2}}{2},\dfrac{\sqrt{6}}{2}\right)$ is

$\sqrt{\left(-\dfrac{\sqrt{2}}{2}-\dfrac{\sqrt{6}}{2}\right)^2+\left(\dfrac{\sqrt{6}}{2}-\dfrac{\sqrt{2}}{2}\right)^2}=\sqrt{2+\sqrt{3}+2-\sqrt{3}}=2$

The points are the solutions to the equation $z^4=a+b\text{i}$ so represent the vertices of a square

Therefore, the solution in part **a** form a square of side length 2 and centre $(0, 0)$

c $\left(\dfrac{\sqrt{6}}{2},\dfrac{\sqrt{2}}{2}\right)\rightarrow\left(\dfrac{\sqrt{6}}{2},\dfrac{3\sqrt{2}}{2}\right)$

$\left(-\dfrac{\sqrt{2}}{2},\dfrac{\sqrt{6}}{2}\right)\rightarrow\left(-\dfrac{\sqrt{2}}{2},\dfrac{\sqrt{6}}{2}+\sqrt{2}\right)$

$\left(-\dfrac{\sqrt{6}}{2},-\dfrac{\sqrt{2}}{2}\right)\rightarrow\left(-\dfrac{\sqrt{6}}{2},\dfrac{\sqrt{2}}{2}\right)$

$\left(\dfrac{\sqrt{2}}{2},-\dfrac{\sqrt{6}}{2}\right)\rightarrow\left(\dfrac{\sqrt{2}}{2},\sqrt{2}-\dfrac{\sqrt{6}}{2}\right)$

6 a $3\sqrt{2}\left(\cos\left(\dfrac{\pi}{4}\right)+\text{i}\sin\left(\dfrac{\pi}{4}\right)\right)$

$3\sqrt{2}\left(\cos\left(\dfrac{13\pi}{20}\right)+\text{i}\sin\left(\dfrac{13\pi}{20}\right)\right)$

$3\sqrt{2}\left(\cos\left(-\dfrac{19\pi}{20}\right)+\text{i}\sin\left(-\dfrac{19\pi}{20}\right)\right)$

$3\sqrt{2}\left(\cos\left(-\dfrac{11\pi}{20}\right)+\text{i}\sin\left(-\dfrac{11\pi}{20}\right)\right)$

$3\sqrt{2}\left(\cos\left(-\dfrac{3\pi}{20}\right)+\text{i}\sin\left(-\dfrac{3\pi}{20}\right)\right)$

b Length of one side of pentagon is given by

$2\times3\sqrt{2}\sin\left(\dfrac{\pi}{5}\right)=6\sqrt{2}\sin\left(\dfrac{\pi}{5}\right)$

So perimeter $=5\times6\sqrt{2}\sin\left(\dfrac{\pi}{5}\right)=30\sqrt{2}\sin\left(\dfrac{\pi}{5}\right)$

7 a 6   **b** $\left(-\dfrac{1}{2},\dfrac{\sqrt{3}}{2}\right)$ **c** $\dfrac{3\sqrt{3}}{2}$

8 a $(0,\sqrt{2})$

$(-1,1)$

$(-\sqrt{2},0)$

$(-1,-1)$

$(0,-\sqrt{2})$

$(1,-1)$

$(\sqrt{2},0)$

b 21.6 cm (3 sf)

Review exercise 6

1 a $3e^{-\frac{\pi}{2}\text{i}}$ **b** $\sqrt{2}e^{\frac{\pi}{4}\text{i}}$

c $5e^0$ **d** $2e^{-\frac{5\pi}{6}\text{i}}$

e $2\sqrt{2}e^{-\frac{\pi}{3}\text{i}}$ **f** $2e^{\frac{2\pi}{3}\text{i}}$

2 a $3e^{\frac{\pi}{7}\text{i}}$ **b** $\sqrt{2}e^{\frac{\pi}{9}\text{i}}$

c $\sqrt{3}\left(\cos\left(-\dfrac{\pi}{8}\right)+\text{i}\sin\left(-\dfrac{\pi}{8}\right)\right)=\sqrt{3}e^{-\frac{\pi}{8}\text{i}}$

d $5\left(\cos\left(\dfrac{\pi}{5}\right)+i\sin\left(\dfrac{\pi}{5}\right)\right)=5e^{\frac{\pi}{5}\text{i}}$

3 a 7i **b** $3+3\sqrt{3}\text{i}$

c $\dfrac{3}{2}-\dfrac{\sqrt{3}}{2}\text{i}$ **d** $-1+\text{i}$

e $-\dfrac{\sqrt{6}}{2}-\dfrac{3\sqrt{2}}{2}\text{i}$ **f** $-3+\sqrt{3}\text{i}$

4 a $|z\omega| = 3 \times 5 = 15$

$\arg(z\omega) = \dfrac{\pi}{5} + \dfrac{\pi}{7} = \dfrac{12}{35}\pi$

b $|z\omega| = \sqrt{2} \times \sqrt{6} = 2\sqrt{3}$

$\arg(z\omega) = -\dfrac{\pi}{8} + \dfrac{\pi}{3} = \dfrac{5}{24}\pi$

c $|z| = 2, \arg(z) = \dfrac{\pi}{3}$

$|\omega| = 3\sqrt{2}, \arg(\omega) = -\dfrac{\pi}{4}$

$|z\omega| = 2 \times 3\sqrt{2} = 6\sqrt{2}$

$\arg(z\omega) = \dfrac{\pi}{3} - \dfrac{\pi}{4} = \dfrac{\pi}{12}$

d $|z\omega| = \dfrac{1}{2} \times 4 = 2$

$\arg(z\omega) = \dfrac{2\pi}{9} + \dfrac{\pi}{3} = \dfrac{5}{9}\pi$

5 a $\left|\dfrac{z}{\omega}\right| = \dfrac{5}{10} = \dfrac{1}{2}$

$\arg\left(\dfrac{z}{\omega}\right) = \dfrac{\pi}{2} - \dfrac{\pi}{4} = \dfrac{\pi}{4}$

b $\left|\dfrac{z}{\omega}\right| = \dfrac{\sqrt{15}}{\sqrt{5}} = \sqrt{3}$

$\arg\left(\dfrac{z}{\omega}\right) = -\dfrac{3\pi}{4} - \dfrac{\pi}{8} = -\dfrac{5\pi}{8}$

c $|z| = 5\sqrt{2}, \arg z = \dfrac{3\pi}{4}$

$|\omega| = 2\sqrt{6}, \arg\omega = -\dfrac{\pi}{3}$

$\left|\dfrac{z}{\omega}\right| = \dfrac{5\sqrt{2}}{2\sqrt{6}} = \dfrac{5}{6}\sqrt{3}$

$\arg\left(\dfrac{z}{\omega}\right) = \dfrac{3\pi}{4} - \dfrac{\pi}{3} = -\dfrac{11\pi}{12}$

d $\left|\dfrac{z}{\omega}\right| = \dfrac{16}{\sqrt{2}} = 8\sqrt{2}$

$\arg\left(\dfrac{z}{\omega}\right) = -\dfrac{2\pi}{11} - \dfrac{5\pi}{11} = -\dfrac{7\pi}{11}$

6 a -64 **b** $-512i$

7 a $8 + 8\sqrt{3}i$ **b** $\dfrac{i}{64}$

8 a $\cos 5\theta + i\sin 5\theta = (\cos\theta + i\sin\theta)^5$

$= \cos^5\theta + 5\cos^4\theta(i\sin\theta) + 10\cos^3\theta(i\sin\theta)^2$

$\quad + 10\cos^2\theta(i\sin\theta)^3 + 5\cos\theta(i\sin\theta)^4 + (i\sin\theta)^5$

$= \cos^5\theta + 5i\cos^4\theta\sin\theta - 10\cos^3\theta\sin^2\theta$

$\quad - 10i\cos^2\theta\sin^3\theta + 5\cos\theta\sin^4\theta + i\sin^5\theta$

$\text{Im}: \sin 5\theta \equiv 5\cos^4\theta\sin\theta - 10\cos^2\theta\sin^3\theta + \sin^5\theta$

$= 5(1 - \sin^2\theta)^2\sin\theta - 10(1 - \sin^2\theta)\sin^3\theta + \sin^5\theta$

$= 5\sin\theta - 10\sin^3\theta + 5\sin^5\theta - 10\sin^3\theta + 10\sin^5\theta + \sin^5\theta$

$= 5\sin\theta - 20\sin^3\theta + 16\sin^5\theta$

b $x = 0, 0.588, 0.951$

9 $\cos^3\theta = \dfrac{1}{8}(2\cos\theta)^3 = \dfrac{1}{8}(e^{i\theta} + e^{-i\theta})^3$

$= \dfrac{1}{8}(e^{3i\theta} + 3e^{i\theta} + 3e^{-i\theta} + e^{-3i\theta})$

$= \dfrac{1}{8}(2\cos 3\theta + 6\cos\theta)$

$= \dfrac{1}{4}(\cos 3\theta + 3\cos\theta), \text{ as required } \left(A = \dfrac{1}{4}\right)$

10 a $k = 0: z = 1$

$k = 1: z = e^{\frac{\pi i}{4}} = \dfrac{\sqrt{2}}{2} + \dfrac{\sqrt{2}}{2}i$

$k = 2: z = e^{\frac{\pi i}{2}} = i$

$k = 3: z = e^{\frac{3\pi i}{4}} = -\dfrac{\sqrt{2}}{2} + \dfrac{\sqrt{2}}{2}i$

$k = 4: z = e^{\pi i} = -1$

$k = 5: z = e^{\frac{5\pi i}{4}} = -\dfrac{\sqrt{2}}{2} - \dfrac{\sqrt{2}}{2}i$

$k = 6: z = e^{\frac{3\pi i}{2}} = -i$

$k = 7: z = e^{\frac{7\pi i}{4}} = \dfrac{\sqrt{2}}{2} - \dfrac{\sqrt{2}}{2}i$

10 b

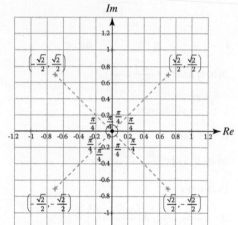

11 a $k = 0: z = \sqrt{2}$

$k = 1: z = \sqrt{2}e^{\frac{\pi i}{4}} = 1 + i$

$k = 2: z = \sqrt{2}e^{\frac{2\pi i}{4}} = \sqrt{2}i$

$k = 3: z = \sqrt{2}e^{\frac{3\pi i}{4}} = -1 + i$

$k = 4: z = \sqrt{2}e^{\frac{4\pi i}{4}} = -\sqrt{2}$

$k = 5: z = \sqrt{2}e^{\frac{5\pi i}{4}} = -1 - i$

$k = 6: z = \sqrt{2}e^{\frac{6\pi i}{4}} = -\sqrt{2}i$

$k = 7: z = \sqrt{2}e^{\frac{7\pi i}{4}} = 1 - i$

b $k = 0: z = e^{\frac{\pi i}{6}} = \dfrac{\sqrt{3}}{2} + \dfrac{1}{2}i$

$k = 1: z = e^{\frac{5\pi i}{6}} = -\dfrac{\sqrt{3}}{2} + \dfrac{1}{2}i$

$k = 2: z = e^{\frac{9\pi i}{6}} = -i$

c $k = 0: z = 3e^{-\frac{\pi i}{4}} = \dfrac{3\sqrt{2}}{2} - \dfrac{3\sqrt{2}}{2}i$

$k = 1: z = 3e^{\frac{3\pi i}{4}} = -\dfrac{3\sqrt{2}}{2} + \dfrac{3\sqrt{2}}{2}i$

d $k = 0: z = \sqrt{5}e^{\frac{\pi i}{6}} = \dfrac{\sqrt{15}}{2} + \dfrac{\sqrt{5}}{2}i$

$k = 1: z = \sqrt{5}e^{\frac{3\pi i}{6}} = \sqrt{5}i$

$k = 2: z = \sqrt{5}e^{\frac{5\pi i}{6}} = -\dfrac{\sqrt{15}}{2} + \dfrac{\sqrt{5}}{2}i$

$k = 3: z = \sqrt{5}e^{\frac{7\pi i}{6}} = -\dfrac{\sqrt{15}}{2} - \dfrac{\sqrt{5}}{2}i$

$k = 4: z = \sqrt{5}e^{\frac{9\pi i}{6}} = -\sqrt{5}i$

$k = 5: z = \sqrt{5}e^{\frac{11\pi i}{6}} = \dfrac{\sqrt{15}}{2} - \dfrac{\sqrt{5}}{2}i$

12 $k = 0: z = \sqrt{2}e^{\frac{-3\pi i}{16}}$

$k = 1: z = \sqrt{2}e^{\frac{5\pi i}{16}}$

$k = 2: z = \sqrt{2}e^{\frac{13\pi i}{16}} = \sqrt{2}e^{-\frac{15\pi i}{16}}$

$k = 3: z = \sqrt{2}e^{\frac{21\pi i}{16}} = \sqrt{2}e^{-\frac{11\pi i}{16}}$

13 $k = 0: z = 32e^{\frac{10\pi i}{6}} = 16 - 16\sqrt{3}i$

$k = 1: z = 32e^{\frac{40\pi i}{6}} = -16 + 16\sqrt{3}i$

Assessment 6

1 a $a = -10, b = 30$

b $5 + i, x^2 = -4 = x = \pm 2i$

c

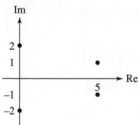

2 a $z = 2\sqrt{3}e^{\frac{-\pi}{6}i}$

b $144\left(\cos\left(\dfrac{-2\pi}{3}\right) + i\sin\left(\dfrac{-2\pi}{3}\right)\right)$

3 a $2k; \dfrac{2\pi}{3}$

b $\dfrac{\sqrt{6}}{2} + \dfrac{\sqrt{2}}{2}i$

4 a $e^{\frac{-\pi}{6}i}, e^{\frac{\pi}{2}i}, e^{\frac{-5\pi}{6}i}$

b

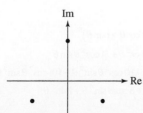

5 a $2\sqrt{2}\left(\cos\left(\dfrac{3\pi}{4}\right) + i\sin\left(\dfrac{3\pi}{4}\right)\right)$

b $\dfrac{1}{32} - \dfrac{1}{32}i$

6 $-8 - 8\sqrt{3}i$

7 $(\cos\theta - i\sin\theta)^n \equiv (\cos(-\theta) + i\sin(-\theta))^n$

since $\sin(-\theta) \equiv -\sin\theta$ and $\cos(-\theta) \equiv \cos\theta$

Therefore

$(\cos\theta - i\sin\theta)^n \equiv \cos(-n\theta) + i\sin(-n\theta)$

by de Moivre's theorem

$\equiv \cos(n\theta) - i\sin(n\theta)$

8 a $\cos(5\theta) + i\sin(5\theta) \equiv (\cos\theta + i\sin\theta)^5$

$\equiv \cos^5\theta + 5i\cos^4\theta\sin\theta$

$-10\cos^3\theta\sin^2\theta$

$-10i\cos^2\theta\sin^3\theta$

$+5\cos\theta\sin^4\theta + i\sin^5\theta$

$\sin(5\theta) \equiv 5\cos^4\theta\sin\theta - 10\cos^2\theta\sin^3\theta + \sin^5\theta$

$\equiv 5(1 - \sin^2\theta)^2\sin\theta - 10(1 - \sin^2\theta)\sin^3\theta + \sin^5\theta$

$\equiv 5\sin\theta - 10\sin^3\theta + 5\sin^5\theta$

$-10\sin^3\theta + 10\sin^5\theta + \sin^5\theta$

$\equiv 5\sin\theta - 20\sin^3\theta + 16\sin^5\theta$

b $6°, 30°, 78°, 102°, 150°, 174°$

9 a $32\cos^5 x \equiv (2\cos x)^5$

$\equiv (e^{ix} + e^{-ix})^5$

$\equiv (e^{ix})^5 + 5(e^{ix})^4(e^{-ix}) + 10(e^{ix})^3(e^{-ix})^2$

$+ 10(e^{ix})^2(e^{-ix})^3 + 5(e^{ix})(e^{-ix})^4 + (e^{-ix})^5$

$\equiv e^{5ix} + 5e^{3ix} + 10e^{ix} + \dfrac{10}{e^{ix}} + \dfrac{5}{e^{3ix}} + \dfrac{1}{e^{5ix}}$

$\equiv 2\cos(5x) + 10\cos(3x) + 20\cos(x)$

b $\dfrac{1}{80}\sin(5x) + \dfrac{5}{48}\sin(3x) + \dfrac{5}{8}\sin(x) + c$

10 a $\cos 3\theta + i\sin 3\theta$

$\equiv (\cos\theta + i\sin\theta)^3$

$\equiv \cos^3\theta + 3i\cos^2\theta\sin\theta - 3\cos\theta\sin^2\theta - i\sin^3\theta$

$\cos 3\theta \equiv \cos^3\theta - 3\cos\theta\sin^2\theta$, as required

b $\sin 3\theta \equiv 3\cos^2\theta\sin\theta - \sin^3\theta$

$\tan 3\theta \equiv \dfrac{\sin 3\theta}{\cos 3\theta}$

$\equiv \dfrac{3\cos^2\theta\sin\theta - \sin^3\theta}{\cos^3\theta - 3\cos\theta\sin^2\theta}$

$\equiv \dfrac{3\tan\theta - \tan^3\theta}{1 - 3\tan^2\theta}$, as required

c $-\dfrac{18}{35}\sqrt{3}$

11 a 6

b $\sqrt{2}\left(\cos\left(\dfrac{-\pi}{6}\right) + i\sin\left(\dfrac{-\pi}{6}\right)\right),$

$$\sqrt{2}\left(\cos\left(\frac{\pi}{6}\right)+\mathrm{i}\sin\left(\frac{\pi}{6}\right)\right),$$

$$\sqrt{2}\left(\cos\left(\frac{\pi}{2}\right)+\mathrm{i}\sin\left(\frac{\pi}{2}\right)\right),$$

$$\sqrt{2}\left(\cos\left(\frac{5\pi}{6}\right)+\mathrm{i}\sin\left(\frac{5\pi}{6}\right)\right),$$

$$\sqrt{2}\left(\cos\left(\frac{-\pi}{2}\right)+\mathrm{i}\sin\left(\frac{-\pi}{2}\right)\right),$$

$$\sqrt{2}\left(\cos\left(\frac{-5\pi}{6}\right)+\mathrm{i}\sin\left(\frac{-5\pi}{6}\right)\right)$$

12 RHS $\equiv 1-2\left(\dfrac{\mathrm{e}^{\mathrm{i}x}-\mathrm{e}^{-\mathrm{i}x}}{2\mathrm{i}}\right)^2$

$$\equiv 1-\dfrac{2(\mathrm{e}^{2\mathrm{i}x}+\mathrm{e}^{-2\mathrm{i}x}-2)}{4\mathrm{i}^2}$$

$$\equiv 1-\dfrac{\mathrm{e}^{2\mathrm{i}x}+\mathrm{e}^{-2\mathrm{i}x}-2}{-2}$$

$$\equiv 1+\dfrac{\mathrm{e}^{2\mathrm{i}x}+\mathrm{e}^{-2\mathrm{i}x}-2}{2}$$

$$\equiv \dfrac{2+\mathrm{e}^{2\mathrm{i}x}+\mathrm{e}^{-2\mathrm{i}x}-2}{2}$$

$$\equiv \dfrac{\mathrm{e}^{2\mathrm{i}x}+\mathrm{e}^{-2\mathrm{i}x}}{2}$$

$$\equiv \cos 2x$$

13 a $\dfrac{3\pi}{4}$ **b** $\dfrac{\pi}{4}$

c $\dfrac{n}{4}\pi$ **d** $\dfrac{\pi}{4}-2\times\dfrac{\pi}{2}-\dfrac{3\pi}{4}$

14 a $\cos 7\theta+\mathrm{i}\sin 7\theta \equiv(\cos\theta+\mathrm{i}\sin\theta)^7$

$\cos 7\theta \equiv \cos^7\theta-21\cos^5\theta\sin^2\theta$
$\qquad +35\cos^3\theta\sin^4\theta-7\cos\theta\sin^6\theta$
$\qquad \equiv \cos^7\theta-21\cos^5\theta(1-\cos^2\theta)+35\cos^3\theta(1-\cos^2\theta)^2$
$\qquad -7\cos\theta(1-\cos^2\theta)^3$
$\qquad \equiv \cos^7\theta-21\cos^5\theta+21\cos^7\theta+35\cos^3\theta-70\cos^5\theta$
$\qquad +35\cos^7\theta-7\cos\theta+21\cos^3\theta-21\cos^5\theta+7\cos^7\theta$
$\qquad \equiv 64\cos^7\theta-112\cos^5\theta+56\cos^3\theta-7\cos\theta$
$\qquad \equiv \cos\theta\left(64\cos^6\theta-112\cos^4\theta+56\cos^2\theta-7\right),$

as required

b $0.975, 0.782, 0.434, -0.434, -0.782, -0.975$

15 $2\mathrm{e}^{\frac{-5}{48}\pi\mathrm{i}}, 2\mathrm{e}^{\frac{7}{48}\pi\mathrm{i}}, 2\mathrm{e}^{\frac{19}{48}\pi\mathrm{i}}, 2\mathrm{e}^{\frac{31}{48}\pi\mathrm{i}}, 2\mathrm{e}^{\frac{43}{48}\pi\mathrm{i}},$

$2\mathrm{e}^{\frac{-17}{48}\pi\mathrm{i}}, 2\mathrm{e}^{\frac{-29}{48}\pi\mathrm{i}}, 2\mathrm{e}^{\frac{-41}{48}\pi\mathrm{i}}$

16 a $(\sin\theta)^4 \equiv\left(\dfrac{\mathrm{e}^{\mathrm{i}\theta}-\mathrm{e}^{-\mathrm{i}\theta}}{2}\right)^4$

$$\equiv \dfrac{1}{16}\left[\begin{array}{l}(\mathrm{e}^{\mathrm{i}\theta})^4-4(\mathrm{e}^{\mathrm{i}\theta})^3(\mathrm{e}^{-\mathrm{i}\theta})+6(\mathrm{e}^{\mathrm{i}\theta})^2(\mathrm{e}^{-\mathrm{i}\theta})^2\\ -4(\mathrm{e}^{\mathrm{i}\theta})(\mathrm{e}^{-\mathrm{i}\theta})^3+(\mathrm{e}^{-\mathrm{i}\theta})^4\end{array}\right]$$

$$\equiv \dfrac{1}{16}\left[\mathrm{e}^{4\mathrm{i}\theta}+\mathrm{e}^{-4\mathrm{i}\theta}-4\left(\mathrm{e}^{2\mathrm{i}\theta}+\mathrm{e}^{-2\mathrm{i}\theta}\right)+6\right]$$

$$\equiv \dfrac{1}{16}\left[2\cos(4\theta)-8\cos(2\theta)+6\right]$$

$$\equiv \dfrac{1}{8}\left(\cos(4\theta)-4\cos(2\theta)+3\right),$$

as required

b $\dfrac{1}{4}\sin(4\theta)-2\sin(2\theta)+3\theta+c$

17 a $1.86\left(\cos\left(\dfrac{-\pi}{12}+\dfrac{n\pi}{2}\right)+\mathrm{i}\sin\left(\dfrac{-\pi}{12}+\dfrac{n\pi}{2}\right)\right)$

$1.86\left(\cos\left(\dfrac{-\pi}{12}\right)+\mathrm{i}\sin\left(\dfrac{-\pi}{12}\right)\right),$

$1.86\left(\cos\left(\dfrac{5\pi}{12}\right)+\mathrm{i}\sin\left(\dfrac{5\pi}{12}\right)\right),$

$1.86\left(\cos\left(\dfrac{-7\pi}{12}\right)+\mathrm{i}\sin\left(\dfrac{-7\pi}{12}\right)\right),$

$1.86\left(\cos\left(\dfrac{11\pi}{12}\right)+\mathrm{i}\sin\left(\dfrac{11\pi}{12}\right)\right)$

b $2.80-0.481\mathrm{i}$
$1.481+1.80\mathrm{i}$
$0.519-1.80\mathrm{i}$
$-0.80+0.481\mathrm{i}$

18 a $\dfrac{1}{z^n}\equiv z^{-n}\equiv(\cos\theta+\mathrm{i}\sin\theta)^{-n}$

$\qquad \equiv \cos(-n\theta)+\mathrm{i}\sin(-n\theta)$
$\qquad \equiv \cos(n\theta)-\mathrm{i}\sin(n\theta)$

$z^n \equiv(\cos\theta+\mathrm{i}\sin\theta)^n \equiv \cos(n\theta)+\mathrm{i}\sin(n\theta)$

$z^n-\dfrac{1}{z^n}\equiv \cos(n\theta)+\mathrm{i}\sin(n\theta)-\left(\cos(n\theta)-\mathrm{i}\sin(n\theta)\right)$

$\qquad \equiv 2\mathrm{i}\sin(n\theta),$ as required

b $4\sin^3\theta\equiv\dfrac{1}{-2\mathrm{i}}\left(2\mathrm{i}\sin\theta\right)^3$

$$\equiv \dfrac{1}{2}\mathrm{i}\left(z-\dfrac{1}{z}\right)^3$$

$$\equiv \dfrac{1}{2}\mathrm{i}\left[(z)^3-3(z)^2\left(\dfrac{1}{z}\right)+3(z)\left(\dfrac{1}{z}\right)^2-\left(\dfrac{1}{z}\right)^3\right]$$

$$\equiv \dfrac{1}{2}\mathrm{i}\left[z^3-\dfrac{1}{z^3}-3\left(z-\dfrac{1}{z}\right)\right]$$

$$\equiv \dfrac{1}{2}\mathrm{i}\left(2\mathrm{i}\sin(3\theta)-6\mathrm{i}\sin\theta\right)$$

$$\equiv 3\sin\theta-\sin(3\theta),\text{ as required}$$

c $\dfrac{7\pi}{6},\dfrac{11\pi}{6}$

19 a

b i -1 **ii** 4

20 a $\cos(5\theta)+\mathrm{i}\sin(5\theta)\equiv(\cos\theta+\mathrm{i}\sin\theta)^5$

$\qquad \equiv \cos^5\theta+5\mathrm{i}\cos^4\theta\sin\theta$
$\qquad -10\cos^3\theta\sin^2\theta-10\mathrm{i}\cos^2\theta\sin^3\theta$
$\qquad +5\cos\theta\sin^4\theta+\mathrm{i}\sin^5\theta$

$$\cos(5\theta) \equiv \cos^5\theta - 10\cos^3\theta\sin^2\theta + 5\cos\theta\sin^4\theta$$
$$\equiv \cos^5\theta - 10\cos^3\theta(1-\cos^2\theta) + 5\cos\theta(1-\cos^2\theta)^2$$
$$\equiv \cos^5\theta - 10\cos^3\theta + 10\cos^5\theta$$
$$+ 5\cos\theta - 10\cos^3\theta + 5\cos^5\theta$$
$$\equiv 16\cos^5\theta - 20\cos^3\theta + 5\cos\theta, \text{ as required}$$

b $\theta = \dfrac{\pi}{2}(2n-1)$ for $n \in \mathbb{Z}$

21 a $\omega^7 = \left(\cos\left(\dfrac{2\pi}{7}\right) + i\sin\left(\dfrac{2\pi}{7}\right)\right)^7$

$\left(e^{\frac{2\pi i}{7}}\right)^7$

$e^{2\pi i}$

$\cos(2\pi) + i\sin(2\pi)$

1, as required

b $\omega^2, \omega^3, \omega^4, \omega^5, \omega^6$

c -1

22 a $\left(\dfrac{1}{2} + \dfrac{\sqrt{3}}{2}i - 1\right)^3 = \left(\dfrac{-1}{2} + \dfrac{\sqrt{3}}{2}i\right)^3$

$= \left(\dfrac{-1}{2}\right)^3 + 3\left(\dfrac{-1}{2}\right)^2\left(\dfrac{\sqrt{3}}{2}i\right) + 3\left(\dfrac{-1}{2}\right)\left(\dfrac{\sqrt{3}}{2}i\right)^2 + \left(\dfrac{\sqrt{3}}{2}i\right)^3$

$-\dfrac{1}{8} + \dfrac{3\sqrt{3}}{8}i + \dfrac{9}{8} - \dfrac{3\sqrt{3}}{8}i$

1, as required

b $\dfrac{1}{2} - \dfrac{\sqrt{3}}{2}i$

2

c

d 1
$(1, 0)$

23 a $n = 0 = z_1 = 2\left(\cos\left(\dfrac{\pi}{4}\right) + i\sin\left(\dfrac{\pi}{4}\right)\right) = \sqrt{2} + \sqrt{2}i$

$n = 1 = z_2 = 2\left(\cos\left(\dfrac{-\pi}{4}\right) + i\sin\left(\dfrac{-\pi}{4}\right)\right) = \sqrt{2} - \sqrt{2}i$

$n = 2 = z_3 = 2\left(\cos\left(\dfrac{-3\pi}{4}\right) + i\sin\left(\dfrac{-3\pi}{4}\right)\right) = -\sqrt{2} - \sqrt{2}i$

$n = 3 = z_4 = 2\left(\cos\left(\dfrac{-5\pi}{4}\right) + i\sin\left(\dfrac{-5\pi}{4}\right)\right) = -\sqrt{2} + \sqrt{2}i$

b 8 square units

24 a $\cos 5\theta + i\sin 5\theta = (\cos\theta + i\sin\theta)^5$
$\cos^5\theta + 5i\cos^4 x\sin\theta - 10\cos^3\theta\sin^2\theta$
$\quad - 10i\cos^2\theta\sin^3\theta + 5\cos\theta\sin^4\theta + i\sin^5\theta$
Imaginary parts give
$\sin 5\theta = 5\cos^4\theta\sin\theta - 10\cos^2\theta\sin^3\theta + \sin^5\theta$

$5(1-\sin^2\theta)^2\sin\theta - 10(1-\sin^2\theta)\sin^3\theta + \sin^5\theta$
$5\sin\theta - 10\sin^3\theta + 5\sin^5\theta - 10\sin^3\theta + 10\sin^5\theta + \sin^5\theta$
$16\sin^5\theta - 20\sin^3\theta + 5\sin\theta, \text{ as required}$

b $\sin\left(\dfrac{\pi}{10}\right), \sin\left(\dfrac{5\pi}{10}\right), \sin\left(\dfrac{9\pi}{10}\right), \sin\left(\dfrac{13\pi}{10}\right), \sin\left(\dfrac{17\pi}{10}\right)$

25 a $k = 0 = z - 1 = 2\left(\cos\left(\dfrac{0}{3}\right) + i\sin\left(\dfrac{0}{3}\right)\right) = 2 = z_1 = 3$

$k = 1 = z - 1 = 2\left(\cos\dfrac{2\pi}{3} + i\sin\dfrac{2\pi}{3}\right)$

$= 2\left(\dfrac{-1}{2} + \dfrac{\sqrt{3}}{2}i\right) = -1 + \sqrt{3}i = z_2 = \sqrt{3}i$

$k = 2 = z - 1 = 2\left(\cos\dfrac{4\pi}{3} + i\sin\dfrac{4\pi}{3}\right)$

$= 2\left(\dfrac{-1}{2} - \dfrac{\sqrt{3}}{2}i\right) = -1 - \sqrt{3}i = z_3 = -\sqrt{3}i$

b

c $3\sqrt{3}$ square units

d $\dfrac{3}{2} + \dfrac{3\sqrt{3}}{2}i \qquad \dfrac{-3}{2} + \dfrac{\sqrt{3}}{2}i \qquad \dfrac{3}{2} - \dfrac{\sqrt{3}}{2}i$

26 a $\left(\dfrac{\frac{1}{2}(1+i) - 1}{\frac{1}{2}(1+i)}\right)^6 = \left(\dfrac{1+i-2}{1+i}\right)^6$

$= \left(\dfrac{i-1}{1+i}\right)^6$

$= \left(\dfrac{(i-1)(1-i)}{(1+i)(1-i)}\right)^6$

$= \left[\dfrac{-1}{2}(i-1)^2\right]^6$

$= \dfrac{1}{64}(i-1)^{12}$

$= \dfrac{1}{64}\left(\sqrt{2}\left(\cos\left(\dfrac{3\pi}{4}\right) + i\sin\left(\dfrac{3\pi}{4}\right)\right)\right)^{12}$

$= \dfrac{64}{64}\left(\cos\left(\dfrac{36\pi}{4}\right) + i\sin\left(\dfrac{36\pi}{4}\right)\right)$

$= (-1 + 0)$

$= -1, \text{ as required}$

b $k = 0 = \dfrac{z-1}{z} = \cos\left(\dfrac{\pi}{6}\right) + i\sin\left(\dfrac{\pi}{6}\right) = \dfrac{\sqrt{3}}{2} + \dfrac{1}{2}i$

$k = 1 = \dfrac{z-1}{z} = \cos\left(\dfrac{\pi}{2}\right) + i\sin\left(\dfrac{\pi}{2}\right) = i$

$k = 2 = \dfrac{z-1}{z} = \cos\left(\dfrac{5\pi}{6}\right) + i\sin\left(\dfrac{5\pi}{6}\right) = \dfrac{-\sqrt{3}}{2} + \dfrac{1}{2}i$

$k = 3 = \dfrac{z-1}{z} = \cos\left(\dfrac{7\pi}{6}\right) + i\sin\left(\dfrac{7\pi}{6}\right) = \dfrac{-\sqrt{3}}{2} - \dfrac{1}{2}i$

$k = 4 = \dfrac{z-1}{z} = \cos\left(\dfrac{3\pi}{2}\right) + i\sin\left(\dfrac{3\pi}{2}\right) = -i$

$k = 5 = \dfrac{z-1}{z} = \cos\left(\dfrac{11\pi}{6}\right) + i\sin\left(\dfrac{11\pi}{6}\right) = \dfrac{\sqrt{3}}{2} - \dfrac{1}{2}i$

$\dfrac{z-1}{z} = \omega = z\omega = z - 1 > z - z\omega = 1$

$> z = \dfrac{1}{1-\omega}$

$z_1 = \dfrac{1}{2} + \dfrac{2+\sqrt{3}}{2}i$

$z_2 = \dfrac{1}{2} + \dfrac{1}{2}i$

$z_3 = \dfrac{1}{2} + \dfrac{2-\sqrt{3}}{2}i$

$z_4 = \dfrac{1}{2} + \dfrac{-2+\sqrt{3}}{2}i$

$z_5 = \dfrac{1}{2} - \dfrac{1}{2}i$

$z_6 = \dfrac{1}{2} + \dfrac{-2-\sqrt{3}}{2}i$

c $\left(\dfrac{1}{4}\left(1+\sqrt{3}\right), \dfrac{1}{4}\left(-1+\sqrt{3}\right)\right)$

27 a $k = 0 = z_1 = 2e^{\frac{-\pi i}{9}}$

$k = 1 = z_2 = 2e^{\frac{5\pi i}{9}}$

$k = -1 = z_3 = 2e^{\frac{-7\pi i}{9}}$

b $3\sqrt{3}$ square units

c Sum of roots $= 0$

So sum of imaginary parts $= 0$

Therefore $\sin\left(\dfrac{-\pi}{9}\right) + \sin\left(\dfrac{5\pi}{9}\right) + \sin\left(\dfrac{-7\pi}{9}\right) = 0$

$> -\sin\left(\dfrac{\pi}{9}\right) + \sin\left(\dfrac{5\pi}{9}\right) - \sin\left(\dfrac{7\pi}{9}\right) = 0$

$> \sin\left(\dfrac{\pi}{9}\right) - \sin\left(\dfrac{5\pi}{9}\right) + \sin\left(\dfrac{7\pi}{9}\right) = 0$

28 First consider the case when n is a positive integer

Let $n = 1$. Then $\left(\cos\theta + i\sin\theta\right)^1 = \cos(1\theta) + i\sin(1\theta)$, so true for $n = 1$

Assume true for $n = k$ and consider when $n = k+1$

$(\cos\theta + i\sin\theta)^{k+1} = (\cos\theta + i\sin\theta)^k(\cos\theta + i\sin\theta)$

$= (\cos(k\theta) + i\sin(k\theta))(\cos\theta + i\sin\theta)$

$= \cos(k\theta)\cos\theta - \sin(k\theta)\sin\theta$
$\quad + i(\cos(k\theta)\sin\theta + \sin(k\theta)\cos\theta)$

$= \cos((k+1)\theta) + i\sin((k+1)\theta)$

So true for $n = k+1$

Hence, since true for $n = 1$ and by assuming it is true for $n = k$ we have proved it is true for $n = k+1$, therefore it is true for all positive integers n.

When $n = 0$, $(\cos\theta + i\sin\theta)^0 = 1$ and $\cos(0) + i\sin(0) = 1$, so

true for $n = 0$

When n is a negative integer, it can be written as $n = -m$, where m is a positive integer

So $(\cos\theta + i\sin\theta)^n = (\cos\theta + i\sin\theta)^{-m}$

$= \dfrac{1}{(\cos\theta + i\sin\theta)^m}$

$= \dfrac{1}{(\cos(m\theta) + i\sin(m\theta))}$ since m is a positive integer and you have just proved that de Moivre's holds for positive integers

$= \dfrac{\cos(m\theta) - i\sin(m\theta)}{(\cos(m\theta) + i\sin(m\theta))(\cos(m\theta) - i\sin(m\theta))}$

$= \dfrac{\cos(m\theta) - i\sin(m\theta)}{\cos^2(m\theta) + \sin^2(m\theta)}$

$= \cos(m\theta) - i\sin(m\theta)$

$= \cos(-m\theta) + i\sin(-m\theta)$

$= \cos(n\theta) + i\sin(n\theta)$, as required

So de Moivre's theorem holds for all integers n

Exercise 7.1A

1 a $\dfrac{2}{x+5} + \dfrac{4}{x-5}$

b $\dfrac{3}{x-3} - \dfrac{7}{x-7}$

c $\dfrac{9}{1-x} + \dfrac{2}{x+6}$

d $\dfrac{1}{2x-5} - \dfrac{6}{3x+4}$

e $\dfrac{1}{x-2} + \dfrac{8}{x-3} + \dfrac{6}{x-4}$

f $\dfrac{2}{x+2} + \dfrac{5}{x-2} - \dfrac{7}{x+3}$

g $\dfrac{4}{2-x} + \dfrac{7}{x+9} - \dfrac{3}{5-x}$

h $\dfrac{9}{3x-1} - \dfrac{7}{4x-3} + \dfrac{2}{3x+5}$

2 a $\dfrac{1}{x+5} + \dfrac{3}{(x+5)^2}$

b $\dfrac{12}{x-3} - \dfrac{2}{(x-7)^2}$

c $\dfrac{7}{x+4} + \dfrac{1}{(x+3)^2}$

d $\dfrac{1}{x-5} + \dfrac{5}{x+4} + \dfrac{3}{(x+4)^2}$

e $\dfrac{3}{x+7} - \dfrac{2}{x-8} + \dfrac{1}{(x-8)^2}$

f $\dfrac{10}{x-4} - \dfrac{12}{x+7} - \dfrac{11}{(x+7)^2}$

g $\dfrac{3}{3-x} + \dfrac{4}{x+11} - \dfrac{3}{(3-x)^2}$

h $\dfrac{4}{3x+1}+\dfrac{5}{2x-7}+\dfrac{1}{(3x+1)^2}$

3 a $\dfrac{n}{n+1}$

b $\dfrac{11}{6}-\dfrac{3n^2+12n+11}{(n+1)(n+2)(n+3)}$

c $\dfrac{2n}{2n+1}$

d $\dfrac{n(n+7)}{12(n+3)(n+4)}$

4 a $\dfrac{n-1}{n}$

b $\dfrac{n}{2(n+2)}$

c $\dfrac{5n^2+13n}{12(n+2)(n+3)}$

d $\dfrac{(3n+2)(n-1)}{4n(n+1)}$

Exercise 7.1B

1 a $r(r+1)-r(r-1)\equiv r^2+r-r^2+r\equiv 2r$

b $\Sigma r(r+1)-r(r-1)\equiv$

$\begin{array}{ccc}
1\times 2 & - & 1\times 0 \\
2\times 3 & - & 2\times 1 \\
3\times 4 & - & 3\times 2 \\
& \downarrow & \\
(n-2)\times(n-1) & - & (n-3)\times(n-2) \\
(n-1)\times n & - & (n-2)\times(n-1) \\
n\times(n+1) & - & (n-1)\times n
\end{array}$

Hence $2\Sigma r\equiv \Sigma r(r+1)-r(r-1)\equiv 1\times 0+n\times(n+1)$

$2\Sigma r\equiv n(n+1)$

$\displaystyle\sum_1^n r\equiv \dfrac{n(n+1)}{2}$

2 a $24r^2+2$

b $\Sigma[(2r+1)^3-(2r-1)^3]=24\Sigma r^2+2n$

$\Sigma[(2r+1)^3-(2r-1)^3]\equiv$

$\begin{array}{ccc}
3^3 & - & 1^3 \\
5^3 & - & 3^3 \\
7^3 & - & 5^3 \\
\downarrow & & \downarrow \\
(2n-3)^3 & - & (2n-5)^3 \\
(2n-1)^3 & - & (2n-3)^3 \\
(2n+1)^3 & - & (2n-1)^3
\end{array}$

$\equiv (2n+1)^3-1$

Hence $24\displaystyle\sum_1^n r^2+2n\equiv (2n+1)^3-1$

$24\displaystyle\sum_1^n r^2\equiv (2n+1)^3-2n-1$

$\equiv 8n^3+12n^2+4n$

$\equiv 4n(n+1)(2n+1)$

Hence $\displaystyle\sum_1^n r^2\equiv \dfrac{n(n+1)(2n+1)}{6}$

3 a $4r^3$

b $\dfrac{1}{4}\left[r^2(r+1)^2\right]$

4 a $3-\dfrac{2(n+1)-2n}{n(n+1)}\equiv \dfrac{3n^2-n-2}{n(n+1)}$

b $\dfrac{3}{4}$

5 a $\dfrac{1}{2(2r-1)}-\dfrac{1}{2(2r+1)}$

b $\dfrac{2n}{2(2n+1)}$

c $\dfrac{1}{2}$

6 $\dfrac{n(n+2)}{(n+1)^2}$; 1

7 Let $\dfrac{2r-1}{r(r+1)(r+2)}\equiv \dfrac{A}{r}+\dfrac{B}{r+1}+\dfrac{C}{r+2}$

$2r-1\equiv A(r+1)(r+2)+Br(r+2)+Cr(r+1)$

When $r=0$, $-1=2A$, so $A=-\dfrac{1}{2}$

When $r=-1$, $-3=-B$, so $B=3$

When $r=-2$, $-5=2C$, so $C=-\dfrac{5}{2}$

Hence $\dfrac{2r-1}{r(r+1)(r+2)}\equiv -\dfrac{1}{2r}+\dfrac{3}{r+1}-\dfrac{5}{2(r+2)}$

$\displaystyle\sum_1^n S\equiv -\dfrac{1}{2}+\dfrac{3}{2}-\dfrac{5}{6}$

$\qquad -\dfrac{1}{4}+\dfrac{3}{3}-\dfrac{5}{8}$

$\qquad -\dfrac{1}{6}+\dfrac{3}{4}-\dfrac{5}{10}$

$\qquad -\dfrac{1}{8}+\dfrac{3}{5}-\dfrac{5}{12}$

$\qquad -\dfrac{1}{10}+\dfrac{3}{6}-\dfrac{5}{14}$

$+\dots$

$\qquad -\dfrac{1}{2n-4}+\dfrac{3}{n-1}-\dfrac{5}{2n}$

$\qquad -\dfrac{1}{2n-2}+\dfrac{3}{n}-\dfrac{5}{2(n+1)}$

$\qquad -\dfrac{1}{2n}+\dfrac{3}{n+1}-\dfrac{5}{2(n+2)}$

Hence $\displaystyle\sum_2^n S\equiv -\dfrac{1}{2}+\dfrac{3}{2}-\dfrac{1}{4}-\dfrac{5}{2(n+1)}+\dfrac{3}{n+1}-\dfrac{5}{2(n+2)}$

$\equiv \dfrac{3}{4}+\dfrac{1}{2(n+1)}-\dfrac{5}{2(n+2)}$

Hence $\displaystyle\sum_2^\infty S=\dfrac{3}{4}+\dfrac{1}{\infty}-\dfrac{5}{\infty}=\dfrac{3}{4}$

8 $\dfrac{1}{4} + \dfrac{1}{n+1} + \dfrac{1}{n+2} - \dfrac{3}{n+3} - \dfrac{3}{n+4}$

9 $\dfrac{3}{4}$

10 a $\dfrac{x-1}{x} - \dfrac{x-2}{x-1} \equiv \dfrac{(x-1)^2 - x(x-2)}{x(x-1)}$

$\equiv \dfrac{(x^2 - 2x + 1 - x^2 + 2x)}{x(x-1)}$

$\equiv \dfrac{1}{x(x-1)}$

b $\dfrac{n-2}{2n}$

c $\dfrac{1}{2}$

11 $\ln\left[\dfrac{k^{n-2}}{(n-1)!}\right]$

12 a $\dfrac{1}{(x-1)!} - \dfrac{1}{x!} \equiv \dfrac{x! - (x-1)!}{x!(x-1)!}$

$\equiv \dfrac{x(x-1)! - (x-1)!}{x!(x-1)!} \equiv \dfrac{(x-1)![x-1]}{x!(x-1)!}$

$\equiv \dfrac{x-1}{x!}$

b $\left(1 - \dfrac{1}{n!}\right)$

c 1

Exercise 7.2A

1

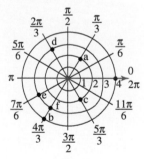

a $x = 2\cos\left(\dfrac{\pi}{3}\right) = 1;\quad y = 2\sin\left(\dfrac{\pi}{3}\right) = \sqrt{3}$

$(1, \sqrt{3})$

b $x = 4\cos\left(\dfrac{4\pi}{3}\right) = -2;\quad y = 4\sin\left(\dfrac{4\pi}{3}\right) = -2\sqrt{3}$

$(-2, -2\sqrt{3})$

c $x = 2\cos\left(\dfrac{5\pi}{3}\right) = 1;\quad y = 2\sin\left(\dfrac{5\pi}{3}\right) = -\sqrt{3}$

$(1, -\sqrt{3})$

d $x = 3\cos\left(\dfrac{2\pi}{3}\right) = -\dfrac{3}{2};\quad y = 3\sin\left(\dfrac{2\pi}{3}\right) = \dfrac{3}{2}\sqrt{3}$

$\left(-\dfrac{3}{2}, \dfrac{3\sqrt{3}}{2}\right)$

e $x = 3\cos\left(-\dfrac{5\pi}{6}\right) = -\dfrac{3\sqrt{3}}{2};\quad y = 3\sin\left(-\dfrac{5\pi}{6}\right) = -\dfrac{3}{2}$

$\left(-\dfrac{3\sqrt{3}}{2}, -\dfrac{3}{2}\right)$

f $x = 3\cos\left(-\dfrac{2\pi}{3}\right) = -\dfrac{3}{2};\quad y = 3\sin\left(-\dfrac{2\pi}{3}\right) = -\dfrac{3\sqrt{3}}{2}$

$\left(-\dfrac{3}{2}, -\dfrac{3\sqrt{3}}{2}\right)$

2 a $\left(4, \dfrac{\pi}{6}\right)$

b $\left(6, \dfrac{2\pi}{3}\right)$

c $\left(8, \dfrac{5\pi}{3}\right)$

d $(13, 4.32)$

3 a $\left(4, -\dfrac{5\pi}{6}\right)$

b $\left(10, -\dfrac{\pi}{3}\right)$

c $(5, 2.21)$

d $(\sqrt{37}, 1.41)$

4 $r = \sqrt{5}$, circle centre origin, radius $\sqrt{5}$

5 Circle, centre origin, radius 4; $x^2 + y^2 = 16$

6 a Sketch should show graph of $y = x$ for positive values of x since $r \geq 0$

$\theta = \dfrac{\pi}{4} \Rightarrow \tan\theta = \tan\dfrac{\pi}{4} = 1 \Rightarrow \dfrac{y}{x} = 1 \Rightarrow y = x$

b Sketch should show graph of $x = 0$ for $x \geq 0$ since $r \geq 0$

$\theta = \dfrac{\pi}{2} \Rightarrow \tan\theta = \tan\dfrac{\pi}{2}$

so gradient is undefined, hence lies on $x = 0$

c Sketch should show $y = -x$ for $x \geq 0$ since $r \geq 0$

$\theta = \dfrac{3\pi}{4} \Rightarrow \tan\theta = \tan\dfrac{3\pi}{4} = -1 \Rightarrow \dfrac{y}{x} = -1 \Rightarrow y = -x$

d Sketch should show $y = -\dfrac{x\sqrt{3}}{3}$ for $x \geq 0$ since $r \geq 0$

$\theta = -\dfrac{\pi}{6} \Rightarrow \dfrac{y}{x} = \tan-\dfrac{\pi}{6} \Rightarrow y = -\dfrac{\sqrt{3}}{3}$

7 a

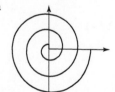

b

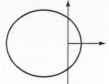

8 **a** **i** Maximum = 1, minimum = 0

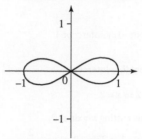

ii Maximum = 1, minimum = −1

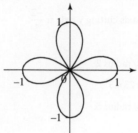

b Maximum = 1, minimum = 0

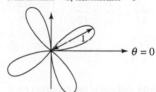

c **i** Maximum = 2, minimum = 0

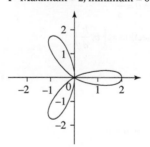

ii Maximum = 2, minimum = −2

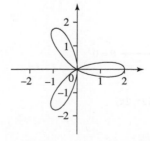

d **i** Maximum = 4, minimum = 0

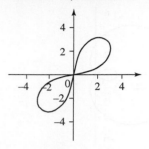

ii Maximum = 4, minimum = −4

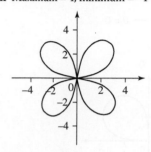

e Max = 2, min = 0

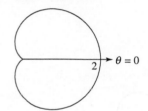

f Max = 5, min = 3

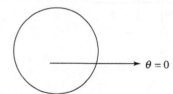

g Max = 5, min = 1

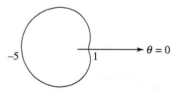

h **i** Maximum = 8, minimum = 0

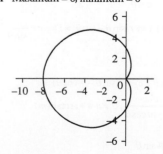

ii Maximum = 8, minimum = −2

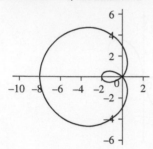

i Max = 4π, min = 0

$\theta = 0$

j **i** Max = 2, min = 0

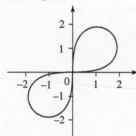

ii Max = 2, min = −2

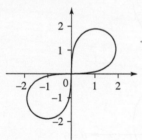

Exercise 7.2B

1 **a** $\theta = \dfrac{\pi}{4}$

b $\theta = \arctan 2$

c $\theta = \dfrac{\pi}{2} \quad r \geq 0$

d $\theta = 0 \quad r \geq 0$

e $r\sin\theta = 5 \Rightarrow r = \dfrac{5}{\sin\theta}, \quad \theta \neq n\pi$

f $r\cos\theta = 3 \Rightarrow r = \dfrac{3}{\cos\theta}, \theta \neq n\pi + \dfrac{\pi}{2}$

g $r\sin\theta = r\cos\theta + 1 \Rightarrow r(\sin\theta - \cos\theta) = 1 \Rightarrow r = \dfrac{1}{\sin\theta - \cos\theta},$

$\theta \neq n\pi + \dfrac{\pi}{4}$

h $r\sin\theta = mr\cos\theta + c \Rightarrow r(\sin\theta - m\cos\theta) = c$

$\Rightarrow r = \dfrac{c}{\sin\theta - m\cos\theta}, \theta \neq n\,\pi + \arctan(m)$

i $r = 5$

j $r = 10\cos\theta + 24\sin\theta$

k $r = 6\cos\theta$

l $r = 8\sin\theta$

m $r = 10(\cos\theta + \sin\theta)$

n $r = \dfrac{\sin\theta}{\cos^2\theta}, \theta \neq n\pi + \dfrac{\pi}{2}$

o $r^2 = \dfrac{5}{1 + \sin 2\theta} \quad \theta \neq n\pi - \dfrac{\pi}{4}$

2 **a** $y = 1 - x$

a straight line, gradient −1, y-intercept 1

b $y = 2x + 5$

a straight line, gradient 2, y-intercept 5

c $r = \dfrac{2}{\cos\theta} \Rightarrow r\cos\theta = 2 \Rightarrow x = 2$

a line parallel to y-axis cutting x-axis at 2

d $r = \dfrac{-2}{3\sin\theta} \Rightarrow 3r\sin\theta = -2 \Rightarrow 3y = -2 \Rightarrow y = -\dfrac{2}{3}$

a line parallel to x-axis cutting y-axis at $\dfrac{-2}{3}$

e $x^2 + (y - 2)^2 = 2^2$

a circle centre $(0, 2)$ radius 2

f $(x - 4)^2 + (y - 3)^2 = 25 = 5^2$

a circle centre $(4, 3)$ radius 5.

3 **a** $\left(2, \dfrac{\pi}{6}\right), \left(2, \dfrac{5\pi}{6}\right)$

b $\left(\dfrac{1}{2}, \pm\dfrac{\pi}{12}\right), \left(\dfrac{1}{2}, \pm\dfrac{5\pi}{12}\right), \left(\dfrac{1}{2}, \pm\dfrac{7\pi}{12}\right), \left(\dfrac{1}{2}, \pm\dfrac{11\pi}{12}\right)$

c $\left(\dfrac{\sqrt{2}}{2}, \dfrac{\pi}{8}\right), \left(\dfrac{\sqrt{2}}{2}, -\dfrac{7\pi}{8}\right)$

4 Spiral has equation $r = a\theta$ and circle has equation $r = b$

Intersect when $a\theta = b \Rightarrow \theta = \dfrac{b}{a}$

So single point of intersection at $\left(b, \dfrac{b}{a}\right)$

5 **a** **i**

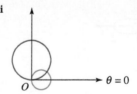
$\theta = 0$

ii $\left(\dfrac{\sqrt{3}}{2}, \dfrac{\pi}{6}\right)$

b **i**

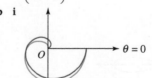
$\theta = 0$

ii $(0, 0), (\pi, \pi), (2\pi, 2\pi)$

c i

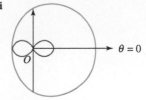

ii $(1, \pi)$

d i

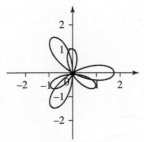

ii $\left(\dfrac{-1+\sqrt{5}}{2}, 0.905 \right), \left(\dfrac{-1+\sqrt{5}}{2}, 5.38 \right)$

e i

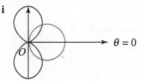

ii $\left(\dfrac{\sqrt{3}}{2}, \dfrac{5\pi}{9} \right), \left(\dfrac{\sqrt{3}}{2}, \dfrac{11\pi}{9} \right), \left(\dfrac{\sqrt{3}}{2}, \dfrac{17\pi}{9} \right)$

f i and ii As in part **e**, but now positive and negative values of r are included, so solutions are:

$\left(\dfrac{\sqrt{3}}{2}, \dfrac{2\pi}{9} \right), \left(\dfrac{\sqrt{3}}{2}, \dfrac{5\pi}{9} \right), \left(\dfrac{\sqrt{3}}{2}, \dfrac{8\pi}{9} \right), \left(\dfrac{\sqrt{3}}{2}, \dfrac{11\pi}{9} \right),$

$\left(\dfrac{\sqrt{3}}{2}, \dfrac{14\pi}{9} \right), \left(\dfrac{\sqrt{3}}{2}, \dfrac{17\pi}{9} \right)$

6 a $-1 \le \cos\theta \le 1$

$-4 \le 4\cos\theta \le 4$

$a - 4 \le a + 4\cos\theta \le a + 4$

So if $a > 4$, then $a + 4\cos\theta > 0$ for all θ. So $r > 0$. To pass through the pole $r = 0$

So curve never passes through pole.

b $r(\theta) = a + 4\cos\theta$

$r(-\theta) = a + 4\cos(-\theta) = a + 4\cos\theta = r(\theta)$

Since $r(-\theta) = r(\theta)$ curve is symmetrical about the initial line.

c i

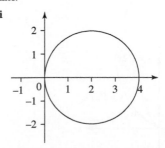

ii

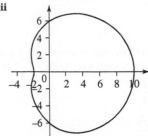

iii

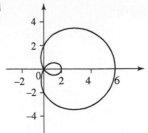

iv

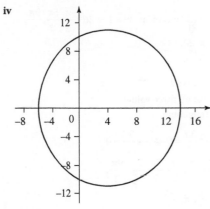

d $a - 4 \le r \le a + 4$

7 $\sec^2\theta = 2\tan\theta$

$\Rightarrow 1 = 2\sin\theta\cos\theta = \sin 2\theta$

$\Rightarrow 2\theta = \dfrac{\pi}{2}, \dfrac{5\pi}{2}$

$\Rightarrow \theta = \dfrac{\pi}{4}, \dfrac{5\pi}{4}$

$\theta = \dfrac{\pi}{4}, r = \sqrt{2}$

$\theta = \dfrac{5\pi}{4}, r = -\sqrt{2}$

So they intersect at $\left(\sqrt{2}, \dfrac{\pi}{4} \right)$ and $\left(-\sqrt{2}, \dfrac{5\pi}{4} \right)$

But these are the same point, so they only intersect at one point.

8 a $(1.39, 0.401), (1.39, 2.74), (0.360, -0.695), (0.360, -2.45)$

Since $1 + \sin\theta$ is always non-negative, extending r to permit negative values does not add any additional solutions.

b $(1.46, 1.01), (1.46, -1.01)$

Since $2 - \cos\theta$ is always non-negative, extending r to permit negative values does not add any additional solutions.

9 $2\sin^2\theta = \cos 2\theta = 1 - 2\sin^2\theta$

$4\sin^2\theta = 1$

$\sin^2\theta = \dfrac{1}{4}$

$\sin\theta = \pm\dfrac{1}{2}$

$\theta = \pm\dfrac{\pi}{6}, \pm\dfrac{5\pi}{6}$

$r = 2\sin^2\theta = \dfrac{1}{2}$ so points of intersection are

$\left(\dfrac{1}{2}, \pm\dfrac{\pi}{6}\right), \left(\dfrac{1}{2}, \pm\dfrac{5\pi}{6}\right)$

The second pair of points are reflections of the first pair through the y-axis, so these form a rectangle, with Cartesian

coordinates $\left(\pm\dfrac{\sqrt{3}}{4}, \pm\dfrac{1}{4}\right)$

$\text{Area} = 4 \times \dfrac{1}{4} \times \dfrac{\sqrt{3}}{4} = \dfrac{\sqrt{3}}{4}$

10 a $r = 1 - \sin 3\theta = 0 \Rightarrow \sin 3\theta = 1 \Rightarrow 3\theta = \dfrac{\pi}{2}, \dfrac{5\pi}{2}, \dfrac{9\pi}{2}, \ldots$

$\Rightarrow \theta = \dfrac{\pi}{6}, \dfrac{5\pi}{6}, \dfrac{3\pi}{2}$

$\dfrac{dr}{d\theta} = -3\cos 3\theta$

$= 0$ at all three values

$\Rightarrow$ no tangents

b $r = \cos^2\dfrac{3\theta}{2} = 0 \Rightarrow \cos\dfrac{3\theta}{2} = 0$

$\Rightarrow \dfrac{3\theta}{2} = \dfrac{\pi}{2}, \dfrac{3\pi}{2}, \dfrac{5\pi}{2}, \dfrac{7\pi}{2}, \dfrac{9\pi}{2}, \ldots$

$\Rightarrow \theta = \dfrac{\pi}{3}, \pi, \dfrac{5\pi}{3}, \ldots$

$y = \sqrt{3}x, \quad y = 0, \quad y = -\sqrt{3}x$

11 a $r = 1 + 2\sin 4\theta = 0 \Rightarrow \sin 4\theta = -\dfrac{1}{2}$

$\Rightarrow 4\theta = \dfrac{5\pi}{6}, \dfrac{7\pi}{6}, \dfrac{17\pi}{6}, \dfrac{19\pi}{6}, \ldots$

$\Rightarrow \theta = \dfrac{5\pi}{24}, \dfrac{7\pi}{24}, \dfrac{17\pi}{24}, \dfrac{19\pi}{24}, \ldots$

b $r = 1 - 2\cos^2\dfrac{5\theta}{3} = 0 \Rightarrow \cos^2\dfrac{5\theta}{3} = \dfrac{1}{2} \Rightarrow \cos\dfrac{5\theta}{3} = \pm\dfrac{\sqrt{2}}{2}$

$\Rightarrow \dfrac{5\theta}{3} = \dfrac{\pi}{4}, \dfrac{3\pi}{4}, \dfrac{5\pi}{4}, \dfrac{7\pi}{4}, \dfrac{9\pi}{4}, \dfrac{11\pi}{4}, \dfrac{13\pi}{4}, \dfrac{15\pi}{4},$

$\dfrac{17\pi}{4}, \dfrac{19\pi}{4}, \ldots$

$\Rightarrow \theta = \dfrac{3\pi}{20}, \dfrac{9\pi}{20}, \dfrac{3\pi}{4}, \dfrac{21\pi}{20}, \dfrac{27\pi}{20}, \dfrac{33\pi}{20}, \dfrac{39\pi}{20},$

$\dfrac{9\pi}{4}, \dfrac{51\pi}{20}, \dfrac{57\pi}{20}, \ldots$

$\Rightarrow \theta = \dfrac{3\pi}{20}, \dfrac{9\pi}{20}, \dfrac{3\pi}{4}, \dfrac{\pi}{20}, \dfrac{7\pi}{20}, \dfrac{13\pi}{20}, \dfrac{19\pi}{20}, \dfrac{\pi}{4}, \dfrac{11\pi}{20}, \dfrac{17\pi}{20}, \ldots$

Exercise 7.3A

1 a 0

 b 1.54 (to 3 sf)

 c $\dfrac{3}{4}$

d $\dfrac{5}{3}$

e 1

f $\dfrac{3}{5}$

2 a
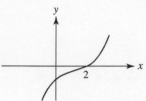

b
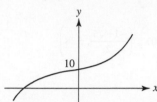

c
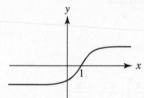

d
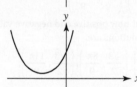

e
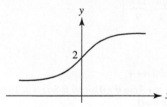

f
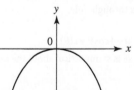

3 a $y = f(x)$

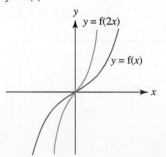

b $\sinh(2x) = 2 \Rightarrow x = 0.722$

4

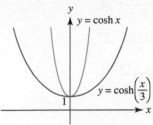

5 a

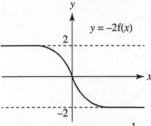

b $-2\tanh(x) = 1 \Rightarrow \tanh(x) = -\dfrac{1}{2}$

$\Rightarrow \tanh(x) = -0.549$

6 a

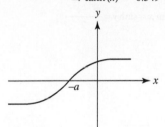

b $y = \pm 1$

7 a

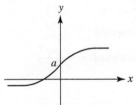

b $y = a + 1; y = a - 1$

8 a

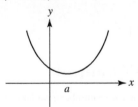

b

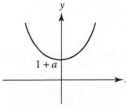

c

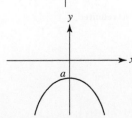

9 a 2.31

b ± 1.32

c -0.549

d $0.763, -2.76$

e 0.698

f 0.549

10 a $x = 0$

b $x = 0$

c $x = \ln(2)$

d $x = \ln(4), \ln\left(\dfrac{1}{4}\right)$

e $x = \ln(2)$

f $x = \ln(9)$

11 a $x = \ln\left(\dfrac{1}{2}\right)$

b $x = 0, \ln\left(\dfrac{1}{5}\right)$

c $x = \ln\left(\dfrac{3 \pm \sqrt{6}}{3}\right)$

d $x = \dfrac{1}{4}\ln 2$

Exercise 7.3B

1 a $x = \ln(2 + \sqrt{5}), -\ln(3 + \sqrt{10})$

b $x = 0, \pm \ln\left(\dfrac{3 + \sqrt{5}}{2}\right)$

c $x = \pm \dfrac{1}{2}\ln(5 + 2\sqrt{6})$

d $x = \pm \dfrac{1}{2}\ln(3 + 2\sqrt{2})$

2 a 0   **b** $\ln(-4 + \sqrt{17}), \ln(1 + \sqrt{2})$

c $0, \ln\left(\dfrac{-2 + \sqrt{13}}{3}\right)$ **d** $\dfrac{1}{2}\ln 3$

e $\ln\dfrac{1}{2}(3 + \sqrt{10})$ **f** $\pm\dfrac{1}{3}\ln(\ln(4 + \sqrt{15}))$

g $\dfrac{1}{6}\ln(3)$

3 $\text{LHS} = \left(\dfrac{e^x + e^{-x}}{2}\right)^2 - \left(\dfrac{e^x - e^{-x}}{2}\right)^2$

$= \dfrac{1}{4}(e^{2x} + 2 + e^{-2x}) - \dfrac{1}{4}(e^{2x} - 2 + e^{-2x})$

$= \dfrac{1}{4}(2 + 2)$

$= 1 = \text{RHS}$

4 a $\sinh(A)\cosh(B) + \sinh(B)\cosh(A)$

$\equiv \left(\dfrac{e^A - e^{-A}}{2}\right)\left(\dfrac{e^B + e^{-B}}{2}\right) + \left(\dfrac{e^B - e^{-B}}{2}\right)\left(\dfrac{e^A + e^{-A}}{2}\right)$

$\equiv \dfrac{e^{A+B} + e^{A-B} - e^{-(A-B)} - e^{-(A+B)}}{4}$

$+ \dfrac{e^{A+B} + e^{-(A-B)} - e^{A-B} - e^{-(A+B)}}{4}$

$$\equiv \frac{2e^{A+B} - 2e^{-(A+B)}}{4}$$

$$\equiv \frac{e^{A+B} - e^{-(A+B)}}{2}$$

$$\equiv \sinh(A+B), \text{ as required}$$

b $\sinh(A)\cosh(B) - \sinh(B)\cosh(A)$

$$= \left(\frac{e^A - e^{-A}}{2}\right)\left(\frac{e^B + e^{-B}}{2}\right) - \left(\frac{e^B - e^{-B}}{2}\right)\left(\frac{e^A + e^{-A}}{2}\right)$$

$$= \frac{e^{A+B} + e^{A-B} - e^{-(A-B)} - e^{-(A+B)}}{4}$$

$$- \frac{e^{A+B} - e^{A-B} + e^{-(A-B)} - e^{-(A+B)}}{4}$$

$$= \frac{2e^{A-B} - 2e^{-(A-B)}}{4}$$

$$= \frac{e^{A-B} - e^{-(A-B)}}{2}$$

$$= \sinh(A-B), \text{ as required}$$

c $2\sinh(x)\cosh(x) = 2\left(\frac{e^x + e^{-x}}{2}\right)\left(\frac{e^x - e^{-x}}{2}\right)$

$$= \frac{(e^{2x} + 1 - 1 - e^{2x})}{2}$$

$$= \frac{(e^{2x} - e^{-2x})}{2}$$

$$= \sinh(2x), \text{ as required}$$

5 a $\tanh x = \dfrac{\sinh x}{\cosh x}$

$$= \frac{\dfrac{e^x - e^{-x}}{2}}{\dfrac{e^x + e^{-x}}{2}}$$

$$= \frac{e^x - e^{-x}}{e^x + e^{-x}}$$

$$= \frac{e^{2x} - 1}{e^{2x} + 1}$$

b $\dfrac{2\tanh x}{1 + \tanh^2 x} \equiv \dfrac{2\left(\dfrac{e^{2x} - 1}{e^{2x} + 1}\right)}{1 + \left(\dfrac{e^{2x} - 1}{e^{2x} + 1}\right)^2}$

$$\equiv \frac{2(e^{2x} - 1)(e^{2x} + 1)}{(e^{2x} + 1)^2 + (e^{2x} - 1)^2}$$

$$\equiv \frac{2(e^{4x} - 1)}{(e^{4x} + 2e^{2x} + 1) + (e^{4x} - 2e^{2x} + 1)}$$

$$\equiv \frac{2(e^{4x} - 1)}{2e^{4x} + 2}$$

$$\equiv \frac{e^{4x} - 1}{e^{4x} + 1}$$

$$\equiv \tanh(2x), \text{ as required}$$

6 a $2\cosh^2 x - 1 \equiv 2\left(\dfrac{e^x + e^{-x}}{2}\right)^2 - 1$

$$\equiv \frac{2(e^{2x} + 2 + e^{-2x})}{4} - 1$$

$$\equiv \frac{e^{2x} + e^{-2x} + 2}{2} - \frac{2}{2}$$

$$\equiv \frac{e^{2x} + e^{-2x}}{2}$$

$$\equiv \cosh(2x), \text{ as required}$$

b $2\cosh^2 x - 1 + \cosh x = 5$

$$2\cosh^2 x + \cosh x - 6 = 0$$

$$\Rightarrow \cosh x = \frac{3}{2}, (-2)$$

$$\Rightarrow \cosh x = \pm\ln\left(\frac{3}{2} + \sqrt{\left(\frac{3}{2}\right)^2 - 1}\right)$$

$$= \pm\ln\left(\frac{3 + \sqrt{5}}{2}\right)$$

7 Let $y = \sinh^{-1} x$ then $x = \sinh y = \dfrac{e^y - e^{-y}}{2}$

$$2x = e^y - e^{-y}$$

$$e^{2y} - 2xe^y - 1 = 0$$

$$(e^y - x^2) = x^2 + 1$$

$$e^y = x \pm \sqrt{x^2 + 1}$$

$e^y > 0$, therefore $x - \sqrt{x^2 + 1}$ not valid as $x^2 + 1 > x^2$

so $\sqrt{x^2 + 1} > x \Rightarrow x - \sqrt{x^2 + 1} < 0$

So $y = \ln(x + \sqrt{x^2 + 1})$, as required

8 a Let $y = \tanh^{-1} x$. Then $x = \tanh y = \dfrac{e^y - e^{-y}}{e^y + e^{-y}}$

$$xe^y + xe^{-y} = e^y - e^{-y}$$

$$(1 + x)e^{-y} = (1 - x)e^y$$

$$\frac{1 + x}{1 - x} = e^{2y}$$

$$2y = \ln\left(\frac{1 + x}{1 - x}\right)$$

So $\text{artanh } x = \dfrac{1}{2}\ln\left(\dfrac{1 + x}{1 - x}\right)$, as required

b Because the log of a negative number is undefined in $\mathbb{R}$

9 $x = \pm\ln\left(2 + \sqrt{3}\right), \pm\ln\left(\dfrac{3 + \sqrt{5}}{2}\right)$

10 a $3\sinh\left(\dfrac{x}{2}\right) - \sinh(x) \equiv 3\sinh\left(\dfrac{x}{2}\right) - 2\sinh\left(\dfrac{x}{2}\right)\cosh\left(\dfrac{x}{2}\right)$

$$\equiv \sinh\left(\frac{x}{2}\right)\left(3 - 2\cosh\left(\frac{x}{2}\right)\right),$$
$$\text{as required}$$

b $x = \pm2\ln\left(\dfrac{3 + \sqrt{5}}{2}\right)$

11 a $\dfrac{1}{\cosh x+1}-\dfrac{1}{\cosh x-1}\equiv\dfrac{(\cosh x-1)-(\cosh x+1)}{(\cosh x+1)\ (\cosh x-1)}$

$$\equiv\dfrac{-2}{\cosh^2(x)-1}$$

$$\equiv-\dfrac{2}{\sinh^2(x)},\text{ as required}$$

b $x=\ln\left(\dfrac{1+\sqrt5}{2}\right),\ \ln\left(\dfrac{-1+\sqrt5}{2}\right)$

12 a $2\sinh x\cosh x=\cosh^2 x$

$\cosh^2 x-2\sinh x\cosh x=0$

$\cosh x\,(\cosh x-2\sinh x)=0$

$\cosh x\neq0$ so $\cosh x-2\sinh x=0\Rightarrow\cosh x=2\sinh x$

$$\Rightarrow\dfrac{1}{2}=\dfrac{\sinh x}{\cosh x}=\tanh x,$$
$$\text{as required}$$

b $x=\dfrac{1}{2}\ln(3)$

13 $x=1$

14 a $\sqrt5$ as $\cosh x>0$

b $\dfrac{2}{\sqrt5}$

15 a $2\sqrt2$ as $\sinh x>0$

b $\dfrac{2\sqrt2}{3}$

16 a $\dfrac{\sqrt3}{3}$

b $\dfrac{2\sqrt3}{3}$

17 a $\dfrac{d}{dx}(\sinh x)=\dfrac{d}{dx}\left(\dfrac{e^x-e^{-x}}{2}\right)=\dfrac{e^x+e^{-x}}{2}=\cosh x$

$\dfrac{d}{dx}(\cosh x)=\dfrac{d}{dx}\left(\dfrac{e^x+e^{-x}}{2}\right)=\dfrac{e^x+e^{-x}}{2}=\sinh x$

b $\dfrac{d}{dx}(\sinh x)_{(x=0)}=\dfrac{e^0+e^0}{2}=\cosh 0=1$

$\dfrac{d}{dx}(\cosh x)_{(x=0)}=\dfrac{e^0-e^0}{2}=\sinh 0=0$

c When $x>0$, $e^{-x}>0\Rightarrow e^x+e^{-x}>e^x>e^x-e^{-x}$

The derivative of $\cosh x$ is $\sinh x$ and the derivative of $\sinh x$ is $\cosh x$

$\Rightarrow e^x+e^{-x}>e^x-e^{-x}\Rightarrow\dfrac{e^x+e^{-x}}{2}>\dfrac{e^x+e^{-x}}{2}$

$\Rightarrow\cosh x>\sinh x$

So the gradient of $y=\sinh x$ is greater than the gradient of $y=\cosh x$ when $x>0$

d 1

Review exercise 7

1 a $\dfrac{1}{2}-\dfrac{1}{n+2}$

b $\dfrac{1}{3}-\dfrac{1}{n+3}$

2 $1-\dfrac{1}{\infty}=1$

3 $\displaystyle\sum_1^k\left(\dfrac{1}{n^2}-\dfrac{1}{(n+1)^2}\right)=\dfrac{1}{1}-\dfrac{1}{4}$

$$+\dfrac{1}{4}-\dfrac{1}{9}$$

$$+\dfrac{1}{9}-\dfrac{1}{16}$$

$$\downarrow$$
$$\downarrow$$

$$+\dfrac{1}{(k-1)^2}-\dfrac{1}{(k)^2}$$

$$+\dfrac{1}{(k)^2}-\dfrac{1}{(k+1)^2}$$

Thus $\displaystyle\sum_1^k\left(\dfrac{1}{n^2}-\dfrac{1}{(n+1)^2}\right)=\dfrac{1}{1}-\dfrac{1}{(k+1)^2}$

$$=\dfrac{(k+1)^2-1}{(k+1)^2}=\dfrac{k^2+2k}{(k+1)^2}=\dfrac{k(k+2)}{(k+1)^2}$$

4 a $\left(\dfrac{5}{2},\dfrac{5\sqrt3}{2}\right)$

b $\left(\dfrac{5}{2},-\dfrac{5\sqrt3}{2}\right)$

5 a $\left(4,\dfrac{\pi}{6}\right)$

b $\left(4,\dfrac{7\pi}{6}\right)$

6 a $r\sin\theta=3r\cos\theta\Rightarrow\tan\theta=3\Rightarrow\theta=\tan^{-1}3$

b $r\sin\theta=2r\cos\theta+1\Rightarrow r=\dfrac{1}{\sin\theta-2\cos\theta}$

c $r=4$

d $r=2(\cos\theta+\sin\theta)$

7 a $x^2+(y-2)^2=4$ centre is $(0,2)$ radius 2

b $x^2+y^2=2\sqrt{x^2+y^2}-4y$

8 a i

ii $y=x$

b i

ii $y=\sqrt3x$

9 a i

ii $\max=3,\ \min=3$

b i

ii $\max=4,\ \min=0$

c i

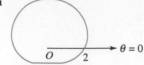

ii max = 3, min = 1

d i

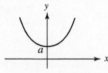

ii max = 6π, min = 0

10 $\left(2, \dfrac{\pi}{3}\right), \left(2, \dfrac{5\pi}{3}\right)$

11 $\left(\dfrac{\sqrt{3}}{2}, \dfrac{\pi}{6}\right), \left(\dfrac{\sqrt{3}}{2}, \dfrac{7\pi}{6}\right)$

12 $r = \sin\dfrac{3\theta}{2} = 0 \Rightarrow \dfrac{3\theta}{2} = 0, \pi, 2\pi, 3\pi \ldots \Rightarrow \theta = 0, \dfrac{2\pi}{3}, \dfrac{4\pi}{3}, \ldots$

$y = 0, \; y = -\sqrt{3}x, \; y = \sqrt{3}x$

13 a $\dfrac{13}{5}$

b $-\dfrac{3}{5}$

14 a

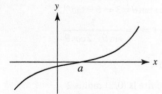

domain $x \in \mathbb{R}$, range $y \in \mathbb{R}$, $y \geq a$

b

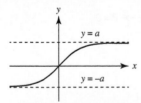

domain $x \in \mathbb{R}$, range $y \in \mathbb{R}$

c

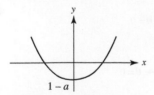

$y = a$

$y = -a$

domain $x \in \mathbb{R}$, range $y \in \mathbb{R}$, $-a < y < a$

d

$1 - a$

domain $x \in \mathbb{R}$, range $y \in \mathbb{R}$, $y \geq 1 - a$

15 $x = \ln(-2 + \sqrt{5}), \ln(1 + \sqrt{2})$

16 a $x = \ln 3$

b $x = \ln(1 + \sqrt{2})$

17 Let $y = \cosh^{-1} 2x$

Then $2x = \cosh y = \dfrac{e^y + e^{-y}}{2}$

$4x = e^y + e^{-y} \Rightarrow e^{2y} - 4xe^y + 1 = 0$

$e^y = 2x + \sqrt{4x^2 - 1} \Rightarrow y = \ln(2x + \sqrt{4x^2 - 1})$

18 a $\ln(2 + \sqrt{5})$

b $\dfrac{1}{2}\ln\left(\dfrac{1}{3}\right)$

Assessment 7

1 a $\dfrac{1}{x} - \dfrac{1}{x+1}$

$= \dfrac{x+1-x}{x(x+1)}$

$= \dfrac{1}{x(x+1)}$

b $1 - \dfrac{1}{n+1}$

2 a $(r+1)^3 - (r-1)^3$

$r^3 + 3r^2 + 3r + 1 - (r^3 - 3r^2 + 3r - 1)$

$6r^2 + 2$, as required

b $\displaystyle\sum_{r=1}^{n} 6r^2 + 2 = 2^3 - 0^3$

$+ 3^3 - 1^3$
$+ 4^3 - 2^3$
$+ 5^3 - 3^3$
$+ \ldots$
$+ n^3 - (n-2)^3$
$+ (n+1)^3 - (n-1)^3$
$= -1 + n^3 + (n+1)^3$
$= -1 + n^3 + n^3 + 3n^2 + 3n + 1$
$= 2n^3 + 3n^2 + 3n$
$= n(2n^2 + 3n + 3)$

So $\displaystyle\sum_{r=1}^{n} 3r^2 + 1 = \dfrac{n}{2}(2n^2 + 3n + 3)$, as required

3 a $\left(2\sqrt{3}, -\dfrac{\pi}{3}\right)$

b i

$\theta = 0$

9

ii

$\dfrac{\pi}{6}$

$\theta = 0$

c i $x^2 + y^2 = 81$ **ii** $y = \dfrac{1}{\sqrt{3}}x$

4 a $(-2\sqrt{3}, -2)$

b i $x^2 + y^2 = 2y$ **ii** $x = 1$

c

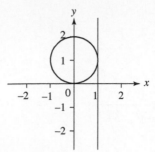

d $\left(\sqrt{2}, \dfrac{\pi}{4}\right)$

5 a $r = \cos\theta$ **b** $r = 2\operatorname{cosec}\theta$

6 a

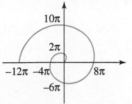

b

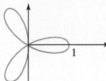

c

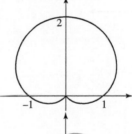

d

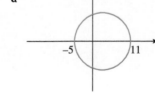

7 $r = \sin^2 4\theta = 0 \Rightarrow \sin 4\theta = 0$

$\Rightarrow 4\theta = 0, \pi, 2\pi, 3\pi, \ldots \Rightarrow \theta = 0, \dfrac{\pi}{4}, \dfrac{\pi}{2}, \dfrac{3\pi}{4}, \ldots$

$y = 0,\ x = 0,\ y = x,\ y = -x$

8 $(2, \dfrac{\pi}{8}),\ (2, \dfrac{5\pi}{8}),\ (2, \dfrac{9\pi}{8}),\ (2, \dfrac{13\pi}{8})$

9 a i, ii

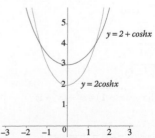

b i $y \geq 2$
  ii $y \geq 3$

10 a $\sinh(\ln 2) = \dfrac{e^{\ln 2} - e^{-\ln 2}}{2}$

$\dfrac{e^{\ln 2} - e^{\ln\left(\frac{1}{2}\right)}}{2}$

$\dfrac{2 - \dfrac{1}{2}}{2}$

$\dfrac{3}{4}$

b $x = 1.8$

11 a

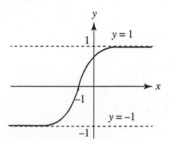

b $y = 1$ $y = -1$

c $\tanh x = \dfrac{\sinh x}{\cosh x}$

$\dfrac{e^x - e^{-x}}{2} \div \dfrac{e^x + e^{-x}}{2}$

$\dfrac{2(e^x - e^{-x})}{2(e^x + e^{-x})}$

$\dfrac{e^x - e^{-x}}{e^x + e^{-x}}$

$\dfrac{e^{2x} - 1}{e^{2x} + 1}$, as required

d $x = \dfrac{1}{2}\ln 3$

12 a $\dfrac{3}{r} - \dfrac{3}{r+1}$

b $\displaystyle\sum_{r=1}^{n} \dfrac{3}{r(r+1)} = 3 - \dfrac{3}{2}$

$+ \dfrac{3}{2} - 1$

$+ 1 - \dfrac{3}{4}$

$+ \dfrac{3}{4} - \dfrac{3}{5}$

$+ \ldots$

$+ \dfrac{3}{n} - \dfrac{3}{n+1}$

$= 3 - \dfrac{3}{n+1}$

$= \dfrac{3(n+1) - 3}{n+1}$

$= \dfrac{3n + 3 - 3}{n+1}$

$= \dfrac{3n}{n+1}$, as required

c $\dfrac{30}{31}$

13 a $\dfrac{1}{r-1}-\dfrac{1}{r+1}$

b $\displaystyle\sum_{r=2}^{n}\dfrac{2}{r^2-1}=1-\dfrac{1}{3}$

$+\dfrac{1}{2}-\dfrac{1}{4}$

$+\dfrac{1}{3}-\dfrac{1}{5}$

$+\dfrac{1}{4}-\dfrac{1}{6}$

$+\ldots$

$+\dfrac{1}{n-1-1}-\dfrac{1}{n-1+1}$

$+\dfrac{1}{n-1}-\dfrac{1}{n+1}$

$=1+\dfrac{1}{2}-\dfrac{1}{n}-\dfrac{1}{n+1}$

$=\dfrac{3}{2}-\dfrac{n+1+n}{n(n+1)}$

$=\dfrac{3(n^2+n)-2(2n+1)}{2n(n+1)}$

$=\dfrac{3n^2-n-2}{2n(n+1)}$

$=\dfrac{(3n+2)(n-1)}{2n(n+1)}$

So $\displaystyle\sum_{r=2}^{n}\dfrac{1}{r^2-1}=\dfrac{(3n+2)(n-1)}{2n(n+1)}\div 2=\dfrac{(3n+2)(n-1)}{4n(n+1)}$,

as required

c $\displaystyle\sum_{r=2}^{\infty}\dfrac{1}{r^2-1}=\lim_{n\to\infty}\dfrac{(3n+2)(n-1)}{4n(n+1)}$

$=\lim_{n\to\infty}\dfrac{3n^2-n-2}{4(n^2+n)}$

$=\lim_{n\to\infty}\dfrac{3-\dfrac{1}{n}-\dfrac{2}{n^2}}{4\left(1+\dfrac{1}{n}\right)}$

$=\dfrac{3}{4}$

14 $r\sin\theta=\dfrac{1}{r\cos\theta}$

$r^2\sin\theta\cos\theta=1$

$r^2\left(\dfrac{1}{2}\sin 2\theta\right)=1$

$r^2=\dfrac{2}{\sin 2\theta}$

$r^2=2\operatorname{cosec}2\theta$

15 a $6xy$

b

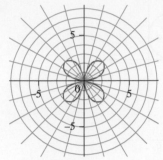

c $r=3$ occurs when $\theta=\dfrac{\pi}{4}$ and $\theta=\dfrac{5}{4}\pi$

16 a $r=7$ is maximum

$r=1$ is minimum

b

17 a

$y=\sinh 2x$

$y=\sinh x$

b $y=\dfrac{e^x-e^{-x}}{2}$

$2y=e^x-e^{-x}$

$e^{2x}-2ye^x-1=0$

$(e^x-y)^2-y^2-1=0$

$(e^x-y)^2=y^2+1$

$e^x-y=\sqrt{y^2+1}$

$e^x=y+\sqrt{y^2+1}$

$x=\ln(y+\sqrt{y^2+1})$, as required

18 a Let $x=\cosh y$ (so $\operatorname{arcosh}x=y$)

$x=\dfrac{e^y+e^{-y}}{2}$

$2x=e^y+e^{-y}$

$e^{2y}-2xe^y+1=0$

$(e^y-x)^2-x^2+1=0$

$e^y-x=\sqrt{x^2-1}$

$e^y=x+\sqrt{x^2-1}$

$y=\ln(x\pm\sqrt{x^2-1})$

$(x+\sqrt{x^2-1})(x-\sqrt{x^2-1})=x^2-(x^2-1)$

$\qquad\qquad\qquad\qquad\qquad 1$

Therefore $x-\sqrt{x^2-1}=\dfrac{1}{x+\sqrt{x^2-1}}$

So $\ln(x-\sqrt{x^2-1})=\ln\dfrac{1}{x+\sqrt{x^2-1}}$

$$\ln(x+\sqrt{x^2-1})^{-1}$$
$$-\ln(x+\sqrt{x^2-1})$$

So $\operatorname{arcosh} x = \pm\ln(x+\sqrt{x^2-1})$, as required

b $x=\pm\ln(3+2\sqrt{2})$

19 a $\dfrac{1}{r!}-\dfrac{1}{(r+1)!}=\dfrac{(r+1)!-r!}{r!(r+1)!}$

$$=\dfrac{r!(r+1)-r!}{r!(r+1)!}$$
$$=\dfrac{r!(r+1-1)}{r!(r+1)!}$$
$$=\dfrac{r}{(r+1)!}, \text{ as required}$$

b $\displaystyle\sum_{r=1}^{n}\dfrac{r}{(r+1)!}=1-\dfrac{1}{2!}$

$$+\dfrac{1}{2!}-\dfrac{1}{3!}$$
$$+\dfrac{1}{3!}-\dfrac{1}{4!}$$
$$+\ldots$$
$$+\dfrac{1}{n!}-\dfrac{1}{(n+1)!}$$
$$=1-\dfrac{1}{(n+1)!}$$

c $\dfrac{(2n+1)!-n!}{n!(2n+1)!}$

20 $\dfrac{n}{30}(n+1)(2n+1)(3n^3+3n-1)$

21

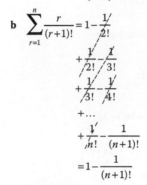

b $(0.599, 2.1957)$ and $(0.987, 5.0640)$

22 a

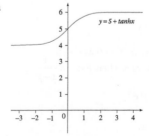

b $y=6 \qquad y=4$

c $y=\operatorname{artanh}(x-5)$ domain is $4<x<6$

Chapter 8

Exercise 8.1A

1 a Not an improper integral; fully defined in interval and limits finite

b An improper integral since $\ln x$ undefined at $x=0$

c An improper integral since one of the limits is ∞

d An improper integral since the limits are $\pm\infty$

e Not an improper integral; fully defined in interval and limits finite

f An improper integral since $\tan x$ undefined at $x=\dfrac{\pi}{2}$

2 a 1 as $a\to\infty$ since $\dfrac{1}{a}\to 0$

b $\dfrac{1}{8}$ as $a\to\infty$ since $\dfrac{1}{a^3}\to 0$

c 1 as $a\to\infty$ since $\dfrac{1}{1-a}\to 0$

d $\dfrac{1}{6}$ as $a\to\infty$ since $\dfrac{1}{3(2-a)}\to 0$

e $\dfrac{1}{8}$ as $a\to\to\infty$ since $\dfrac{1}{(a+2)^2}\to 0$

f $\dfrac{1}{3}$ as $a\to-\infty$ since $\dfrac{1}{(a-2)^3}\to 0$

g $\dfrac{1}{4}$ as $a\to\infty$ since $ae^{-2a}\to 0$ and $e^{-2a}\to 0$

h 1 as $a\to\infty$ since $\dfrac{1}{a}\to 0$ and $a^{-1}\ln a\to 0$

3 $\dfrac{1}{e^3}$ as $a\to\infty$ since $2e^{-a}\to 0$ and $ae^{-a}\to 0$

4 2

5 a 6 as $a\to 0$ since $\sqrt{a}\to 0$

b $\dfrac{27}{2}$ as $a\to 0$ since $a^{\frac{2}{3}}\to 0$

c $2\sqrt{2}$ as $a\to 2$ since $\sqrt{a-2}\to 0$

d 3 as $a\to 3$ since $\sqrt{9-a^2}\to 0$

e -4 as $a\to 0$ since $2a^{\frac{1}{2}}\ln a\to 0$ and $4a^{\frac{1}{2}}\to 0$

f 2 as $a\to 0$ since $e^a\to 1$ so $\sqrt{e^a-1}\to 0$

g 2 as $a\to\dfrac{\pi}{2}$ since $\cos a\to 0$

h $\dfrac{\sqrt{2}}{2}$ as $a\to 0$ since $\sin 2a\to 0$

6 a $\dfrac{e^2}{4}$ as $a\to 0$ since $\dfrac{1}{2}a^2\ln a\to 0$ and $\dfrac{1}{4}a^2\to 0$

b $\dfrac{2}{9}e^3$ as $a\to 0$ since $\dfrac{1}{3}a^3\ln a\to 0$ and $\dfrac{1}{9}a^3\to 0$

7 By parts $u=\ln(1-x) \qquad \dfrac{dv}{dx}=1$

$$\dfrac{du}{dx}=-\dfrac{1}{1-x};\quad v=x$$

$$\int_{1-e}^{a}\ln(1-x)dx=\Big[x\ln(1-x)\Big]_{1-e}^{a}+\int_{1-e}^{a}\dfrac{x}{1-x}dx$$

$$=a\ln(1-a)-(1-e)+\int_{1-e}^{a}\left(\dfrac{1}{1-x}-1\right)dx$$

$$=a\ln(1-a)-(1-e)+\Big[-\ln(1-x)-x\Big]_{1-e}^{a}$$

$$=a\ln(1-a)-(1-e)-\ln(1-a)-a+1+(1-e)$$

$$=(a-1)\ln(1-a)-a+1$$

$$\to 0 \text{ as } a\to 1$$

8 Function is symmetrical about $x = 0$

$$\Rightarrow \int_{-e}^{e} x^2 \ln|x| \, dx = 2\int_{0}^{e} x^2 \ln x \, dx = 2 \times \frac{2e^3}{9} = \frac{4e^3}{9}$$

as in question 6b

Exercise 8.1B

1 a As $a \to 0$, $\frac{1}{a^3} \to \infty$, so integral diverges

b As $a \to \infty$, $\frac{1}{a^3} \to 0$, so integral converges to the value $\frac{1}{3}$

c $-\ln\left(\frac{\sqrt{2}}{2}\right) = \ln\sqrt{2}$

d As $a \to \frac{\pi}{2}$, $\cos a \to 0$, so $\ln \cos a \to -\infty$

So integral diverges

e As $a \to \infty$, $\sin a$ does not converge, so integral diverges

f As $a \to \infty$, $\ln a \to \infty$, so integral diverges

g As $a \to -\infty$, $\ln(3 - a) \to \infty$, so integral diverges

h $2\sqrt{7}$ as $a \to 7$ since $\sqrt{7 - a} \to 0$

i As $a \to 7$, $\frac{1}{7 - a} \to \infty$, so integral diverges

j $\ln\frac{1}{2} - \ln\frac{1}{3} = \ln\frac{3}{2}$

k As $a \to \infty$, $\frac{\frac{1}{a}}{1 + \frac{1}{a^2}} \to 0$, so $\ln\frac{a}{a^2 + 1} \to -\infty$

So integral diverges

l So integral converges to the value given by

$$\frac{1}{2}\ln\frac{1}{2} - \frac{1}{2}\ln\frac{2}{3} = \frac{1}{2}\ln\frac{3}{4} \text{ or } \ln\left(\frac{\sqrt{3}}{2}\right)$$

2 a $\displaystyle\int \frac{x}{x^2 + 3} - \frac{2}{2x + 3} \, dx = \frac{1}{2}\ln(x^2 + 3) - \ln(2x + 3) + c$

$$= \frac{1}{2}\ln(x^2 + 3) - \frac{1}{2}\ln(2x + 3)^2 + c$$

$$= \frac{1}{2}\ln\frac{x^2 + 3}{(2x + 3)^2} + c$$

$$= \frac{1}{2}\ln\frac{x^2 + 3}{4x^2 + 12x + 9} + c$$

b $\displaystyle\int_{0}^{a} \frac{x}{x^2 + 3} - \frac{2}{2x + 3} \, dx = \frac{1}{2}\left[\ln\frac{x^2 + 3}{4x^2 + 12x + 9}\right]_0^a$

$$= \frac{1}{2}\ln\frac{a^2 + 3}{4a^2 + 12a + 9} - \frac{1}{2}\ln\frac{3}{9}$$

$$= \frac{1}{2}\ln\frac{1 + \frac{3}{a^2}}{4 + \frac{12}{a} + \frac{9}{a^2}} - \frac{1}{2}\ln\frac{1}{3}$$

$$\to \frac{1}{2}\ln\frac{1}{4} - \frac{1}{2}\ln\frac{1}{3} \text{ since } \frac{1 + \frac{3}{a^2}}{4 + \frac{12}{a} + \frac{9}{a^2}} \to \frac{1}{4} \text{ as } a \to \infty$$

$$= \frac{1}{2}\ln\left(\frac{\frac{1}{4}}{\frac{1}{3}}\right)$$

$$= \frac{1}{2}\ln\left(\frac{3}{4}\right)$$

$$= \ln\left(\frac{\sqrt{3}}{2}\right) \quad \left(k = \frac{\sqrt{3}}{2}\right)$$

3 a $\displaystyle\int_{a}^{0} \frac{6}{3x - 2} - \frac{2x}{x^2 + 4} \, dx = \left[2\ln(3x - 2) - \ln(x^2 + 4)\right]_a^0$

$$= \left[\ln\frac{(3x - 2)^2}{x^2 + 4}\right]_a^0$$

$$= \ln 1 - \ln\frac{(3a - 2)^2}{a^2 + 4}$$

$$= -\ln\frac{9a^2 - 12a + 4}{a^2 + 4}$$

$$\frac{9a^2 - 12a + 4}{a^2 + 4} = \frac{9 - \frac{12}{a} + \frac{4}{a^2}}{1 + \frac{4}{a^2}}$$

As $a \to -\infty$, $\dfrac{9 - \frac{12}{a} + \frac{4}{a^2}}{1 + \frac{4}{a^2}} \to 9$

So integral converges to $-\ln 9$ or $\ln\frac{1}{9}$

b $\displaystyle\int_{a}^{0} \frac{3 - x}{x^2 - 6x + 1} - \frac{8}{5 - 8x} \, dx$

$$= \left[-\frac{1}{2}\ln(x^2 - 6x + 1) + \ln(5 - 8x)\right]_a^0$$

$$= \left[\frac{1}{2}\ln(5 - 8x)^2 - \frac{1}{2}\ln(x^2 - 6x + 1)\right]_a^0$$

$$= \left[\frac{1}{2}\ln\frac{(5 - 8x)^2}{x^2 - 6x + 1}\right]_a^0$$

$$= \frac{1}{2}\ln 25 - \frac{1}{2}\ln\frac{(5 - 8a)^2}{a^2 - 6a + 1}$$

$$\frac{(5 - 8a)^2}{a^2 - 6a + 1} = \frac{25 - 80a + 64a^2}{a^2 - 6a + 1}$$

$$= \frac{\frac{25}{a^2} - \frac{80}{a} + 64}{1 - \frac{6}{a} + \frac{1}{a^2}} \to 64 \text{ as } a \to -\infty$$

So integral converges to the value given by

$$\frac{1}{2}\ln 25 - \frac{1}{2}\ln 64 = \frac{1}{2}\ln\frac{25}{64} = \ln\frac{5}{8}$$

4 $\displaystyle\int_{a}^{1} x^{-p} \, dx = \left[\frac{x^{-p+1}}{-p + 1}\right]_a^1$ if $p \neq 1$

$$= \frac{1}{1 - p}(1 - a^{1-p})$$

If $p < 1$, then $1 - p > 0$, so $a^{1-p} \to 0$ as $a \to 0$

Therefore integral converges when $p < 1$ to $\dfrac{1}{1 - p}$

5 $\displaystyle\int_{1}^{a} x^{-p} \, dx = \left[\frac{x^{-p+1}}{-p + 1}\right]_1^a$ if $p \neq 1$

$$= \frac{1}{1 - p}(a^{1-p} - 1)$$

If $p > 1$, then $1 - p < 0$, so $a^{1-p} \to 0$ as $a \to \infty$

Therefore integral converges when $p > 1$ to

$$\frac{1}{1 - p} \times (-1) = \frac{1}{p - 1}$$

Exercise 8.2A

1 a $\dfrac{45}{4}$ or 11.25 square units

b $\dfrac{15}{4}$ or 3.75

2 a

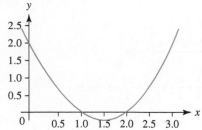

b $\dfrac{1}{6}$ (0.17) square units

c i $-\dfrac{1}{6}$ **ii** $\dfrac{5}{6}$

3 a

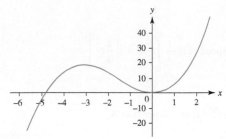

b $\dfrac{625}{12}$

c i $\dfrac{125}{12}$ **ii** $\dfrac{26}{3}$ (8.67)

4 a 256

b 121

c 11

d $\dfrac{1}{16}$

5 a 9

b $\dfrac{113}{3}$

c 5

d $\dfrac{23}{3}$ (7.67)

6 a $-\dfrac{325}{2}$ (−162.5)

b $-\dfrac{99}{2}$ (−49.5)

c 0

d $\dfrac{245}{2}$ (122.5)

7 $\dfrac{1}{5-2}\displaystyle\int_{2}^{5} x^{-2}\,dx = \dfrac{1}{3}\Big[-x^{-1}\Big]_{2}^{5}$

$= \dfrac{1}{3}\left(-\dfrac{1}{5}\right) - \dfrac{1}{3}\left(-\dfrac{1}{2}\right)$

$= \dfrac{1}{10}$ (0.1)

8 $\dfrac{1}{4-1}\displaystyle\int_{1}^{4} x^{\frac{1}{2}}\,dx = \dfrac{1}{3}\left[\dfrac{2}{3}x^{\frac{3}{2}}\right]_{1}^{4}$

$= \dfrac{1}{3}\left(\dfrac{16}{3}\right) - \dfrac{1}{3}\left(\dfrac{2}{3}\right)$

$= \dfrac{14}{9}$ (1.56)

9 $\dfrac{1}{9-4}\displaystyle\int_{4}^{9} x^{-\frac{1}{2}}\,dx = \dfrac{1}{5}\left[2x^{\frac{1}{2}}\right]_{4}^{9}$

$= \dfrac{1}{5}(6) - \dfrac{1}{5}(4)$

$= \dfrac{2}{5}$ (0.4)

10 a $f(x) = \dfrac{2\sqrt{x}+3x}{2x} = x^{-\frac{1}{2}} + \dfrac{3}{2},\ A=1,\ B=\dfrac{3}{2},\ c=-\dfrac{1}{2}$

b $\dfrac{1}{9-1}\displaystyle\int_{1}^{9} x^{-\frac{1}{2}} + \dfrac{3}{2}\,dx = \dfrac{1}{8}\left[2x^{\frac{1}{2}} + \dfrac{3}{2}x\right]_{1}^{9}$

$= \dfrac{1}{8}\left(6 + \dfrac{27}{2}\right) - \dfrac{1}{8}\left(2 + \dfrac{3}{2}\right)$

$= 2$

11 a $g(x) = \dfrac{3x-x^2}{5\sqrt{x}} = \dfrac{3}{5}x^{\frac{1}{2}} - \dfrac{1}{5}x^{\frac{3}{2}}\ \ A=\dfrac{3}{5},\ B=\dfrac{1}{5},\ c=\dfrac{1}{2},\ d=\dfrac{3}{2}$

b $\dfrac{8}{25} - \dfrac{6}{25}\sqrt{2}$

12 a $-\dfrac{13}{3}$ (−4.33)

b $-\dfrac{229}{200}$ (−1.145)

c $\dfrac{1}{12}$ (0.0833)

d $\dfrac{28}{5}$ (5.6)

13 a $1 - 2T + \dfrac{4}{3}T^2$

b $4T^2 + 4T + \dfrac{7}{3}$

14 $\dfrac{26}{9}\sqrt{2}$

15 $\dfrac{45}{4}a^3 - \dfrac{3}{8a^3}$

16 $-\dfrac{1}{6}X^4$

Exercise 8.2B

1 a $3.6\,\text{ms}^{-1}$

b $\dfrac{1}{2}\left[\dfrac{2}{5}t^3 - \dfrac{1}{5}t\right]_{1}^{3} = \dfrac{1}{2}\left(\dfrac{54}{5} - \dfrac{3}{5}\right) - \dfrac{1}{2}\left(\dfrac{2}{5} - \dfrac{1}{5}\right)$

$= 5\,\text{ms}^{-2}$

2 a $\dfrac{1}{5-0}\displaystyle\int_{0}^{5} \dfrac{3t^2-5}{5}\,dt = \dfrac{1}{25}\left[t^3 - 5t\right]_{0}^{5}$

$= \dfrac{1}{25}(125-25) - 0$

$= 4\ \text{ms}^{-1}$, as required

b $3\,\text{ms}^{-1}$

3 $s = 2t^{\frac{3}{2}}$

$v = 3t^{\frac{1}{2}}$

$\left(a = \dfrac{3}{2}t^{-\frac{1}{2}}\right)$

Mean acceleration $= \dfrac{1}{1-\dfrac{1}{2}}\displaystyle\int_{1}^{4}\dfrac{3}{2}t^{-\frac{1}{2}}\,dt$

$= 2\left[3t^{\frac{1}{2}}\right]_{0.5}^{1}$

$= 2\left(3-\dfrac{3\sqrt{2}}{2}\right)$

$= 6-3\sqrt{2}\ \text{m s}^{-2}$

4 a $\dfrac{1}{3-1}\displaystyle\int_{1}^{3}\dfrac{t}{10}-10t^{-3}\,dt = \dfrac{1}{2}\left[\dfrac{t^2}{20}+5t^{-2}\right]_{1}^{3}$

$= \dfrac{1}{2}\left(\dfrac{9}{20}+\dfrac{5}{9}\right)-\dfrac{1}{2}\left(\dfrac{1}{20}+5\right)$

$= -\dfrac{91}{45}\ \text{m s}^{-2}$

b $\dfrac{11}{6}$ (1.83) m s^{-1}

5 Mean value $= \dfrac{1}{b-a}\displaystyle\int_{a}^{b}mx+c\,dx$

$= \dfrac{1}{b-a}\left[\dfrac{mx^2}{2}+cx\right]_{a}^{b}$

$= \dfrac{1}{b-a}\left(\dfrac{mb^2}{2}+cb\right)-\dfrac{1}{b-a}\left(\dfrac{ma^2}{2}+ca\right)$

$= \dfrac{1}{b-a}\left(\dfrac{m}{2}(b^2-a^2)+c(b-a)\right)$

$= \dfrac{m(b+a)(b-a)}{2(b-a)}+\dfrac{c(b-a)}{b-a}$

$= \dfrac{m(b+a)}{2}+c$, as required

6 Mean value $= \dfrac{1}{a-0}\displaystyle\int_{0}^{a}x^2\,dx$

$= \dfrac{1}{a}\left[\dfrac{x^3}{3}\right]_{0}^{a}$

$= \dfrac{1}{a}\dfrac{a^3}{3}$

$= \dfrac{a^2}{3}$, as required

7 a

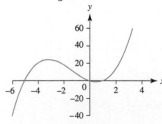

b $\dfrac{443}{6}$ (73.8) square units

c i $\dfrac{175}{12}$ **ii** $-\dfrac{11}{12}$

8 $\dfrac{49}{36}$

9 -8

10 7

11 6, -9

12 $b=4$
$\Rightarrow a=-1$

13 $\dfrac{1}{3-1}\displaystyle\int_{1}^{3}x^4-2x^3+3x-5\,dx = \dfrac{1}{2}\left[\dfrac{x^5}{5}-\dfrac{x^4}{2}+\dfrac{3}{2}x^2-5x\right]_{1}^{3}$

$= \dfrac{1}{2}\left(\dfrac{243}{5}-\dfrac{81}{2}+\dfrac{27}{2}-15\right)-\dfrac{1}{2}\left(\dfrac{1}{5}-\dfrac{1}{2}+\dfrac{3}{2}-5\right)$

$= \dfrac{26}{5}$

$\dfrac{1}{-1--2}\left[\dfrac{x^5}{5}-\dfrac{x^4}{2}+\dfrac{3}{2}x^2-5x\right]_{-2}^{-1}$

$= \left(-\dfrac{1}{5}-\dfrac{1}{2}+\dfrac{3}{2}+5\right)-\left(-\dfrac{32}{5}-8+6+10\right)$

$= \dfrac{21}{5}$

So the mean value in the range [1, 3] is bigger by 1

14 Mean speed of A: $\dfrac{1}{\dfrac{1}{2}}\displaystyle\int_{0}^{1.5}t^2+t\ dt = 2\left[\dfrac{t^3}{3}+\dfrac{t^2}{2}\right]_{0}^{0.5}$

$= \dfrac{1}{3}$

Mean speed of B: $\dfrac{1}{\dfrac{1}{2}}\displaystyle\int_{0}^{0.5}t^{\frac{1}{2}}\,dt = 2\left[\dfrac{2}{3}t^{\frac{3}{2}}\right]_{0}^{0.5}$

$= \dfrac{\sqrt{2}}{3}$

B is $\dfrac{1}{3}(\sqrt{2}-1)\ \text{m s}^{-1}$ faster than A

15 $\dfrac{1}{3-\dfrac{1}{2}}\displaystyle\int_{\frac{1}{2}}^{3}\ln x\,dx = \dfrac{2}{5}\left[x\ln x-x\right]_{\frac{1}{2}}^{3}$

$= \dfrac{2}{5}(3\ln 3-3)-\dfrac{2}{5}\left(\dfrac{1}{2}\ln\dfrac{1}{2}-\dfrac{1}{2}\right)$

$= \dfrac{2}{5}\left(\ln 3^3-3-\ln\left(\dfrac{1}{2}\right)^{\frac{1}{2}}+\dfrac{1}{2}\right)$

$= \dfrac{2}{5}\left(\ln 27-\ln\left(\dfrac{1}{\sqrt{2}}\right)-\dfrac{5}{2}\right)$

$= \dfrac{2}{5}\ln(27\sqrt{2})-1$

16 $\dfrac{1}{8}\ln\left(\dfrac{4k^2+8k+1}{4k^2-3}\right)$

17 $2+\dfrac{1}{2}\ln 3$

18 a $\dfrac{3}{2\pi}$

b 0 (or can use graph)

c $\dfrac{1}{3}(e^3-1)$

d $\dfrac{4}{3}e^3+\dfrac{2}{3}$

Exercise 8.3A

1 a $\dfrac{\pi}{4}\ln\left(\dfrac{5}{2}\right)$ cubic units

b $\dfrac{2}{3}\sqrt{3}\pi$ cubic units

c $\dfrac{127}{14}\pi$ cubic units

d $\frac{9}{4}\pi\left(1-\frac{1}{e}\right)$ cubic units or $\frac{9\pi(e-1)}{4e}$ cubic units

2 $\frac{\pi^2}{2}$ (4.93) cubic units

3 **a** $=\frac{\pi}{18}(24\ln 2-7)$ cubic units

b $\frac{\pi}{2}(e^1-2)$ cubic units

4 $\frac{9\pi}{20}$ cubic units

5 $\frac{\pi}{12}(3\sqrt{3}+2\pi)$ cubic units

6 **a** $\frac{288}{7}\pi$ cubic units

b $\frac{48}{5}\pi$ cubic units

7 $\pi\left(\frac{15}{2}-8\ln 2\right)$ cubic units

8 3π cubic units

9 $\frac{\pi}{2}\int_0^1 (\sqrt{\tan y})^2 dy = \frac{\pi}{2}\int_0^1 \tan y\, dy$

$= \frac{\pi}{2}\left[\ln\sec y\right]_{\frac{\pi}{6}}^{\frac{\pi}{3}}$

$= \frac{\pi}{2}\left(\ln 2 - \ln\left(\frac{2\sqrt{3}}{3}\right)\right)$

$= \frac{\pi}{4}\ln 3 \qquad \left(A = \frac{1}{4}\right)$

10 $\frac{\sqrt{3}}{4}\pi$ cubic units

Exercise 8.3B

1 **a** $\frac{27}{40}\pi$ cubic units

b $\frac{27}{16}\pi$ cubic units

2 **a** $\frac{375}{2}\pi$ cubic units

b $\frac{25}{2}\pi(8-\sqrt{2})$ cubic units

3 **a** $\frac{\pi}{12}$ cubic units

b $\frac{\pi}{21}$ cubic units

4 **a** $\sin^3\theta = \sin\theta(\sin^2\theta)$
$= \sin\theta(1-\cos^2\theta)$
$= \sin\theta - \sin\theta\cos^2\theta$, as required

b $\frac{2}{3}\pi$ cubic units

5 **a** **i** $\frac{4\sqrt{2}}{5}\pi$ cubic units

ii $\frac{9}{4}\pi$ cubic units

b **i** $y=2\cos^2 t-1$

$\Rightarrow y=2\left(\frac{x}{3}\right)^2-1$

$\Rightarrow y=\frac{2}{9}x^2-1$

When $y=0$, $x^2=\frac{9}{2}$

$\Rightarrow x=\pm\frac{3}{\sqrt{2}}$

$V=2\times\frac{\pi}{2}\int_0^{\frac{3}{\sqrt{2}}}\left(\frac{2}{9}x^2-1\right)^2 dx$

$=\pi\int_0^{\frac{3}{\sqrt{2}}}\left(\frac{4}{81}x^4-\frac{4}{9}x^2+1\right)dx$

$=\pi\left[\frac{4}{405}x^5-\frac{4}{27}x^3+x\right]_0^{\frac{3}{\sqrt{2}}}$

$=\pi\left(\frac{3}{10}\sqrt{2}-\sqrt{2}+\frac{3\sqrt{2}}{2}\right)-0$

$=\frac{4\sqrt{2}}{5}\pi$, as required

ii $y=\frac{2}{9}x^2-1\Rightarrow x=\sqrt{\frac{9}{2}(y+1)}$

$V=\pi\int_{-1}^{0}\left(\sqrt{\frac{9}{2}(y+1)}\right)^2 dy$

$=\frac{9}{2}\pi\int_{-1}^{0}(y+1)dy$

$=\frac{9}{2}\pi\left[\frac{y^2}{2}+y\right]_{-1}^{0}$

$=0-\frac{9}{2}\pi\left(\frac{1}{2}-1\right)$

$=\frac{9}{4}\pi$, as required

6 **a** $\int\cos^3\theta\,d\theta=\int\cos\theta(1-\sin^2\theta)d\theta$

$=\int\cos\theta-\sin^2\theta\cos\theta\,d\theta$

$=\sin\theta-\frac{1}{3}\sin^3\theta+c$

b 3π cubic units

Exercise 8.4A

1 **a** $-\frac{1}{\sqrt{1-x^2}}$

b $\frac{1}{1+x^2}$

c $\frac{2}{\sqrt{1-4x^2}}$

d $-\frac{5}{\sqrt{1-25x^2}}$

e $\frac{1}{x^2-2x+2}$

f $\frac{2}{\sqrt{1-x^2}}$

g $-\frac{3}{\sqrt{9-x^2}}$

h $-\frac{3}{\sqrt{4x-x^2-3}}$

i $-\dfrac{2x}{\sqrt{1-x^4}}$

j $\dfrac{x}{\sqrt{1-x^2}}+\arcsin x$

2 Let $y=\operatorname{arcsec}x$

$\Rightarrow x=\sec y$

$=(\cos y)^{-1}$

$\dfrac{dx}{dy}=-(\cos y)^{-2}(-\sin y)$

$=\dfrac{\sin y}{\cos^2 y}$

$=\tan y\sec y$

$=\sec y\sqrt{\sec^2 y-1}$

$=x\sqrt{x^2-1}$

$\dfrac{dy}{dx}=\dfrac{1}{x\sqrt{x^2-1}}$, as required

3 Let $y=\operatorname{arccosec}x$

$\Rightarrow x=\operatorname{cosec}y$

$=(\sin y)^{-1}$

$\dfrac{dx}{dy}=-(\sin y)^{-2}(\cos y)$

$=-\dfrac{\cos y}{\sin^2 y}$

$=-\cot y\operatorname{cosec}y$

$=-\operatorname{cosec}y\sqrt{\operatorname{cosec}^2 y-1}$

$=-x\sqrt{x^2-1}$

$\dfrac{dy}{dx}=-\dfrac{1}{x\sqrt{x^2-1}}$, as required

4 Let $y=\operatorname{arccot}x$

$\Rightarrow x=\cot y$

$=(\tan y)^{-1}$

$\dfrac{dx}{dy}=-(\tan y)^{-2}(\sec^2 y)$

$=-\dfrac{\sec^2 y}{\tan^2 y}$

$=-\operatorname{cosec}^2 y$

$=-(1+\cot^2 y)$

$=-(1+x^2)$

$\dfrac{dy}{dx}=-\dfrac{1}{1+x^2}$, as required

5 a $e^x\arctan x+\dfrac{e^x}{1+x^2}$

b $-\dfrac{6x}{\sqrt{6x^2-9x^4}}$

c $-\dfrac{2\sin x}{\sqrt{1-(2x)^2}}+\cos x\arccos 2x$

d $\dfrac{2(\arcsin x)}{\sqrt{1-x^2}}$

e $\dfrac{e^x}{\sqrt{1-e^{2x}}}$

f $\dfrac{2}{1+4x^2}e^{\arctan 2x}$

6 $x=\cos u\Rightarrow\dfrac{dx}{du}=-\sin u$

$\displaystyle\int\dfrac{1}{\sqrt{1-x^2}}dx=\int\dfrac{1}{\sqrt{1-\cos^2 u}}(-\sin u)du$

$\displaystyle=-\int\dfrac{\sin u}{\sin u}du$

$\displaystyle=-\int 1du$

$=-u+c$

$=-\arccos x+c$, as required

7 $x=\tan u\Rightarrow\dfrac{dx}{du}=\sec^2 u$

$\displaystyle\int\dfrac{1}{1+x^2}dx=\int\dfrac{1}{1+\tan^2 u}(\sec^2 u)du$

$\displaystyle=\int\dfrac{\sec^2 u}{\sec^2 u}du$

$\displaystyle=\int 1du$

$=u+c$

$=\arctan x+c$, as required

8 $\dfrac{1}{3}\arcsin 3x+c$

9 $5\arcsin\left(\dfrac{x}{5}\right)+c$

10 $\dfrac{1}{3}\arctan\left(\dfrac{x}{3}\right)+c$

Exercise 8.4B

1 a $\arcsin\left(\dfrac{x}{3}\right)+c$

b $\arcsin\left(\dfrac{x}{10}\right)+c$

c $-2\arcsin\left(\dfrac{x}{6}\right)+c$

d $-4\arcsin\left(\dfrac{x}{\sqrt{8}}\right)+c$

e $\dfrac{1}{5}\arctan\left(\dfrac{x}{5}\right)+c$

f $\dfrac{1}{7}\arctan\left(\dfrac{x}{7}\right)+c$

g $-\dfrac{3}{\sqrt{2}}\arctan\left(\dfrac{x}{\sqrt{2}}\right)+c$

h $\sqrt{6}\arctan\left(\dfrac{x}{\sqrt{6}}\right)+c$

i $\dfrac{1}{2}x\sqrt{1-x^2}+\dfrac{1}{2}\arcsin x+c$

j $\dfrac{1}{2}x\sqrt{16-x^2}+8\arcsin\left(\dfrac{x}{4}\right)+c$

2 a $8\arctan\left(\dfrac{x}{8}\right)+c$

b $5\arctan(5x)+c$

c $\sqrt{2}\arctan\left(\dfrac{x}{\sqrt{8}}\right)+c$

d $\dfrac{2\sqrt{3}}{3}\arctan\left(\dfrac{x}{\sqrt{3}}\right)+c$

e $3\arcsin\left(\dfrac{x}{3}\right)+c$

f $\dfrac{1}{2}\arcsin\left(\dfrac{x}{2}\right)+c$

g $-6\arcsin\left(\dfrac{x}{6}\right)+c$

h $\arcsin\left(\dfrac{x}{\sqrt{2}}\right)+c$

i $\dfrac{1}{18}x\sqrt{81-x^2}+\dfrac{9}{2}\arcsin\left(\dfrac{x}{9}\right)+c$

j $\dfrac{3}{2}x\sqrt{4-x^2}+6\arcsin\left(\dfrac{x}{2}\right)+c$

3 a $\dfrac{1}{2}\arctan\left(\dfrac{x+1}{2}\right)+c$

b $\arctan(x-3)+c$

c $\dfrac{1}{2}\arctan\left(\dfrac{x-7}{2}\right)+c$

d $\sqrt{2}\arctan\left(\dfrac{x+6}{\sqrt{2}}\right)+c$

e $\arcsin\left(\dfrac{x+4}{6}\right)+c$

f $\arcsin\left(\dfrac{x-1}{\sqrt{2}}\right)+c$

g $-2\arcsin\left(\dfrac{x-4}{4}\right)+c$

h $\arcsin\left(\dfrac{x+2}{\sqrt{8}}\right)+c$

i $\dfrac{x-2}{2}\sqrt{12+4x-x^2}+8\arcsin\left(\dfrac{x-2}{4}\right)+c$

j $\dfrac{1}{2}(x+3)\sqrt{16-6x-x^2}+\dfrac{25}{2}\arcsin\left(\dfrac{x+3}{5}\right)+c$

4 $\dfrac{\pi}{12}$

Exercise 8.5A

1 a

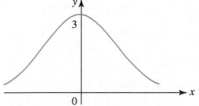

Domain: $x\in\mathbb{R}$, $x\neq 1$, range: $y\in\mathbb{R}$, $y\neq 0$

b

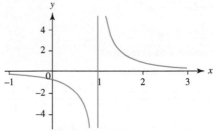

Domain: $x\in\mathbb{R}$, range: $y\in\mathbb{R}$, $0<y\leq 3$

c

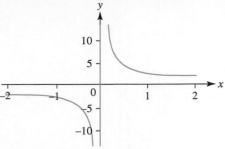

Domain: $x\in\mathbb{R}$, $x\neq 0$, range: $y\in\mathbb{R}$, $y<-2, y>2$

d

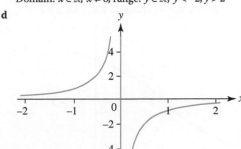

Domain: $x\in\mathbb{R}$, $x\neq 0$, range: $y\in\mathbb{R}$, $y\neq 0$

e

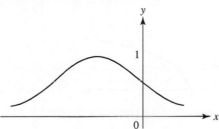

Domain: $x\in\mathbb{R}$, range: $y\in\mathbb{R}$, $0<y\leq 1$

f

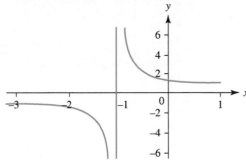

Domain: $x\in\mathbb{R}$, $x\neq -1$, range: $y\in\mathbb{R}$, $y<-1, y>1$

g

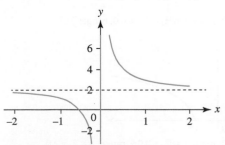

Domain: $x\in\mathbb{R}$, $x\neq 0$, range: $y\in\mathbb{R}$, $y\neq 2$

h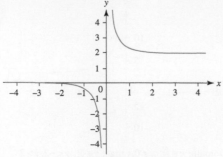

Domain: $x \in \mathbb{R}$, $x \neq 0$, range: $y \in \mathbb{R}$, $y < 0$, $y > 2$

2 a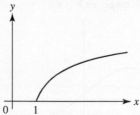

Domain: $x \in \mathbb{R}$, range: $y \in \mathbb{R}$

b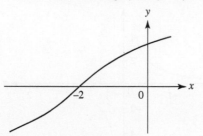

Domain: $x \in \mathbb{R}$, $x \geq 1$ range: $y \in \mathbb{R}$, $y \geq 0$

c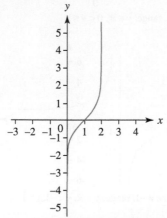

Domain: $x \in \mathbb{R}$, $-2 < x < 0$, range: $y \in \mathbb{R}$

d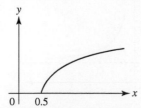

Domain: $x \in \mathbb{R}$, $x \geq 0.5$, range: $y \in \mathbb{R}$, $y \geq 0$

e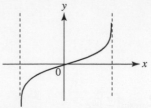

Domain: $x \in \mathbb{R}$, $-4 < x < 4$, range: $y \in \mathbb{R}$

f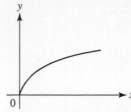

Domain: $x \in \mathbb{R}$, $x \geq 0$, range: $y \in \mathbb{R}$, $y \geq 0$

3 a $\cosh^2 x + \sinh^2 x = \left(\dfrac{e^x + e^{-x}}{2}\right)^2 + \left(\dfrac{e^x - e^{-x}}{2}\right)^2$

$= \dfrac{1}{4}(e^{2x} + 2 + e^{-2x}) + \dfrac{1}{4}(e^{2x} - 2 + e^{-2x})$

$= \dfrac{1}{4}(2e^{2x} + 2e^{-2x})$

$= \dfrac{1}{2}(e^{2x} + e^{-2x})$

$= \cosh 2x$, as required

b $\sinh A \cosh B + \sinh B \cosh A$

$= \left(\dfrac{e^A - e^{-A}}{2}\right)\left(\dfrac{e^B + e^{-B}}{2}\right) + \left(\dfrac{e^B - e^{-B}}{2}\right)\left(\dfrac{e^A + e^{-A}}{2}\right)$

$= \dfrac{1}{4}(e^{A+B} + e^{A-B} - e^{-(A-B)} - e^{-(A+B)})$

$\quad + \dfrac{1}{4}(e^{A+B} + e^{-(A-B)} - e^{A-B} - e^{-(A+B)})$

$= \dfrac{1}{4}(2e^{A+B} - 2e^{-(A+B)})$

$= \dfrac{1}{2}(e^{A+B} - e^{-(A+B)})$

$= \sinh(A + B)$, as required

c $\cosh A \cosh B - \sinh B \sinh A$

$= \left(\dfrac{e^A + e^{-A}}{2}\right)\left(\dfrac{e^B + e^{-B}}{2}\right) - \left(\dfrac{e^B - e^{-B}}{2}\right)\left(\dfrac{e^A - e^{-A}}{2}\right)$

$= \dfrac{1}{4}(e^{A+B} + e^{A-B} + e^{-(A-B)} + e^{-(A+B)})$

$\quad - \dfrac{1}{4}(e^{A+B} - e^{-(A-B)} - e^{A-B} + e^{-(A+B)})$

$= \dfrac{1}{4}(2e^{A-B} + 2e^{-(A-B)})$

$= \dfrac{1}{2}(e^{A-B} + e^{-(A-B)})$

$= \cosh(A - B)$, as required

4 a $\pm\ln(3 + 2\sqrt{2})$

b $\ln(4 \pm \sqrt{15})$

c $x = \dfrac{1}{2}\ln(3 + 2\sqrt{2})$ or $x = \dfrac{1}{2}\ln(3 - 2\sqrt{2})$

d $x = \frac{1}{2}\ln\left(\frac{1+\frac{1}{2}}{1-\frac{1}{2}}\right) = \frac{1}{2}\ln 3$

e $\ln\left(\frac{1}{2}\right)$

5 $\cosh x = \frac{1}{2}(e^x + e^{-x})$

So $\dfrac{d(\cosh x)}{dx} = \frac{1}{2}(e^x - e^{-x})$

$\qquad = \sinh x$

6 $\tanh x = \dfrac{e^x - e^{-x}}{e^x + e^{-x}}$

So $\dfrac{d(\tan x)}{dx} = \dfrac{(e^x + e^{-x})(e^x + e^{-x}) - (e^x - e^{-x})(e^x - e^{-x})}{(e^x + e^{-x})^2}$

$\qquad = \dfrac{(e^{2x} + 2 + e^{-2x}) - (e^{2x} - 2 + e^{-2x})}{(e^x + e^{-x})^2}$

$\qquad = \dfrac{4}{(e^x + e^{-x})^2}$

$\qquad = \left(\dfrac{2}{e^x + e^{-x}}\right)^2$

$\qquad = \dfrac{1}{\sinh^2 x}$

$\qquad = \operatorname{sech}^2 x$

7 a $-\coth x \operatorname{cosech} x$

b $-\operatorname{cosech}^2 x$

c $2\cosh 2x$

d $\frac{1}{3}\sinh\left(\frac{x}{3}\right)$

e $\operatorname{sech}^2(x-2)$

f $2x\cosh x^2$

g $2\sinh(2x-3)$

h $\sinh 2x$

i $\tanh x + x\operatorname{sech}^2 x$

j $\sqrt{\operatorname{sech} x} + \frac{1}{2}x\tanh x\sqrt{\operatorname{sech} x}$

8 a $\tanh x + c$

b $-\operatorname{cosech} x$

c $\frac{1}{2}\sinh 2x + c$

d $3\cosh\left(\frac{x}{3}\right) + c$

e $\frac{1}{3}\sinh^3 x + c$

f $\frac{1}{4}\sinh 2x + \frac{1}{2}x + c$

g $\frac{1}{4}\ln\sinh 4x + c$

h $-\frac{1}{5}\coth 5x + c$

9 Let $y = \operatorname{arsinh} x$. Then $x = \sinh y$

$\dfrac{dx}{dy} = \cosh y$

$\dfrac{dy}{dx} = \dfrac{1}{\cosh y}$

$\qquad = \dfrac{1}{\pm\sqrt{1 + \sinh^2 y}}$

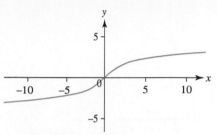

Gradient always positive

$\Rightarrow \dfrac{dy}{dx} = \dfrac{1}{\sqrt{1 + x^2}}$, as required

10 Let $y = \operatorname{artanh} 2x$. Then $x = \frac{1}{2}\tanh y$

$\dfrac{dx}{dy} = \frac{1}{2}\operatorname{sech}^2 y$

$\dfrac{dy}{dx} = \dfrac{2}{\operatorname{sech}^2 y}$

$\qquad = \dfrac{2}{1 - \tanh^2 y}$

$\qquad = \dfrac{2}{1 - (2x)^2}$

$\qquad = \dfrac{2}{1 - 4x^2}$, as required

11 Let $y = \operatorname{arcoth} x$ then $x = \coth y$

$\dfrac{dx}{dy} = -\operatorname{cosech}^2 y$

$\dfrac{dy}{dx} = \dfrac{1}{-\operatorname{cosech}^2 y}$

$\qquad = \dfrac{1}{-(\coth^2 y - 1)}$

$\qquad = \dfrac{1}{-(x^2 - 1)}$

$\qquad = \dfrac{1}{1 - x^2}$, as required

12 Let $y = \operatorname{arcosech} x$. Then $x = \operatorname{cosech} y$

$\dfrac{dx}{dy} = -\coth y \operatorname{cosech} y$

$\dfrac{dy}{dx} = \dfrac{1}{-\coth y \operatorname{cosech} y}$

$\qquad = \dfrac{1}{\pm\sqrt{\operatorname{cosech}^2 y + \operatorname{cosech} y}} = \pm\dfrac{1}{x\sqrt{x^2 + 1}}$

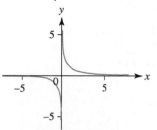

Gradient always negative

$\Rightarrow \dfrac{dy}{dx} = -\dfrac{1}{|x|\sqrt{x^2 + 1}}$, as required

13 a $\dfrac{4}{\sqrt{16x^2 - 1}}$

b $\dfrac{1}{\sqrt{x^2 + 25}}$

c $\dfrac{2x}{\sqrt{x^4-1}}$

d $\dfrac{e^x}{\sqrt{e^{2x}+1}}$

e $\dfrac{\cosh x}{\sqrt{\sinh^2 x-1}}$

f $e^x \operatorname{arsinh}(x^2-1)+\dfrac{2xe^x}{\sqrt{x^4-2x^2+2}}$

Exercise 8.5B

1 a $\operatorname{arsinh}\left(\dfrac{x}{7}\right)+c$

b $\operatorname{arcosh}\left(\dfrac{x}{9}\right)+c$

c $\operatorname{arsinh}\left(\dfrac{x}{4}\right)+c$

d $-3\operatorname{arcosh}\left(\dfrac{x}{3}\right)+c$

e $\operatorname{arsinh}\left(\dfrac{x+3}{4}\right)+c$

f $\operatorname{arsinh}(x-5)+c$

g $\operatorname{arcosh}\left(\dfrac{x+7}{5}\right)+c$

h $3\operatorname{arcosh}\left(\dfrac{x-12}{10}\right)+c$

2 a 0.398

b 23.7

3 a $\ln(1+\sqrt{2})$

b $\dfrac{1}{\sqrt{3}}\ln 2$ as integral is positive

4 a $\dfrac{1}{9}\sqrt{16+9x^2}+\dfrac{1}{3}\operatorname{arsinh}\left(\dfrac{3x}{4}\right)+c$

b $6\operatorname{arcosh}\left(\dfrac{x}{\sqrt{12}}\right)-2\sqrt{x^2-12}+c$

c $\sqrt{8}\arcsin\left(\dfrac{x}{\sqrt{8}}\right)-\dfrac{1}{2}\sqrt{16-2x^2}+c$

d $-7\ln\left(49+x^2\right)-5\arctan\left(\dfrac{x}{7}\right)+c$

5 Let $v=\operatorname{arsinh} u$, $\dfrac{dv}{dx}=\dfrac{1}{\sqrt{x^2+1}}$, $\dfrac{du}{dx}=1$, $u=x$

$\displaystyle\int \operatorname{arsinh} x\,dx = x\operatorname{arsinh} x-\int \dfrac{x}{\sqrt{x^2+1}}\,dx$

$\qquad\qquad = x\operatorname{arsinh} x-\sqrt{x^2+1}+c$

6 a $x\operatorname{arcosh} x-\sqrt{x^2-1}+c$

b $x\operatorname{arcoth} x+\dfrac{1}{2}\ln\sqrt{x^2-1}+c$

7 A: $\left[\operatorname{arcosh}\left(\dfrac{x}{a}\right)\right]_t^{2a} = \operatorname{arcosh}(2)-\operatorname{arcosh}\left(\dfrac{t}{a}\right)$

$\qquad\qquad = \ln(2+\sqrt{4-1})-\ln\left(\dfrac{t}{a}+\sqrt{\left(\dfrac{t}{a}\right)^2-1}\right)$

As $t\to a$, $\dfrac{t}{a}\to 1$, so $\dfrac{t}{a}+\sqrt{\left(\dfrac{t}{a}\right)^2-1}\to 1$

so $\ln\left(\dfrac{t}{a}+\sqrt{\left(\dfrac{t}{a}\right)^2-1}\right)\to\ln 1=0$

So the integral exists and is equal to $\ln(2+\sqrt{3})$

B: $\left[\operatorname{arcosh}\left(\dfrac{x}{a}\right)\right]_{2a}^t = \operatorname{arcosh}\left(\dfrac{t}{a}\right)-\operatorname{arcosh}(2)$

$\qquad = \ln\left(\dfrac{t}{a}+\sqrt{\left(\dfrac{t}{a}\right)^2-1}\right)-\ln(2+\sqrt{3})$

As $t\to\infty$, $\dfrac{t}{a}\to\infty$, so $\dfrac{t}{a}+\sqrt{\left(\dfrac{t}{a}\right)^2-1}\to\infty$

So $\ln\left(\dfrac{t}{a}+\sqrt{\left(\dfrac{t}{a}\right)^2-1}\right)\to\infty$

So the integral does not exist

Exercise 8.6A

1 a $5x+\ln(x-2)+3\ln(x+4)+c$

b $-3x+2\ln(x+1)-3\ln(x-5)+c$

c $\dfrac{5}{2}x^2+x+2\ln(x+3)-3\ln(x-3)+c$

d $8x+3\ln(x+3)-2\ln(x+8)+3\ln x+c$

e $\dfrac{x^3}{3}+\dfrac{x^2}{2}-3x+\ln(x-9)-2\ln(x+1)+c$

f $\dfrac{9}{2}x^2+27x+\ln(x+1)+80\ln(x-2)-\dfrac{48}{x-2}+c$

g $-\dfrac{x^2}{3}-x+\dfrac{5}{3}\ln(3x+4)+\dfrac{3}{4}\ln(3-4x)+c$

2 a $3\ln\left(\dfrac{x+3}{\sqrt{x^2+1}}\right)+2\arctan x+c$

b $\ln\left(\dfrac{x-3}{\sqrt{x^2+25}}\right)+\arctan\left(\dfrac{x}{5}\right)+c$

c $\ln\left(\dfrac{\sqrt{x^2+36}}{6-x}\right)+\dfrac{5}{6}\arctan\left(\dfrac{x}{6}\right)+c$

d $\ln\dfrac{x^2+2}{(1-x)^2}+\dfrac{5}{\sqrt{2}}\arctan\left(\dfrac{x}{\sqrt{2}}\right)+c$

e $\ln\dfrac{(x+1)(x-2)^2}{(x^2+9)\sqrt{x^2+9}}-\arctan\left(\dfrac{x}{3}\right)+c$

f $\ln\dfrac{x^2+16}{(x-1)^2}-\dfrac{3}{x-1}-\dfrac{1}{4}\arctan\left(\dfrac{x}{4}\right)+c$

3 a $\ln\dfrac{x}{\sqrt{x^2+49}}+\dfrac{3}{7}\arctan\left(\dfrac{x}{7}\right)+c$

b $3\ln\dfrac{\sqrt{x^2+64}}{x}+\dfrac{1}{8}\arctan\left(\dfrac{x}{8}\right)+c$

c $\ln\dfrac{3x-1}{\sqrt{x^2+4}}+\dfrac{3}{2}\arctan\dfrac{x}{2}+c$

4 a $\dfrac{1}{x-3}+\dfrac{1}{(x-3)^2}-\dfrac{x+4}{x^2+5}$

b $\ln\dfrac{x-3}{\sqrt{x^2+5}}-\dfrac{1}{x-3}-\dfrac{4\sqrt{5}}{5}\arctan\dfrac{\sqrt{5}}{5}x+c$

1 a $\ln\dfrac{4}{3}$

b $\ln\left(\dfrac{1}{6}\right)$ or $-\ln 6$

c $2+\ln\left(\dfrac{5}{2}\right)$

d $\dfrac{1}{2}\ln 10$ or $-\dfrac{1}{2}\ln\dfrac{1}{10}$

2 $\dfrac{2x+12}{2x^3-x^2+6x-3}=\dfrac{2(x+6)}{x^2(2x-1)+3(2x-1)}$

$=\dfrac{2(x+6)}{(x^2+3)(2x-1)}=\dfrac{Ax+B}{x^2+3}+\dfrac{C}{2x-1}$

$2(x+6)=(Ax+B)(2x-1)+C(x^2+3)$

Let $x=\dfrac{1}{2}$. Then $13=\dfrac{13}{4}C\Rightarrow C=4$

$x^2:0=2A+C\Rightarrow A=-2$

$1:12=-B+3C\Rightarrow B=0$

A: $\displaystyle\int_1^a \dfrac{4}{2x-1}-\dfrac{2x}{x^2+3}\,dx=\left[2\ln(2x-1)-\ln(x^2+3)\right]_1^a$

$=\left[\ln\dfrac{(2x-1)^2}{x^2+3}\right]_1^a$

$=\ln\dfrac{(2a-1)^2}{a^2+3}-\ln\dfrac{1}{4}$

$\dfrac{(2a-1)^2}{a^2+3}=\dfrac{4a^2-4a+1}{a^2+3}\to 4$ as $a\to\infty$

So integral exists and is $2\ln 4$

B: Point of discontinuity at $x=\dfrac{1}{2}$

$\dfrac{(2a-1)^2}{a^2+3}\to 0$ as $a\to\dfrac{1}{2}\Rightarrow \ln\dfrac{(2a-1)^2}{a^2+3}\to\infty$ as $a\to\dfrac{1}{2}$

Hence integral does not exist.

3 a $\dfrac{1}{x}+\dfrac{2-x}{x^2+a}$

b $\displaystyle\int_0^a \dfrac{1}{x}\,dx$ does not converge for any value of a, and $\displaystyle\int \dfrac{2-x}{x^2+a}$

converges for any value of a therefore $\displaystyle\int_0^a \dfrac{2x+a}{x^3+ax}\,dx$ does not exist.

4 $\dfrac{3-a}{(x+a)(x+3)}=\dfrac{A}{x+a}+\dfrac{B}{(x+3)}\Rightarrow A(x+3)+B(x+a)$

$=3-a\Rightarrow A=1,\ B=-1$

$\displaystyle\int \dfrac{3-a}{(x+a)(x+3)}\,dx=\int \dfrac{1}{x+a}-\dfrac{1}{(x+3)}\,dx$

$=\ln(x+a)-\ln(x+3)+c$

$=\ln\left(\dfrac{x+a}{x+3}\right)+c$

a $\dfrac{1}{(x+3)}$ is discontinuous at $x=-3$

so consider value when $x\to-3,\ \dfrac{x+a}{x+3}\to 0$

$\ln\left(\dfrac{x+a}{x+3}\right)\to\infty$, so undefined for all values of $a\neq-3$

b If $a<0$, then $\dfrac{x+a}{x+3}=0$ for a positive value of x

$\Rightarrow\ln\dfrac{x+a}{x+3}$ does not exist for a positive value of x,

so the integral does not exist.

If $a>0$, then $\ln\left(\dfrac{x+a}{x+3}\right)\to\ln 1=0$ as $x\to\infty$, so the

integral exists and is equal to 0

1 a i Max = 2, min = 0

ii

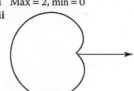

b i Max = 3, min = 1

ii

c i Max = 4, min = 0

ii

d i Max = 9, min = 3

ii

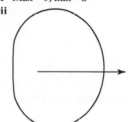

e i Max = 6, min = 4

ii

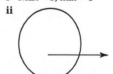

f i Max = 1, min = 0

ii

g i Max = a, min = 0

ii

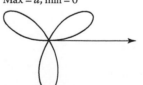

h i Max = b, min = 0

ii

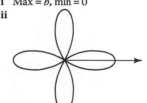

i **i** Max $= c$, min $= 0$

   ii

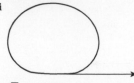

2 $\dfrac{4}{3}\pi + \sqrt{3}$ square units

3 $\dfrac{\pi^3}{48}$ square units

4 $\dfrac{\pi}{3}$ square units

5 $2\sqrt{2}$ square units

6 a $\dfrac{19}{2}\pi$ square units

  b 27π square units

  c $\dfrac{3}{2}\pi$ square units

  d 11π square units

7 a $\dfrac{\pi}{4}$ square units

  b $\dfrac{\pi}{4}$ square units

  c 4π square units

  d $\dfrac{25}{4}\pi$ square units

  e 2 square units

  f 5π square units

Exercise 8.7B

1 a $\cos\theta + \sin\theta = 2 + \sin\theta$

    $\Rightarrow \cos\theta = 2$ which has no solutions so they do not intersect

  b 4π square units

2 a $(0, 0)$ and $\left(\dfrac{\pi}{3}, \dfrac{\sqrt{3}}{2}\right)$

  b $\dfrac{3\sqrt{3}}{16} - \dfrac{\pi}{12}$ square units

3 $\dfrac{3\pi}{4} + \dfrac{3\sqrt{3}}{16}$ square units

4 a $\dfrac{5}{24}\pi - \dfrac{\sqrt{3}}{4}$ square units

  b $\dfrac{9}{4}\pi - 3\sqrt{3}$ square units

  c $\dfrac{7}{4}\pi$ square units

  d $\dfrac{\pi}{2} - \dfrac{3}{4}\sqrt{3}$ square units

  e $\dfrac{59}{3}\pi - \dfrac{19}{2}\sqrt{3}$ square units

5 a 2.514 units

  b 8.473 square units

6 $\dfrac{9\pi}{8}$ square units

7 $\dfrac{3\pi}{4}$ square units

8 $\dfrac{3\pi - 8}{2}$ square units

9 $\pi + \dfrac{3\sqrt{3}}{4}$ square units

10 $\dfrac{\pi - 2}{4}$ square units

11 $\pi - 3$ square units

Exercise 8.8A

1 a $\dfrac{(-x)^r}{r!}$ or $\dfrac{(-1)^r x^r}{r!}$

  b $(-1)^{r+1}\dfrac{(-2x)^r}{r}$ or $(-1)^r\dfrac{2^r x^r}{r}$

  c $(-1)^r\dfrac{\left(\dfrac{x}{5}\right)^{2r+1}}{(2r+1)!}$ or $(-1)^r\dfrac{x^{2r+1}}{5^{2r+1}(2r+1)!}$

  d $\dfrac{n(n-1)\ldots(n-1+r)(3x)^r}{r!}$ or $\dfrac{n(n-1)\ldots(n-1+r)(3^r)(x)^r}{r!}$

2 $f(x) = \sin x \rightarrow f(0) = \sin 0 = 0$

    $f'(x) = \cos x \rightarrow f'(0) = \cos 0 = 1$

    $f''(x) = -\sin x \rightarrow f''(0) = -\sin 0 = 0$

    $f'''(x) = -\cos x \rightarrow f'''(0) = -\cos 0 = -1$

    Series then goes back to $\sin x$ etc.

    Hence $\sin x \equiv x - \dfrac{x^3}{3!} + \dfrac{x^5}{5!} - \dfrac{x^7}{7!} + \ldots$

    The general term is $(-1)^r\dfrac{x^{2r+1}}{(2r+1)!}$

3 Since the differentiation of cos goes $\cos \rightarrow -\sin \rightarrow -\cos \rightarrow$ $\sin \rightarrow \cos$ etc. the expansion can be done in the same way

    and you get $\cos x \equiv 1 - \dfrac{x^2}{2!} + \dfrac{x^4}{4!} - \dfrac{x^6}{6!} + \ldots$

    The general term is $(-1)^r\dfrac{x^{2r}}{(2r)!}$

4 $-4x - \dfrac{16x^3}{3} - \dfrac{64}{5}x^5$

5 $1 - x^2 + \dfrac{x^4}{3} - \ldots$

6 $1 + \dfrac{3x^2}{2} - \dfrac{9x^4}{8} - \ldots$

7 $x + 2x^2 + \dfrac{11}{6}x^3 \ldots$

8 $x + \dfrac{x^2}{2} + \dfrac{x^3}{3} \ldots$

9 a 1

  b -1

  c 3

  d $\dfrac{1}{2}$

  e $-\dfrac{1}{4}$

  f $\dfrac{1}{3}$

  g -1

  h -11

10 a 1.5

  b $\dfrac{1}{3}$

  c -3

  d $-\dfrac{5}{8}$

  e 2

  f $\dfrac{\sqrt{2}}{2}$

1 a $4 - 16x + 64x^2 - 256x^3 + \dots$

  b $4x - 8x^2 + \dfrac{64}{3}x^3 - 64x^4 + \dots$

2 a -6

  b $\dfrac{1}{2}$

  c 8

  d -1

3 a 2

  b 0

  c 0

  d $-\dfrac{1}{5}$

  e $\dfrac{1}{4}$

4 a $x + \dfrac{x^2}{2} + \dfrac{5x^3}{6} + \dfrac{7x^4}{12} \dots$

  b $\sqrt{2}\left(1 - \dfrac{9}{4}x + \dfrac{143}{32}x^2 - \dfrac{1145}{128}x^3 + \dots\right)$

5 a $\dfrac{9}{(x+1)} - \dfrac{15}{(x+2)}$

  b $\dfrac{3}{2} - \dfrac{21x}{4} + \dfrac{57x^2}{8} - \dfrac{129x^3}{16} + \dots$

  c $\lim\limits_{x \to 0}\left(\dfrac{3(1-2x)}{(x+2)(1+x)}\right) = \dfrac{3}{2 \times 1} = \dfrac{3}{2}$

  When $x = 0$, $\dfrac{3}{2} - \dfrac{21x}{4} + \dfrac{57x^2}{8} - \dfrac{129x^3}{16} + \dots = \dfrac{3}{2}$

6 -1

7 $e^2(2 - 3x + 2x^2 - \dots); 2e^2$

8 $1 - \dfrac{x}{2} + \dfrac{x^2}{12} + \dots$

  Hence $\lim\limits_{x \to 0} \dfrac{x}{e^x - 1} = \lim\limits_{x \to 0} 1 - \dfrac{x}{2} + \dfrac{x^2}{12} + \dots = 1$

9 a $2 - x + \dfrac{x^2}{2} + \dots$

  b $x + \dfrac{5x^2}{8}$

  c $\dfrac{1}{4}$

10 a $2x - \dfrac{4x^3}{3} + \dfrac{4x^5}{15}$

   b $-\dfrac{2x^2}{3} - \dfrac{4x^4}{45}$

   c $-\dfrac{2}{3}$

11 $x + \dfrac{x^3}{3!} + \dfrac{x^5}{5!} + \dots; 3$

12 $\ln 2 + \dfrac{3x}{2} + \dfrac{3x^2}{8} + \dfrac{3x^3}{8} + \dots$

13 a $\ln(\cos x) = \ln\left(1 - \dfrac{x^2}{2!} + \dfrac{x^4}{4!} - \dfrac{x^6}{6!} + \dots\right)$

   $= -\dfrac{x^2}{2!} + \dfrac{x^4}{4!} - \dfrac{x^6}{6!} - \dfrac{1}{2}\left(-\dfrac{x^2}{2!} + \dfrac{x^4}{4!} - \dfrac{x^6}{6!}\right)^2$

   $+ \dfrac{1}{3}\left(-\dfrac{x^2}{2!} + \dfrac{x^4}{4!} - \dfrac{x^6}{6!}\right)^3 \dots$

   $= -\dfrac{x^2}{2!} + \dfrac{x^4}{4!} - \dfrac{x^6}{6!} - \dfrac{1}{2}\left(\dfrac{x^4}{4} - \dfrac{x^6}{24} + \dots\right) + \dfrac{1}{3}\left(-\dfrac{x^6}{8} \dots\right) \dots$

   $= -\dfrac{x^2}{2} - \dfrac{x^4}{12} - \dfrac{x^6}{45} - \dots$

   b $\dfrac{1}{2}$

14 0

1 A is an improper integral as the integrand is undefined at $x = 1$ which is within the limits.
  B is not an improper integral as it is fully defined within the limits.
  C is an improper integral as one of its limits is ∞

2 a $1 - \dfrac{2}{a} \to 1$ as $a \to \infty$

  b $12 - 4\sqrt{3 - a} \to 12$ as $a \to 3$

3 $\displaystyle\int_a^3 \ln x \, dx = [x \ln x - x]_a^3$

   $= (3\ln 3 - 3) - (a \ln a - a)$

   $\to 3\ln 3 - 3$ as $a \to 0$ since $a \ln a \to 0$

4 $\displaystyle\int_1^a \dfrac{2}{x}\, dx = [2\ln x]_1^a$

   $= 2\ln a - 2\ln 1$

   $\to \infty$ as $a \to \infty$ so integral does not exist

5 $\dfrac{241}{3}$

6 $\dfrac{1}{8-2}\displaystyle\int_2^8 \dfrac{1+x}{\sqrt{x}}\, dx = \dfrac{1}{6}\int_2^8 x^{-\frac{1}{2}} + x^{\frac{1}{2}}\, dx$

   $= \dfrac{1}{6}\left[2x^{\frac{1}{2}} + \dfrac{2}{3}x^{\frac{3}{2}}\right]_2^8$

   $= \dfrac{1}{6}\left(4\sqrt{2} + \dfrac{32}{3}\sqrt{2}\right) - \dfrac{1}{6}\left(2\sqrt{2} + \dfrac{4}{3}\sqrt{2}\right)$

   $= \dfrac{17}{9}\sqrt{2}$

7 $\dfrac{75}{8}$

8 $\dfrac{1}{1-a}\displaystyle\int_a^1 x^{-3} - \dfrac{1}{2}x^{-2}\, dx = \dfrac{1}{1-a}\left[-\dfrac{1}{2}x^{-2} + \dfrac{1}{2}x^{-1}\right]_a^1$

   $= \dfrac{1}{1-a}\left(-\dfrac{1}{2} + \dfrac{1}{2}\right) - \dfrac{1}{1-a}\left(-\dfrac{1}{2a^2} + \dfrac{1}{2a}\right)$

   $= \dfrac{1}{1-a}\dfrac{(1-a)}{2a^2}$

   $= \dfrac{1}{2a^2}$, as required

9 a $12\,\text{ms}^{-1}$

  b $15.4\,\text{ms}^{-2}$

10 $\dfrac{9}{2}\pi \ln 3$ cubic units

11 $\dfrac{\pi}{16}(4 - \pi)$ cubic units

12 $\dfrac{\pi^2}{4} - \dfrac{\pi}{3}$ cubic units

13 a 3π cubic units

   b $\dfrac{49\pi}{4}$ cubic units

14 a $\dfrac{73\pi}{30}$ cubic units

   b $\dfrac{49\pi}{15}$ cubic units

15 Let $y = \arccos x$. Then $x = \cos y \Rightarrow \dfrac{dx}{dy} = -\sin y$

$$\dfrac{dy}{dx} = -\dfrac{1}{\sin y}$$

$$= -\dfrac{1}{\pm\sqrt{1 - \cos^2 y}}$$

$$= -\dfrac{1}{\sqrt{1 - x^2}} \text{ as gradient is always negative}$$

16 a $2x \arctan x + \dfrac{x^2}{1 + x^2}$

b $\dfrac{x}{\sqrt{1 - \left(\dfrac{x^2}{2}\right)^2}} = \dfrac{2x}{\sqrt{4 - x^4}}$

17 a $\arcsin\left(\dfrac{x}{2}\right) + c$

b $\dfrac{1}{4}\arctan\left(\dfrac{x}{4}\right) + c$

18 a $\sinh A \cosh B - \sinh B \cosh A = \left(\dfrac{e^A - e^{-A}}{2}\right)\left(\dfrac{e^B + e^{-B}}{2}\right)$

$$-\left(\dfrac{e^B - e^{-B}}{2}\right)\left(\dfrac{e^A + e^{-A}}{2}\right)$$

$$= \dfrac{1}{4}(e^{A+B} + e^{A-B} - e^{-(A-B)} - e^{-(A+B)})$$

$$-\dfrac{1}{4}(e^{A+B} + e^{-(A-B)} - e^{(A-B)} - e^{-(A+B)})$$

$$= \dfrac{1}{4}(2e^{A-B} - 2e^{-(A-B)})$$

$$= \dfrac{1}{2}(e^{A-B} - e^{-(A-B)})$$

$$= \sinh(A - B), \text{ as required}$$

b $\cosh^2 x = \left(\dfrac{e^x + e^{-x}}{2}\right)^2$

$$= \dfrac{1}{4}(e^{2x} + 2 + e^{-2x})$$

$$= \dfrac{1}{2} + \dfrac{1}{4}(e^{2x} + e^{-2x})$$

$$= \dfrac{1}{2}\left(1 + \dfrac{1}{2}(e^{2x} + e^{-2x})\right)$$

$$= \dfrac{1}{2}(1 + \cosh 2x), \text{ as required}$$

19 a $2\sinh 2x$

b $2x\sinh x + x^2 \cosh x$

20 Let $y = \operatorname{artanh} x$. Then $x = \tanh y \Rightarrow \dfrac{dx}{dy} = \operatorname{sech}^2 y$

$$\dfrac{dy}{dx} = \dfrac{1}{\operatorname{sech}^2 y}$$

$$= \dfrac{1}{1 - \tanh^2 y}$$

$$= \dfrac{1}{1 - x^2}, \text{ as required}$$

21 a $\dfrac{4x}{\sqrt{4x^4 + 1}}$

b $\operatorname{arcosh}(x - 1) + \sqrt{\dfrac{x}{x - 2}}$

22 a $\ln(1 + \sqrt{2})$

b $\dfrac{\sqrt{2}}{2}\ln(3 + 2\sqrt{2})$

23 a $2x - 3 + \dfrac{4}{x - 2} - \dfrac{1}{x + 3}$

b $4 + \ln\left(\dfrac{96}{7}\right)$

24 $\ln 640 + 3\arctan(3)$ or 10.2

25 a i

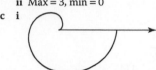

ii Max $= 8$, min $= 0$

b i

ii Max $= 3$, min $= 0$

c i

ii Max $= \dfrac{\pi}{2}$, min $= 0$

d i

ii Max $= 12$, min $= 2$

26 a $\dfrac{19}{2}\pi$ square units

b π square units

27 $\dfrac{7\pi}{4}$ square units

28 $2x + \dfrac{5x^3}{3} - 2x^4 \ldots$

29 a $\dfrac{\left(\dfrac{x}{3}\right)^r}{r!} = \dfrac{x^r}{3^r r!}$

b $((-1)^r)\dfrac{(x^2)^r}{r} = ((-1)^r)\dfrac{x^{2r}}{r}$

c $((-1)^r)\dfrac{\left(\dfrac{x}{3}\right)^{2r}}{(2r)!} = ((-1)^r)\dfrac{(x)^{2r}}{(3)^{2r}(2r)!}$

d $((-1)^r)\dfrac{(4x + 5)^{(2r + 1)}}{(2r + 1)!}$

e $\dfrac{n(n - 1)(n - 2)(n - 3)\ldots(n - r + 1)}{r!}\left(-\dfrac{x}{6}\right)^r$

f $\dfrac{x^{(r + 1)}}{r!}$

30 a $-\sqrt{2} < x < \sqrt{2}$

b $-2 \leq x \leq 2$

31 a $-6 < x < 6$

b $-\dfrac{1}{3} \leq x < \dfrac{1}{3}$

c $-\dfrac{1}{3} \leq x < \dfrac{1}{3}$

32 a $x + 2x^2 + \dfrac{11x^3}{6} + \ldots$

b $4 - 8x + 16x^2 + \ldots$

c $\dfrac{4}{\ln 2} - \dfrac{2}{(\ln 2)^2}x + \dfrac{2 + \ln 2}{2(\ln 2)^3}x^3$

d $1 - 2x - 2x^2$

e $3^x = 1 + (\ln 3)x + \dfrac{(\ln 3)^2}{2}x^2 + \ldots$

f $\dfrac{2}{5} - \dfrac{2x}{25} - \dfrac{23x^2}{125} + \ldots$

Assessment 8

1 a $\dfrac{1}{2}x - \dfrac{3}{2}x^{-\frac{3}{2}}$

   b $\dfrac{3}{4}$

2 $\dfrac{9}{5}\sqrt{3} + \dfrac{4}{5}$

3 6

4 a 4 square units

   b $2\pi\,(e^2-1)$ cubic units

5 177π cubic units

6 a

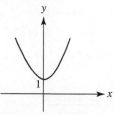

   b $\dfrac{\pi}{32}(e^4 + 8 - e^{-4})$ cubic units

7 $2\pi - \dfrac{\pi^2}{2}$ cubic units

8 $\dfrac{2}{3}$

9 $\ln\left(\dfrac{e^x - 3}{e^x + 3}\right) + c$

10 a $P\left(\dfrac{3}{2}, \dfrac{\pi}{3}\right)$

   b $\dfrac{\pi}{2} - \dfrac{9\sqrt{3}}{16}$ square units

11 a $\dfrac{\pi}{2}$

   b $\cosh^{-1}2$

12 a

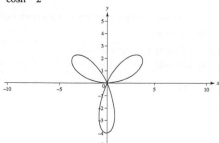

   b $\dfrac{4\pi}{3}$ square units

13 a $\sinh y = x$

$$\cosh y \dfrac{dy}{dx} = 1$$

$$\sqrt{1 + \sinh^2 y}\,\dfrac{dy}{dx} = 1$$

$$\dfrac{dy}{dx} = \dfrac{1}{\sqrt{1 + x^2}}$$

   b $\displaystyle\int_0^2 1 \times \sinh^{-1} x\,dx$

$$= \left[x\sinh^{-1}x\right]_0^2 - \int_0^2 \dfrac{x}{\sqrt{1+x^2}}\,dx$$

$$= \left[x\sinh^{-1}x - \sqrt{1+x^2}\right]_0^2$$

$$= 2\sinh^{-1}2 - (5) - (-1)$$

$$= 2\ln(2+\sqrt{5}) + 1 - \sqrt{5}$$

14 $x^2 - 2x + 10 \equiv (x-1)^2 + 9$

$$\int_1^4 \dfrac{2x+1}{\sqrt{x^2 - 2x + 10}}\,dx = \int_1^4 \dfrac{2x+1}{\sqrt{(x-1)^2 + 9}}\,dx$$

$$x - 1 = 3\sinh\theta$$

$$\dfrac{dx}{d\theta} = 3\cosh\theta$$

Limits of 0 and $\sinh^{-1}1$

$$\int_0^{\sinh^{-1}1} \dfrac{6\sinh\theta + 3}{\sqrt{9\sinh^2\theta + 9}}\,3\cosh\theta\,d\theta$$

$$\int_0^{\sinh^{-1}1} 6\sinh\theta + 3\theta$$

$$= \left[6\cosh\theta + 3\theta\right]_0^{\sinh^{-1}1}$$

$$= \left[6\sqrt{1 + \sinh^2\theta} + 3\theta\right]_0^{\sinh^{-1}1}$$

$$= 6\sqrt{2} + 3\sinh^{-1}1 - 6$$

$$= 6\sqrt{2} + 3\ln(1+\sqrt{2}) - 6$$

15 a $\dfrac{1}{3}\tan^{-1}\left(\dfrac{x-2}{3}\right) + c$

   b 2

16 a $\left(\dfrac{\sqrt{3}}{2}, \dfrac{\pi}{6}\right)$

   b $\dfrac{1}{2}\displaystyle\int_0^{\frac{\pi}{6}} (\sin 2\theta)^2\,d\theta$

$$\dfrac{1}{2}\int_0^{\frac{\pi}{6}} \dfrac{1 - \cos 4\theta}{2}\,d\theta$$

$$= \left[\dfrac{\theta}{4} - \dfrac{\sin 4\theta}{16}\right]_0^{\frac{\pi}{6}}$$

$$= \dfrac{\pi}{24} - \dfrac{\sqrt{3}}{32}$$

$$= \dfrac{1}{2}\int_{\frac{\pi}{6}}^{\frac{\pi}{2}} (\cos\theta)^2\,d\theta$$

$$= \dfrac{1}{2}\int_{\frac{\pi}{6}}^{\frac{\pi}{2}} \dfrac{1 + \cos 2\theta}{2}\,d\theta$$

$$= \left[\dfrac{\theta}{4} + \dfrac{\sin 2\theta}{8}\right]_{\frac{\pi}{6}}^{\frac{\pi}{2}}$$

$$= \dfrac{\pi}{12} - \dfrac{\sqrt{3}}{16}$$

Total area $= \dfrac{\pi}{8} - \dfrac{3\sqrt{3}}{32}$

17 a $\ln\dfrac{3}{2}$

b $\dfrac{3x-x^2}{(x+1)(x^2+3)} \equiv \dfrac{A}{x+1} + \dfrac{Bx+C}{x^2+3}$

$3x-x^2 \equiv A(x^2+3) + (Bx+C)(x+1)$

$-3-1 = 4A \implies A = -1$

$-1 = A+B \implies B = 0$

$0 = 3A+C \implies C = 3$

$\displaystyle\int_1^3 -\dfrac{1}{x+1} + \dfrac{3}{x^2+3} \, dx$

$= \left[-\ln(x+1) + \dfrac{3}{\sqrt{3}}\tan^{-1}\left(\dfrac{x}{\sqrt{3}}\right) \right]_1^3$

$= -\ln 4 + \dfrac{3}{\sqrt{3}}\tan^{-1}(\sqrt{3}) - \left(-\ln 2 + \dfrac{3}{\sqrt{3}}\tan^{-1}\left(\dfrac{1}{\sqrt{3}}\right) \right)$

$= -\ln\left(\dfrac{4}{2}\right) + \dfrac{3}{\sqrt{3}}\left(\dfrac{\pi}{3} - \dfrac{\pi}{6}\right)$

$= \dfrac{\pi}{2\sqrt{3}} - \ln 2$

18 a $\left(\dfrac{3}{2}, \dfrac{\pi}{3}\right)$ $\left(\dfrac{3}{2}, -\dfrac{\pi}{3}\right)$

b

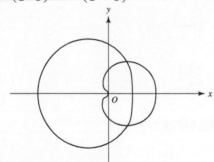

c $3\sqrt{3} - \pi$ square units

19 a $\dfrac{\pi}{4}$ square units

b $\dfrac{\pi}{4}\ln\left(\dfrac{2+\sqrt{2}}{2-\sqrt{2}}\right)$ cubic units

c $2\pi(2-\sqrt{2})$ cubic units

20 a $f(x) = \cos 3x$ $\quad f(0) = 1$

$f'(x) = -3\sin 3x$ $\quad f'(0) = 0$

$f''(x) = -9\cos x$ $\quad f''(0) = -9$

$f'''(x) = 27\sin x$ $\quad f'''(0) = 0$

$f''''(x) = 81\cos x$ $\quad f''''(0) = 81$

So $\cos x = 1 + 0x - \dfrac{9}{2!}x^2 + 0x^3 + \dfrac{81}{4!}x^4$

$= 1 - \dfrac{9}{2}x^2 + \dfrac{27}{8}x^4 - \ldots$

b $\dfrac{(-1)^r (3x)^{2r}}{(2r)!}$

21 a $1 - \dfrac{3}{2}x - \dfrac{9}{8}x^2 - \dfrac{27}{16}x^3 + \ldots$

b $\dfrac{4}{3}$ as $x \to 0$

22 $1 + \dfrac{1}{2}x^2$

23 $1 + 2x + 2x^2 + \ldots$

1 a $e^{3x}y = \dfrac{e^{2x}}{2} + c$ or $y = \dfrac{e^{-x}}{2} + ce^{-3x}$

b $e^{x^2}y = 4e^{x^2} + c$ or $y = 4 + ce^{-x^2}$

c $\sec xy = \dfrac{\sin^3 x}{3} + c$ or $y = \dfrac{\sin^3 x \cos x}{3} + c\cos x$

d $\dfrac{y}{x} = \ln A(x-5)$ or $y = x\ln A(x-5)$

e $x^{\frac{1}{3}}y = \dfrac{2}{5}x^{\frac{5}{6}} + c$ or $y = \dfrac{2}{5}x^{\frac{1}{2}} + cx^{-\frac{1}{3}}$

f $\dfrac{y}{x} = x\ln x - x + c$ or $y = x^2\ln x - x^2 + cx$

g $y\cos x = 4\tan x + c$ or $y = 4\sec x\tan x + c\sec x$

h $e^{(x+1)^2}y = \dfrac{1}{2}e^{(2x^2+2)} + c$ or $y = \dfrac{1}{2}e^{(x^2-2x+1)} + ce^{-(x+1)^2}$

2 a $e^{-2x}y = 2xe^x - 2e^x + 1$ or $y = 2xe^{3x} - 2e^{3x} + e^{2x}$

b $x^3y = 2x^2 + 3$ or $y = \dfrac{2}{x} + \dfrac{3}{x^3}$

c $y\sin x = -\cos^4 x + \dfrac{11}{16}$ or $y = \dfrac{11 - 16\cos^4 x}{16\sin x}$

d $ye^{-\sin 2x} = 2x + 1$ or $y = e^{\sin 2x}(2x+1)$

e $(x+1)^2 y = 14x + 31$ or $y = \dfrac{14x+31}{(x+1)^2}$

f $y\sin x = 2\cos^3 x + \dfrac{1}{\sqrt{2}}$ or $y = \dfrac{\sqrt{2} + 4\cos^3 x}{2\sin x}$

g $y = \dfrac{x}{4}(2x^2\ln x - x^2 + 9)$

h $y\cosh x = e^{5x} + 3$ or $y = \dfrac{3 + e^{5x}}{\cosh x}$

3 a $y\sec^2 x = \tan x - x + c$

b $y\sec^2 x = \tan x - x + \dfrac{\pi}{4}$

4 $y\cos x = x^3 + 2$ or $y = \dfrac{x^3 + 2}{\cos x}$

5 $\dfrac{y}{x^2} = x\ln x - x + 6$ or $y = x^3\ln x - x^3 + 6x^2$

6 a $ye^{2x} = e^{2x}(2\sin x - \cos x) + c$ or $y = 2\sin x - \cos x + ce^{-2x}$

b $y = 2\sin x - \cos x + 2e^{-2x}$

7 a $y\sec^2 x = e^x + c$ or $y = \cos^2 x(e^x + c)$

b $y = \cos^2 x(e^x + 4)$

8 a $y\sin x = 8\sin^3 x + c$

b $2\sin\left(\dfrac{\pi}{4}\right) = 8\sin^3\left(\dfrac{\pi}{4}\right) + c$

$\sqrt{2} = 2\sqrt{2} + c \implies c = -\sqrt{2}$

$y\sin x = 8\sin^3 x - \sqrt{2}$

9 a $x^3 e^x - 3x^2 e^x + 6xe^x - 6e^x + c$

b $x\dfrac{dy}{dx} + (x+2)y = 2x^2$

i.e. $\dfrac{dy}{dx} + \left(1 + \dfrac{2}{x}\right)y = 2x$

integrating factor

$e^{\int\left(1 + \frac{2}{x}\right)dx} = e^{x + 2\ln x} = e^x e^{2\ln x} = x^2 e^x$

so $x^2 e^x \dfrac{dy}{dx} + e^x(x^2 + 2x)y = 2x^3 e^x$

$\dfrac{d}{dx}[x^2 e^x y] = 2x^3 e^x$

$x^2 e^x y = 2\displaystyle\int x^3 e^x dx$

$x^2 e^x y = 2e^x(x^3 - 3x^2 + 6x - 6) + c$

or $x^2 y = 2(x^3 - 3x^2 + 6x - 6) + ce^{-x}$

10 a $\int \sec^3 x \, dx = \int \sec x \sec^2 x \, dx$

$$= \sec x \tan x - \int (\sec x \tan x) \tan x \, dx$$

$$= \sec x \tan x - \int \sec x (\sec^2 - 1) \, dx$$

$$= \sec x \tan x - \int \sec^3 x \, dx + \int \sec x \, dx$$

$$= \sec x \tan x - \int \sec^3 x \, dx + \ln(\sec x + \tan x) + c$$

So $2 \int \sec^3 x \, dx = \sec x \tan x + \ln(\sec x + \tan x) + c$

$$\int \sec^3 x \, dx = \frac{1}{2}[\sec x \tan x + \ln(\sec x + \tan x)] + c$$

b $x\dfrac{dy}{dx} + (1 - x \tan x)y = 2\sec^4 x$

i.e. $\dfrac{dy}{dx} + \left(\dfrac{1}{x} - \tan x\right)y = \dfrac{2\sec^4 x}{x}$

Integrating factor $e^{\int \left(\frac{1}{x} - \tan x\right) dx} = e^{(\ln x + \ln \cos x)} = x \cos x$

So $x \cos x \dfrac{dy}{dx} + (\cos x - x \sin x)y = 2\sec^3 x$

$\dfrac{d}{dx}[y(x \cos x)] = 2\sec^3 x$

$y(x \cos x) = \sec x \tan x + \ln(\sec x + \tan x) + c$

$y = \dfrac{\sec x \tan x + \ln(\sec x + \tan x) + c}{x \cos x}$

11 $x = \dfrac{e^{2t} + c}{\ln t}$

12 a $\dfrac{3\ln(1+t) - 1}{9} + \dfrac{c}{(1+t)^3}$

b $\dfrac{8}{9} = -\dfrac{1}{9} + c \Rightarrow c = 1$

$x = \dfrac{3\ln(1+t) - 1}{9} + \dfrac{1}{(1+t)^3}$

13 a By the chain rule $\dfrac{du}{dx} = \dfrac{du}{dy} \times \dfrac{dy}{dx} = -\dfrac{2}{y^3} \times \dfrac{dy}{dx}$

so $\dfrac{dy}{dx} = -\dfrac{y^3}{2}\dfrac{du}{dx}$

Substituting for $\dfrac{dy}{dx}$ gives $-\dfrac{y^3}{2}\dfrac{du}{dx} = y + 2xy^3$

$\Rightarrow -\dfrac{1}{2}\dfrac{du}{dx} = \dfrac{1}{y^2} + 2x$ or $-\dfrac{1}{2}\dfrac{du}{dx} = u + 2x$ or $\dfrac{du}{dx} + 2u = -4x$

b $\dfrac{1}{y^2} = 1 - 2x + 3e^{-2x}$ or $y^2(1 - 2x + 3e^{-2x}) = 1$

Exercise 9.1B

1 a $\int \dfrac{dT}{T - 20} = \int -\dfrac{1}{8} dt$

$\ln(T - 20) = -\dfrac{1}{8}t + c$

$t = 0$, $T = 100$ (temperature of boiling water), so $c = \ln 80$

and $\ln(T - 20) = -\dfrac{1}{8}t + \ln 80$

$\ln\left(\dfrac{T - 20}{80}\right) = -\dfrac{1}{8}t \Rightarrow T - 20 = 80e^{-\frac{1}{8}t}$

or $T = 20\left(1 + 4e^{-\frac{1}{8}t}\right)$

b 82.3°C

c 22.2 minutes

2 a $\int \dfrac{12\,000}{N(6000 - N)} dN = \int 1 \, dt$

$\int \left(\dfrac{2}{N} + \dfrac{2}{6000 - N}\right) dN = \int 1 \, dt$

$2\ln N - 2\ln(6000 - N) = t + c$

$2\ln\left(\dfrac{N}{6000 - N}\right) = t + c$

$t = 0$, $N = 1000$, gives $2\ln\left(\dfrac{1}{5}\right) = c$

$2\ln\left(\dfrac{N}{6000 - N}\right) = t + 2\ln\left(\dfrac{1}{5}\right)$

$2\ln\left(\dfrac{5N}{6000 - N}\right) = t$

$\left(\dfrac{5N}{6000 - N}\right) = e^{\frac{1}{2}t}$

$5N = 6000e^{\frac{1}{2}t} - Ne^{\frac{1}{2}t}$

$N\left(5 + e^{\frac{1}{2}t}\right) = 6000e^{\frac{1}{2}t}$

$N = \dfrac{6000e^{\frac{1}{2}t}}{5 + e^{\frac{1}{2}t}}$

b 2836

c $N = \dfrac{6000e^{\frac{1}{2}t}}{5 + e^{\frac{1}{2}t}}$, dividing throughout by $e^{\frac{1}{2}t}$ gives

$N = \dfrac{6000}{5e^{-\frac{1}{2}t} + 1}$

As $t \to \infty$, $e^{-\frac{1}{2}t}$ decreases from 1 towards 0, so N increases from 1000 towards 6000

3 a $\int \dfrac{dP}{P} = \int 0.02(t + 1) \, dt$

$\ln P = 0.01(t^2 + 2t) + c$

$t = 0$, $P = 10\,000$ gives $c = \ln(10\,000)$

so $\ln\left(\dfrac{P}{10\,000}\right) = 0.01(t^2 + 2t)$

$P = 10\,000e^{0.01(t^2 + 2t)}$

b 16 160 (to 4 sf)

4 a The integrating factor is $e^{\int 2 \, dt} = e^{2t}$

$e^{2t}\dfrac{dI}{dt} + 2e^{2t}I = 6e^{2t}$

so $\dfrac{d}{dt}\left[e^{2t}I\right] = 6e^{2t}$

$e^{2t}I = 3e^{2t} + c$ or $I = 3 + ce^{-2t}$

$t = 0$, $I = 8$ gives $8 = 3 + c$, so $c = 5$

Therefore $I = 3 + 5e^{-2t}$

b $e^{-2t} > 0$, so $I > 3$

c

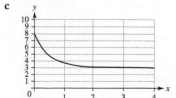

5 a $v = \dfrac{10t(1+t)}{1+2t}$

b $t = 4.4\,\text{s}$

c In this model, as $t \to \infty, v \to \infty$. In practice, the hailstone will approach a terminal velocity.

6 a Integrating factor $e^{\int \frac{1}{20+t}\,dt} = e^{\ln(20+t)} = (20+t)$

$(20+t)\dfrac{dC}{dt} + C = 4(20+t)$

So $\dfrac{d}{dt}[(20+t)C] = 4(20+t)$

$(20+t)C = 2(20+t)^2 + K$

$t = 0$, $C = 10$ gives $200 = 800 + K$, so $K = -600$

so $(20+t)C = 2(20+t)^2 - 600$

or $C = 2(20+t) - \dfrac{600}{20+t}$

b 10 minutes

7 a $x = \dfrac{1}{2}e^{-\frac{1}{2}t}(4 - \sin t)$

b $0.2\,\text{m}$

8 a Multiplying throughout by $(1+2x)$ gives

$2v\dfrac{dv}{dx}(1+2x) + 2v^2 = -4(1+2x)$

Notice that $\dfrac{d}{dx}\left[v^2(1+2x)\right] = 2v\dfrac{dv}{dx}(1+2x) + 2v^2$

so $\dfrac{d}{dx}\left[v^2(1+2x)\right] = -4(1+2x)$

Integrating gives $v^2(1+2x) = -(1+2x)^2 + c$

$x = 0$, $v = 13$, gives $13^2 = 1 + c$, so $c = 170$

$v^2(1+2x) = 170 - (1+2x)^2$

b $6.02\,\text{m}$

Exercise 9.2A

1 a $y = Ae^{2x} + Be^{4x}$

b $y = (Ax + B)e^{-4x}$

c $y = e^{2x}(A\cos x + B\sin x)$

d $y = A + Be^{-3x}$

e $y = Ae^{3x} + Be^{-4x}$

f $y = A\cos x + B\sin x$

g $y = Ae^{\frac{1}{2}x} + Be^{-3x}$

h $y = e^{-\frac{1}{2}x}(A\cos x + B\sin x)$

2 a $y = Ae^{3x} + Be^{-5x} + 2x - 1$

b $y = (Ax + B)e^{3x} + 3e^{2x}$

c $y = e^{-x}(A\cos 4x + B\sin 4x) + \cos 3x$

d $y = Ae^{4x} + Be^{-4x} - 2x + 3$

e $y = Ae^x + Be^{-4x} - e^{-3x}$

f $y = (Ax + B)e^{-x} + x^2 + 5$

g $y = Ae^{-\frac{1}{3}x} + Be^{3x} - \dfrac{5}{7}e^{2x}$

h $y = (Ax + B)e^{-\frac{1}{2}x} + 5x - 2$

3 a $\dfrac{d^2y}{dx^2} - 2a\dfrac{dy}{dx} + a^2y = \dfrac{d^2y}{dx^2} - a\dfrac{dy}{dx} - a\dfrac{dy}{dx} + a^2y$

$= \dfrac{d}{dx}\left[\dfrac{dy}{dx} - ay\right] - a\left[\dfrac{dy}{dx} - ay\right]$

Let $u = \dfrac{dy}{dx} - ay$. Then $\dfrac{du}{dx} - au = 0$ (1)

b $u = Ae^{ax}$

c $u = \dfrac{dy}{dx} - ay$, so $\dfrac{dy}{dx} - ay = Ae^{ax}$ (2)

Multiplying (2) throughout by the integrating factor of e^{-ax}

gives $e^{-ax}\dfrac{dy}{dx} - ae^{-ax}y = A$

$\dfrac{d}{dx}\left[ye^{-ax}\right] = A$

$ye^{-ax} = Ax + B$

$y = (Ax + B)e^{ax}$

4 $Ae^{(m+in)x} + Be^{(m-in)x} = Ae^{mx}e^{inx} + Be^{mx}e^{-inx}$

$= Ae^{mx}(\cos nx + i\sin nx)$

$+ Be^{mx}(\cos nx - i\sin nx)$

$= e^{mx}\left[(A+B)\cos nx + (Ai - Bi)\sin nx\right]$

Let $A + B = \alpha$ and $Ai - Bi = \beta$

Then $Ae^{(m+in)x} + Be^{(m-in)x} = e^{mx}(\alpha\cos nx + \beta\sin nx)$

Exercise 9.2B

1 a $y = 2e^x + e^{6x}$

b $y = (4x+1)e^{-2x}$

c $y = 3\cos 5x + 6\sin 5x$

d $y = 3e^{2x} - 2e^{-3x}$

e $y = (5 - 12x)e^{3x}$

f $y = 6(e^{4x} + e^{-3x})$

g $y = 12e^{\frac{2}{3}x}$

h $y = 2e^{\frac{1}{2}x}(3\cos x - 2\sin x)$

2 a $y = 15e^{2x} - 5e^{5x} + 3e^x$

b $y = (1 - 8x)e^{6x} + 2$

c $y = 2(\cos x + \sin x) + e^{-2x}(3\cos x + 2\sin x)$

d $y = \dfrac{1}{6}(13e^{2x} + 11e^{-4x}) + 3x - 3$

e $y = 6e^{\frac{2}{3}x} - 9e^{-\frac{2}{3}x} - x^2 - 5$

f $y = e^{2x}(8\cos x + 4\sin x) + \cos 2x$

g $y = 3e^{\frac{3}{2}x} + 7e^{-\frac{3}{2}x} + x^2 - 3x - 2$

h $y = (x+7)e^{\frac{1}{4}x} + 3e^{-\frac{1}{4}x}$

3 $a = -2$

The general solution is $y = Ae^{4x} + Be^{-2x} - 2xe^{-2x}$

4 $a = 3$

The general solution is $y = (Ax + B)e^x + 3x^2e^x$

5 $y = 2e^x - 2e^{-6x} + 3xe^x - 2$

6 $y = (2x+3)e^{3x} + 17x^2e^{3x}$

7 $y = 3e^{2x} - e^{3x} + 2e^{4x}$

8 a Given $x = e^u$, $\dfrac{dx}{du} = e^u = x$

By the chain rule $\dfrac{dy}{dx} = \dfrac{dy}{du} \times \dfrac{du}{dx}$

$\dfrac{dy}{dx} = \dfrac{dy}{du} \times \dfrac{1}{e^u} = \dfrac{1}{x}\dfrac{dy}{du}$

$\dfrac{d^2y}{dx^2} = -\dfrac{1}{x^2}\dfrac{dy}{du} + \dfrac{1}{x}\dfrac{d^2y}{du^2} \times \dfrac{du}{dx}$

$\dfrac{d^2y}{dx^2} = -\dfrac{1}{x^2}\dfrac{dy}{du} + \dfrac{1}{x^2}\dfrac{d^2y}{du^2}$

Substituting $\dfrac{dy}{dx} = \dfrac{1}{x}\dfrac{dy}{du}$ and $\dfrac{d^2y}{dx^2} = -\dfrac{1}{x^2}\dfrac{dy}{du} + \dfrac{1}{x^2}\dfrac{d^2y}{du^2}$

into $x^2\dfrac{d^2y}{dx^2} - 4x\dfrac{dy}{dx} + 6y = 12$ gives

$x^2\left(-\dfrac{1}{x^2}\dfrac{dy}{du} + \dfrac{1}{x^2}\dfrac{d^2y}{du^2}\right) - 4x\left(\dfrac{1}{x}\dfrac{dy}{du}\right) + 6y = 12$

$$-\frac{dy}{du} + \frac{d^2y}{du^2} - 4\frac{dy}{du} + 6y = 12$$

$$\frac{d^2y}{du^2} - 5\frac{dy}{du} + 6y = 12$$

b $y = 7x^2 - 2x^3 + 2$

Exercise 9.3A

1
  a $x = 7\cos 5t + 5\sin 5t$
  b $x = 8\sin 2t - 4\cos 2t$
  c $x = e^{3t}(9\sin 2t - 5\cos 2t)$
  d $x = e^{-5t}(\sin 4t - \cos 4t)$
  e $x = e^{-\frac{5}{2}t}\left(6\cos\frac{1}{2}t + 14\sin\frac{1}{2}t\right)$
  f $x = 2e^{\frac{1}{3}t}\left(4\cos\frac{2}{3}t + \sin\frac{2}{3}t\right)$

2
  a $x = 2\sin 4t - 3\cos 4t + 3$
  b $x = 10\cos 3t + 6\sin 3t + t^2 + 4t + 2$
  c $x = 2\cos 2t + 4\sin 2t + \cos 4t - \sin 4t$
  d $x = e^{-5t}(8\cos 2t - 3\sin 2t) + t - 2$
  e $x = e^{-2t}(12\cos t - 5\sin t) + 4$
  f $x = e^{-t}(7\cos t - 6\sin t) + \sin t$

3 $x = 4\cos 5t + 3\sin 5t$

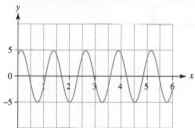

4 $x = 5\cos 3t - 12\sin 3t + 2$

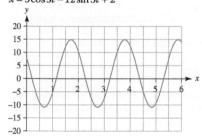

5 $x = 4\cos 2t + 3\sin 2t + 2t$
6 $x = 4 + e^{-t}\sin 2t$

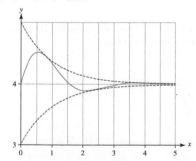

7
  a $x = (2 - \cos t - \sin t)e^{-t}$
  b $\cos t + \sin t = \sqrt{2}\cos\left(t - \frac{\pi}{4}\right)$

  So $x = \sqrt{2}e^{-t}\left(\sqrt{2} - \cos\left(t - \frac{\pi}{4}\right)\right)$

  $\cos\left(t - \frac{\pi}{4}\right) < 1$, so $\sqrt{2}\cos\left(t - \frac{\pi}{4}\right) > 0$ and $e^{-t} > 0$

Hence $x > 0$ for all t

$$\frac{dx}{dt} = -(2 - \cos t - \sin t)e^{-t} + e^{-t}(\sin t - \cos t)$$

$$= e^{-t}(2\sin t - 2) \le 0 \text{ for all } t$$

c

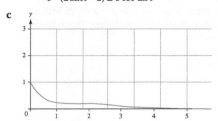

Exercise 9.3B

1
  a $x = 2e^{-\frac{3}{2}t}(5\sin t - \cos t) + 4$
  b 5.74 metres

2
  a $\theta = \frac{\pi e^{-3t}}{8}(2\cos 2t + 3\sin 2t)$
  b

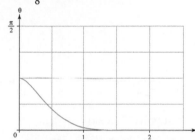

  c As $t \to \infty$, $\theta \to 0 \Rightarrow$ the door settles down towards a closed position.

3
  a $y = \frac{1}{4}(2t + 1 - \cos 2t - \sin 2t)$
  b $t = \frac{3\pi}{4}$
  c

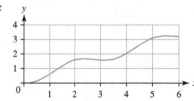

4
  a $y = 4\cos\left(\frac{t}{2}\right) + 3\sin\left(\frac{t}{2}\right) + 7$
  b 12 metres
  c 6.17 am

5
  a $x = \frac{4}{3}e^{-2t} - \frac{4}{3}e^{-\frac{t}{2}} + 2\sin t$
  b 0.0120 cm below A

6
  a $I = 2\sin 4t + 3\sin 2t$
  b 1.21 s

7
  a $x = e^{-\frac{1}{2}t}\left(3 + 2\cos\frac{1}{2}t + 2\sin\frac{1}{2}t\right)$
  b $x = e^{-\frac{1}{2}t}\left(3 + 2\cos\frac{1}{2}t + 2\sin\frac{1}{2}t\right)$
    $= e^{-\frac{1}{2}t}\left(3 + 2\sqrt{2}\sin\left(\frac{1}{2}t + \frac{\pi}{4}\right)\right) > 0$ for all values of t since
    $3 > 2\sqrt{2}$ and $\sin\left(\frac{1}{2}t + \frac{\pi}{4}\right) > -1$
    So the particle never reaches the origin.
  c 7.98 s

8 a $-\dfrac{1}{12}$

b $V = \dfrac{1}{72}\sin 6t - \dfrac{1}{16}t\cos 6t$

c As t becomes large, the voltage oscillates with an amplitude that increases with time. This is known as resonance.

9 a $x = e^{-t}(\cos t + \sin t)$

b $t = \pi,\ 2\pi,\ 3\pi,\ .....$

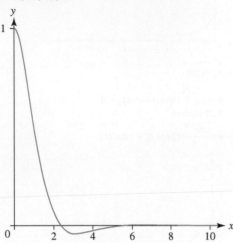

c $x(\pi) = -e^{-\pi},\ x(2\pi) = e^{-2\pi},\ x(3\pi) = -e^{-3\pi},\ ....$

so total distance $= 1 + 2(e^{-\pi} + e^{-2\pi} + e^{-3\pi} + ...\ )$

$$= 1 + \dfrac{2e^{-\pi}}{1 - e^{-\pi}}\ \text{(summing an infinite}$$
$$\text{geometric progression)}$$

$$= \dfrac{1 + e^{-\pi}}{1 - e^{-\pi}}$$

$$= \dfrac{e^{\frac{\pi}{2}} + e^{-\frac{\pi}{2}}}{e^{\frac{\pi}{2}} - e^{-\frac{\pi}{2}}}$$

$$= \coth\left(\dfrac{\pi}{2}\right)$$

10 a (i) After t seconds the end A has moved a displacement $3t$ metres, and box B has moved a displacement y metres. Since the spring is initially in equilibrium, the extension of the spring, x metres, is given by $x = 3t - y$, i.e. $x + y = 3t$

(ii) Differentiating $x + y = 3t$ gives $\dfrac{dx}{dt} + \dfrac{dy}{dt} = 3$ (1)

$$\text{and}\ \dfrac{d^2x}{dt^2} + \dfrac{d^2y}{dt^2} = 0\quad(2)$$

Applying $F = ma$ to B gives $\dfrac{90x}{2} - 30v = 5\dfrac{d^2y}{dt^2}$

So $45x - 30\dfrac{dy}{dt} = 5\dfrac{d^2y}{dt^2}$ i.e. $9x - 6\dfrac{dy}{dt} = \dfrac{d^2y}{dt^2}$

Substituting for $\dfrac{dy}{dt}$ and $\dfrac{d^2y}{dt^2}$ from (1) and (2)

gives $9x - 6\left(3 - \dfrac{dx}{dt}\right) = -\dfrac{d^2y}{dt^2}$

So $\dfrac{d^2x}{dt^2} + 6\dfrac{dx}{dt} + 9x = 18$

(iii) Auxiliary equation is $m^2 + 6m + 9 = 0$ so

$(m + 3)^2 = 0 \Rightarrow m = -3$

Complementary function is $x = (At + B)e^{-3t}$

Particular integral is x = 2, so general solution is

$x = (At + B)e^{-3t} + 2$

$\dfrac{dx}{dt} = Ae^{-3t} - 3(At + B)e^{-3t}$

At $t = 0$, $x = y = 0$ and $\dfrac{dy}{dt} = 0$. Using (1) gives $\dfrac{dx}{dt} = 3$

$t = 0$, $x = 0$ and $\dfrac{dx}{dt} = 3$ give $B + 2 = 0$ and

$A - 3B = 3 \Rightarrow B = -2$ and $A = -3$

So particular solution is $x = 2 - (2 + 3t)e^{-3t}$

b For large values of t the oscillations damp down, and the box moves with a near constant velocity of 3 ms^{-1}, and the spring stretched by a near constant 2 metres.

Exercise 9.4A

1 a $\displaystyle\int -\dfrac{100}{h^2}\,dh = \int 1\,dt$

$\dfrac{100}{h} = t + c$

$t = 0$, $h = 50$ gives $\dfrac{100}{50} = c \Rightarrow c = 2$

$\dfrac{100}{h} = t + 2 \Rightarrow h = \dfrac{100}{t + 2}$

b

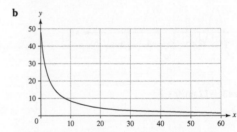

c Under this model the tank is never empty.

2 a $Q = 5\left(4 - 3e^{-\frac{1}{4}}\right)$

b As $t \to \infty$, $Q \to 20$, so it doesn't matter how long you charge it, you can never get more than 20 ampere-hours.

c 10.83 hrs = 10 hrs 50 mins

3 $N = \dfrac{3}{2e^{-\frac{t}{2}} + 1}$

$e^{-\frac{t}{2}} > 0$ so $N < 3$

In a lake there is a limited amount of food, so the lake can only sustain a limited number of fish.

4 a $P = 10000\,e^{-\sin 2t}$

b 13 224

c $P_{max} = 27\,183, P_{min} = 3679$

5 a $T = 10\sqrt{\dfrac{4t + 1}{t + 1}}$

b 17.3 °C

c As $t \to \infty$, T increases towards $T = 10\sqrt{4} = 20$°C

6 a $t = 0$, $x = 8$, so $\dfrac{dx}{dt} = -8 + 4 \Rightarrow \dfrac{dx}{dt} = -4 < 0$, so at $t = 0$ the population is declining.

b $x = t + 3 + 5e^{-t}$

c 5 609 000

d As $t \to \infty$, $x \to \infty \Rightarrow$ the population increases without limit.

Exercise 9.4B

1 a $\dfrac{dV}{dt}=-kV$

$$\int \dfrac{1}{V}dV = \int -k\ dt$$

$\ln(cV)=-kt$

$t=0$, $V=2000$ gives $\ln(2000c)=0$, so $c=\dfrac{1}{2000}$

$\ln\left(\dfrac{V}{2000}\right)=-kt$

$V=2000e^{-kt}$

b $V=1000$ when $t=4$ gives $1000=2000e^{-4k}$

$\ln\left(\dfrac{1}{2}\right)=-4k$

$k=\dfrac{\ln 2}{4}$

c 500

2 a $\dfrac{dT}{dt}$ is the rate of change of temperature.

$(T-20)$ is the difference between the water temperature and the room temperature. Cooling gives the minus sign as the temperature is falling. And k is the constant of proportionality.

b $20\left(1+4e^{-\frac{t(\ln 5)}{10}}\right)$

c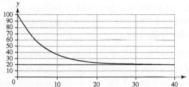

d 27 minutes

3 a $\dfrac{dx}{dt}=kx$

$$\int \dfrac{1}{x}dx = \int k\ dt$$

$\ln(cx)=kt$

$t=0$, $x=100$ gives $\ln(100c)=0$, so $c=\dfrac{1}{100}$

$\ln\left(\dfrac{x}{100}\right)=kt$

$x=100e^{kt}$

b $x=500$ when $t=10$ gives $500=100e^{10k}$

$\ln 5=10k$

$k=\dfrac{\ln 5}{10}$

c 12 500

4 a Let $V\,\text{m}^3$ be the volume of water in the tank at a time t minutes

Then $\dfrac{dV}{dt}=2-0.4x$ (1)

Also $V=16x$

so $\dfrac{dV}{dt}=16\dfrac{dx}{dt}$ (2)

$16\dfrac{dx}{dt}=2-0.4x$

$40\dfrac{dx}{dt}=5-x$

or $\dfrac{dx}{dt}=\dfrac{5-x}{40}$

b $x=5(1-e^{-\frac{t}{40}})$

c 36 mins 39 secs

d $x=5-5e^{-\frac{t}{40}}$, so, whatever the value of t, the depth will always be less than 5 m.

5 a Let $V\,\text{m}^3$ be the volume of water in the tank at a time t minutes

Then $\dfrac{dV}{dt}=1000-C\sqrt{x}$ (1)

Also $V=500x$

so $\dfrac{dV}{dt}=500\dfrac{dx}{dt}$ (2)

$500\dfrac{dx}{dt}=1000-C\sqrt{x}$

Relabel $C=500k$

$500\dfrac{dx}{dt}=1000-500k\sqrt{x}$

$\dfrac{dx}{dt}=2-k\sqrt{x}$

b When $x=25$, $500k\sqrt{x}=500$, so $500k\sqrt{25}=500 \Rightarrow k=\dfrac{1}{5}$

So $\dfrac{dx}{dt}=2-\dfrac{1}{5}\sqrt{x}$

$5\dfrac{dx}{dt}=10-\sqrt{x}$

$\dfrac{dx}{dt}=\dfrac{10-\sqrt{x}}{5}$

$$\int \left(\dfrac{5}{10-\sqrt{x}}\right)dx = \int 1\ dt$$

c $-10\sqrt{x}-100\ln(10-\sqrt{x})=t-100\ln 10$

d 19.3 secs

6 a Using $F=ma$ gives $1600-100v=200v\dfrac{dv}{dx}$

$16-v=2v\dfrac{dv}{dx}$

b 20.4 m

7 a 7.2 s

b 76.2 m

8 a $A=20$, $B=20$, $C=-1$

b Applying $F=ma$ to the van gives $\dfrac{40000}{v}-25v=600v\dfrac{dv}{dx}$

So $\dfrac{1600}{v}-v=24v\dfrac{dv}{dx}$

Therefore $1600-v^2=24v^2\dfrac{dv}{dx}$ or $\dfrac{dv}{dx}=\dfrac{1600-v^2}{24v^2}$

c $480\ln\left(\dfrac{40+v}{120-3v}\right)+480-24v$

d 413 metres.

9 a $\dfrac{dx}{dt}=-kx$

$$\int \dfrac{1}{x}dx = \int -k\ dt$$

$\ln(cx)=-kt$

$t=0$, $x=x_0$ gives $\ln(cx_0)=0$, so $c=\dfrac{1}{x_0}$

$\ln\left(\dfrac{x}{x_0}\right)=-kt$

$x=x_0e^{-kt}$

b Using the sum to infinity of a geometric progression,

$x_0+x_0e^{-kT}+x_0e^{-2kT}+x_0e^{-3kT}+\ldots=\dfrac{x_0}{1-e^{-kT}}$

10 a Using $F=ma$ gives $-13x-12v=4a$

$-13x-12\dfrac{dx}{dt}=4\dfrac{d^2x}{dt^2}$ or $4\dfrac{d^2x}{dt^2}+12\dfrac{dx}{dt}+13x=0$

b $x = 2e^{-\frac{3}{2}t}(\cos t + 3\sin t)$

c $0.266\,\text{s}$

Exercise 9.5A

1 a $y = Ae^{2t} + Be^{3t}$

b $y = A\cos 3t + B\sin 3t$

c $y = Ae^{\frac{16}{3}t} + Be^{t}$

d $y = Ae^{3t} + Be^{-2t} + \frac{4}{3}$

e $y = e^{-5t}(A\cos t + B\sin t)$

f $y = e^{-t}(A\cos 2t + B\sin 2t)$

2 a $y = 2e^{3t} + e^{-2t},\ x = 8e^{3t} - e^{-2t}$

b $y = e^{t} + 3e^{-3t} - 5,\ x = e^{t} + \frac{3}{5}e^{-3t} - 4$

c $y = e^{-t}(3\cos t - 2\sin t),\ x = e^{-t}(4\cos t - 7\sin t)$

d $y = 3te^{-t} - 5t + 9,\ x = 9t - 11 - \frac{3}{2}e^{-t}(2t+1)$

e $y = e^{-2t}(5\cos t + 12\sin t),\ x = \frac{e^{-2t}}{2}(7\cos t - 17\cos t) - 1$

f $y = \left(2t + \frac{5}{3}\right)e^{t} + \frac{1}{3}e^{-2t},\ x = 3(t+1)e^{t}$

3 $x = e^{t} + 2e^{3t},\ y = -e^{t} + 2e^{3t}$

4 $x = 17 - 7t - (4t+9)e^{-t},\ y = 11(t-2) + (4t+11)e^{-t}$

5 a From (2): $\dfrac{d^2y}{dt^2} = \dfrac{dy}{dt} - \dfrac{dz}{dt}$,

so $\dfrac{d^2y}{dt^2} = \dfrac{dy}{dt} + x$ (4)

From (1): $\dfrac{d^2x}{dt^2} = \dfrac{dx}{dt} + 2\dfrac{dy}{dt}$ (5)

Differentiating (5): $\dfrac{d^3x}{dt^3} = \dfrac{d^2x}{dt^2} + 2\dfrac{d^2y}{dt^2}$

From (4): $\dfrac{d^3x}{dt^3} = \dfrac{d^2x}{dt^2} + 2\left(\dfrac{dy}{dt} + x\right)$,

so $\dfrac{d^3x}{dt^3} = \dfrac{d^2x}{dt^2} + 2\dfrac{dy}{dt} + 2x$

Substituting $2\dfrac{dy}{dt} = \dfrac{d^2x}{dt^2} - \dfrac{dx}{dt}$ from (5):

$\dfrac{d^3x}{dt^3} = \dfrac{d^2x}{dt^2} + \dfrac{d^2x}{dt^2} - \dfrac{dx}{dt} + 2x$

$\Rightarrow \dfrac{d^3x}{dt^3} - 2\dfrac{d^2x}{dt^2} + \dfrac{dx}{dt} - 2x = 0$

b $x = e^{2t} + 2\cos t - \sin t$

Exercise 9.5B

1 a $y = 20e^{-t} - 20e^{-3t},\ x = 60e^{-t} - 20e^{-3t}$

b $t = \frac{1}{2}\ln 3$

2 a $y = 3e^{-t} + 2e^{-5t},\ x = 9e^{-t} - 2e^{-5t}$

b $x - y = 9e^{-t} - 2e^{-5t} - (3e^{-t} + 2e^{-5t})$

$= 6e^{-t} - 4e^{-5t}$

$= 2e^{-5t}(3e^{4t} - 2)$

for $t > 0,\ 3e^{4t} - 2 > 0$, so $x > y$, as required.

3 a $y = 20e^{2t} + 10e^{6t},\ x = 60e^{2t} - 10e^{6t}$

b X becomes extinct when $t = \frac{1}{4}\ln 6$

4 a $x = \dfrac{5e^{-2t} - 4e^{-7t} + 5}{2},\ y = 5e^{-2t} + 6e^{-7t} + 3$

b As $t \to \infty,\ x \to \frac{5}{2}$ and $y \to 3$ so $x : y \to 5 : 6$

5 a $L = -220e^{2t} + 300e^{4t},\ G = 220e^{2t} - 100e^{4t}$

b 0.4 years

6 a $R = 30e^{3t} + 70\,e^{5t},\ F = 90e^{3t} + 70e^{5t}$

b $\dfrac{F}{R} = \dfrac{90e^{3t} + 70e^{5t}}{30e^{3t} + 70e^{5t}} = \dfrac{960e^{-2t} + 70}{30e^{-2t} + 70} \to \dfrac{70}{70} = 1$

as $t \to \infty$

So, over time, the number of foxes will be approximately the same as the number of rabbits.

7 a $P = \dfrac{1}{10}(10 + 7t)\,e^{\frac{1}{10}t},\ B = \dfrac{e^{\frac{1}{10}t}}{30}(60 - 7t)$

b $8\frac{4}{7}$ years

8 a $\dfrac{d^2y}{dt^2} = 0.4\dfrac{dy}{dt} + 0.1\dfrac{dx}{dt}$

$= 0.4\dfrac{dy}{dt} + 0.1(-0.2y + 0.2x)$

$= 0.4\dfrac{dy}{dt} - 0.02y + 0.2 \times 0.1x$

$= 0.4\dfrac{dy}{dt} - 0.02y + 0.2\left(\dfrac{dy}{dt} - 0.4y\right)$

So $\dfrac{d^2y}{dt^2} - 0.6\dfrac{dy}{dt} + 0.1y = 0$

b $x = 4e^{0.3t}(250\cos 0.1t - 251\sin 0.1t)$ and
$y = 4e^{0.3t}(\cos 0.1t - 251\sin 0.1t)$

c 7.8 years

d The lake is not a closed region. Other animals may feed off the fish in the lake, and the bears might also go elsewhere for their food.

9 a $\dfrac{dH}{dt}$ is the rate of growth of the hare population. The term $1.2H$ represents a growth equal of 20% of the hare population. The term $-1.15L$ represents a decline equal to 115% of the lynx population.

b $\dfrac{d^2H}{dt^2} = 1.2\dfrac{dH}{dt} - 1.15\dfrac{dL}{dt}$

So $\dfrac{d^2H}{dt^2} = 1.2\dfrac{dH}{dt} - 1.15(0.05H)$

$\dfrac{d^2H}{dt^2} - 1.2\dfrac{dH}{dt} + 0.0575H = 0$

c $H = 460e^{1.15t} + 40e^{0.05t}$ and $L = 20e^{1.15t} + 40e^{0.05t}$

d As $t \to \infty$ this model says that both populations increase without limit, which is not realistic.

10 a $\dfrac{dZ}{dt} = 1.3Z - 1.1L$

So $\dfrac{d^2Z}{dt^2} = 1.3\dfrac{dZ}{dt} - 1.1\dfrac{dL}{dt}$

$\dfrac{d^2Z}{dt^2} = 1.3\dfrac{dZ}{dt} - 1.1(0.2Z)$

$\dfrac{d^2Z}{dt^2} - 1.3\dfrac{dZ}{dt} + 0.22Z = 0$

b $Z = 100(11e^{1.1t} - 2e^{0.2t})$ and $L = 100(2e^{1.1t} - 2e^{0.2t})$

c $\dfrac{L}{Z} = \dfrac{(2e^{1.1t} - e^{0.2t})}{(11e^{1.1t} - e^{0.2t})} \to \dfrac{2}{11}$ as $t \to \infty$

11 a $x = 100e^{-0.2t}$

b $y = -200e^{-0.2t} + 200e^{-0.1t}$

c $z = 100e^{-0.2t} - 200e^{-0.1t} + 100$

d $x + y + z = 100e^{-0.2t} - 200e^{-0.2t} + 200e^{-0.1t} + 100e^{-0.2t} - 200e^{-0.1t} + 100$

$= 100$, as required.

Review exercise 9

1 $y \sin x = -\dfrac{1}{4}\cos^4 x + 1$ or $y = \dfrac{4 - \cos^4 x}{4 \sin x}$

2 a $S = \dfrac{t(300 - t)}{600}$

   b 37.5 kg

3 a $\dfrac{dT}{dt} = -kT$

$$\int \dfrac{1}{T} dT = \int -k \, dt$$

$\ln(cT) = -kt$

$t = 0$, $T = 100$ giving $\ln(100c) = 0$, so $c = \dfrac{1}{100}$

$\ln\left(\dfrac{T}{100}\right) = -kt$

$T = 100e^{-kt}$

   b $T = 25$ when $t = 6$ gives $25 = 100e^{-6k}$

$\ln\left(\dfrac{1}{4}\right) = -6k$

$k = \dfrac{\ln 4}{6} = \dfrac{\ln 2}{3}$

   c 13.0 minutes

4 a Using $F = ma$ gives $600 - 6v^2 = 120\dfrac{dv}{dt} \Rightarrow 100 - v^2 = 20\dfrac{dv}{dt}$

   b 2.94 seconds

   c 16.6 metres

5 a $y = e^{-2x}(A\cos 3x + B\sin 3x) + 2e^{4x}$

   b $y = e^{-2x}(4\cos 3x - \sin 3x) + 2e^{4x}$

6 a $y = -\dfrac{1}{8}\sin 4t + \dfrac{1}{2}t$

   b $\dfrac{\pi}{2}$ seconds

   c

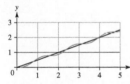

7 a Using $F = ma$ gives $-9x - 12v = 4a$

$\Rightarrow -9x - 12\dfrac{dx}{dt} = 4\dfrac{d^2x}{dt^2}$ or $4\dfrac{d^2x}{dt^2} + 12\dfrac{dx}{dt} + 9x = 0$

   b $4m^2 + 12m + 9 = 0 \Rightarrow (2m+3)^2 = 0$, so $m = -\dfrac{3}{2}$

$x = (At + B)e^{-\frac{3}{2}t}$

$\dfrac{dx}{dt} = -\dfrac{3}{2}e^{-\frac{3}{2}t}(At+B) + Ae^{-\frac{3}{2}t}$

$x = 6$ and $\dfrac{dx}{dt} = -11$ when $t = 0$ gives $B = 6$ and

$-\dfrac{3}{2}B + A = -11$, so $A = -2$ and the particular solution is

$x = e^{-\frac{3}{2}t}(6 - 2t)$

   c 3 s

   d $0.0222 \, \text{m s}^{-1}$

8 a $\dfrac{d^2x}{dt^2} = \dfrac{dx}{dt} - \dfrac{dy}{dt} - 3 = \dfrac{dx}{dt} - y + 4x$

$= \dfrac{dx}{dt} + \dfrac{dx}{dt} - x + 3t + 4x = 2\dfrac{dx}{dt} + 3x + 3t$

$\Rightarrow \dfrac{d^2x}{dt^2} - 2\dfrac{dx}{dt} - 3x = 3t$

   b $x = \dfrac{1}{12}e^{3t} + \dfrac{5}{4}e^{-t} - t + \dfrac{2}{3}$

$y = \dfrac{5}{2}e^{-t} - \dfrac{1}{6}e^{3t} - 4t + \dfrac{5}{3}$

Assessment 9

1 $-e^{-y} = \dfrac{e^x(\cos x + \sin x)}{2}$

2 $y = -\ln(x \ln x - x + 2)$

3 a $y^2 = x^2 \ln x - \dfrac{x^2}{2} + 2c$

   b $y^2 = x^2 \ln x - \dfrac{x^2}{2} + \dfrac{33}{2}$

4 a $P = \dfrac{500e^{500(kt+c)}}{1 + e^{500(kt+c)}}$

   b $P = \dfrac{500e^{(1.45t - 3.89)}}{1 + e^{(1.45t - 3.89)}}$

   c Year 8

5 $y = \dfrac{x^2}{4} + \dfrac{c}{x^2}$

6 $4x^2 \tan x + 8x - 8\tan x + \dfrac{1}{\cos x}$

7 a $\ln x - 2 + \dfrac{c}{\sqrt{x}}$

   b $\ln x - 2 + \dfrac{1}{\sqrt{x}}$

8 a $y = Ae^{-x} + Be^{-3x}$

   b $y = Ae^{-2x} + Bxe^{-2x}$

   c $y = e^{-2x}(A\cos x + B\sin x)$

9 $y = 2e^{\frac{1}{2}x} + 3e^{-2x}$

10 $y = 6e^{\frac{x}{3}} + 2xe^{\frac{x}{3}}$

11 $y = e^{-x}(10\cos 2x + 6\sin 2x)$

12 $y = Ae^{3x} + Be^{-4x} - x - \frac{1}{6}$

13 $y = e^{2x}(4\cos 3x - 4\sin 3x) + 3\sin x + \cos x$

14 $y = -\dfrac{14}{3}e^{2x} + \dfrac{5}{3}e^{5x} + 3e^{x}$

15 a $\dfrac{\ln(T - 19) - \ln 16}{\ln\left(\dfrac{15}{16}\right)} = t$

   b The victim died 2.25 hours before the body was found.

16 a $\dfrac{dP}{dt} = k(1000 - P)$

   b $P = 1000 - e^{-kt - c}$

   c $P = 1000 - 995\left(\dfrac{900}{995}\right)^{\frac{t}{10}}$

   d 186

17 a $\dfrac{dA}{dt} = kA$; $A(0) = 1000$; $A(1) = 1005$

   b $\ln A = \ln\left(1000\left(\dfrac{1005}{1000}\right)^t\right)$

   c The account has £1025.25 after 5 years.

   d The amount will exceed £2000 after 138.98 years.

18 a $\dfrac{dv}{dt} = g - kv$

$\dfrac{dv}{g - kv} = dt$

$\displaystyle\int \dfrac{dv}{g - kv} = \int dt$

Let $u = g - kv$

$du = -kdv$

$-\dfrac{du}{k} = -\dfrac{k}{dv}$

$\therefore \displaystyle\int \dfrac{dv}{g - kv} = -\dfrac{1}{k}\int \dfrac{1}{u}\,du$

$-\dfrac{1}{k}\ln|g - kv| = t + c$

$\ln|g - kv| = -k(t + c)$

$g - kv = e^{-k(t+c)}$

$v = \dfrac{g}{k} - \dfrac{1}{k}e^{-kt}e^{-kc}$

When $t = 0$, $v = 0$, so $e^{-kc} = g$

$\therefore v = \dfrac{g}{k}(1 - e^{-kt})$

Therefore, as $t \to \infty$, $v \to \dfrac{g}{k}$

b

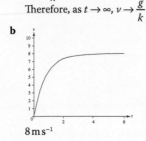

$8\,\mathrm{m\,s^{-1}}$

19 $x = 4\sin\frac{1}{2}t$

20 a The discriminant is $b^2 - 4ac$

which is $0.3^2 - 4 \times 0.15^2 = 0$

Therefore the auxiliary equation has equal roots so the damping is critical. The oscillations will die away quickly.

b $x = 0.5e^{-0.15t} + 0.5te^{-0.15t}$ or $x = 0.5e^{-0.15t}(1 + t)$

c

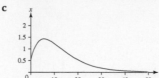

21 a $m\dfrac{d^2x}{dt^2} = -5m\dfrac{dx}{dt} - 6mx + 3m\sin 2t$

$\dfrac{d^2x}{dt^2} + 5\dfrac{dx}{dt} + 6x = 3\sin 2t$

b $x = Ae^{-2t} + Be^{-3t}$

c $x = -\dfrac{15}{52}\cos 2t + \dfrac{3}{52}\sin 2t$

d $x = \dfrac{7}{4}e^{-2t} - \dfrac{19}{13}e^{-3t} + \dfrac{3}{52}(\sin 2t - 5\cos 2t)$

22 a $x = 9(2e^{-t} - e^{-3t} - 1)$

$y = \dfrac{1}{2}(9e^{-3t} - 9e^{-t} + 2)$

b As $t \to \infty$, e^{-t} and $e^{-3t} \to 0$ so $x \to -9$ and $y \to 1$
The system settles around the point $(-9, 1)$

Index

A
acceleration 146
arccos/arcsin/arctan 82–5, 94
arcosh/arsinh/artanh 53–4, 87, 89
areas 21, 95–9
Argand diagrams 16, 21
argument 2–3, 6–7, 16–17
auxiliary equations 127–32, 134–5, 137–9

B
binomial expansions 8, 10

C
calculators
    complex numbers 3
    definite integrals 72
    differential equations 135, 139
    simple harmonic motion 139
Cartesian coordinates 21, 35–6, 42
Cartesian equations 42–3, 45–6
chain rule 65, 86, 87, 100, 146
coefficients 8, 31, 91, 129, 131
common ratios 11, 19
complementary functions 128–32, 135, 137, 139, 150
completing the square 84–5, 90
complex numbers 1–28
    de Moivre's theorem 6–15
    exponential form 2–5, 11–13, 16–17, 20
    roots of unity 16–22
complex roots 127–8, 131, 135, 137, 139
constant of integration 119, 122, 144
convergent integrals 64–6, 68, 70
convergent series 11, 100
coordinates 21, 35–47
cosecant (cosec) 86
cosine (cos)
    complex numbers 2, 4–5, 6–15
    de Moivre's theorem 6–15
    Euler's formula 2, 4, 6
    hyperbolic functions 48–55, 86–90
    inverse function 82–5
    polar coordinates 36–9, 42–6
    polar graphs 43–6, 97–8
cotangent (cot) 86
coupled equations 149–54
cube roots of unity 16
cuboids 145

D
damped harmonic motion 135
decay rates 144
definite integrals 64–70, 72
de Moivre's theorem 6–15
differences method 30–4
differential equations 117–60
    coupled equations 149–54
    first order 118–25
    modelling systems 142–8
    second order 126–33, 150, 152
    simple harmonic motion 134–41
    uses of 134, 142
differentiation 63–116
    chain rule 65, 86, 87, 100, 146

hyperbolic functions 87
inverse trigonometric functions 82
Maclaurin series 100, 102
volumes of revolution 79–80
discontinuity points 67, 93–4
displacement 146
divergent integrals 64–6, 69–70

E
equating coefficients 91, 129, 131
equations
    Cartesian 42–3, 45–6
    complex solutions 19–20
    coupled 149–54
    differential 117–60
    polar 37–40, 42–6
    quadratic 50, 53–4
    simultaneous 73, 75, 92, 94, 97–8, 132, 150, 152
equilateral triangles 16, 20–1
Euler's formula 2, 4, 6
exact equations 118–20, 123–4
exponentials
    complex numbers 2–5, 11–13, 16–17, 20
    differential equations 122, 144
    hyperbolic functions 48, 52, 86

F
factorisation 74–5, 92, 93
factors 30, 92–3
first order differential equations 118–25

G
general solutions
    coupled equations 150, 152
    first order differential equations 119–20, 122–4
    second order differential equations 127–32
    simple harmonic motion 134–5, 137–9
general terms
    functions 30
    series 101–2, 106
geometric series 11–13, 19
graphs 37–40, 43–6, 49, 95–9
growth rates 144

H
half-lines 95, 97–8
hexagons 17
homogeneous equations 126–7
Hooke's law 137
hyperbolas 48
hyperbolic functions 48–55, 86–90

I
identities
    hyperbolic 52
    trigonometric 8–9, 77, 82–4, 87, 89, 96, 97
imaginary parts of complex numbers 9, 12
improper fractions 91
improper integrals 64–70, 93–4
index laws 4, 6, 10–11, 16, 19, 52, 72
induction, proof by 6, 100

infinite series 11, 100
infinity 64–6, 95, 103, 105
initial line 35
integrands 64, 66, 69, 84, 89, 93
integrating factors 118, 120, 122–4, 126
integration 63–116
    by parts 67, 70
    by substitution 84–5, 89–90, 92
    constant of 119, 122, 144
    first order differential equations 118–20
    hyperbolic functions 86–90
    improper integrals 64–70, 93–4
    inverse trigonometric functions 82–5, 94
    mean value theorem 71–6
    polar graphs 95–9
    volumes of revolution 77–81
intersections of curves 43–4, 97–8
inverse hyperbolic functions 48, 53–4, 87, 89
inverse trigonometric functions 82–5, 94

L
limits 64–70
    common 66–7
    Maclaurin series 102–3, 105–6
    partial fractions 93–4
    polar graphs 95–8
    volumes of revolution 79–80
line symmetry 49
logarithms
    differential equations 122, 144
    hyperbolic functions 48, 50, 52–3, 87
    improper integrals 67–8, 70
    Maclaurin series 101–2
    volumes of revolution 78

M
Maclaurin series 100–7
many-to-one functions 53–4
mean value theorem 71–6
method of differences 30–4
modelling systems 142–8
modulus-argument 2–3, 6–7, 16–17
de Moivre's theorem 6–15

N
Newton's second law 137–8, 146
nth roots 16–17, 19
nth terms 101–2

O
one-to-one functions 54
oscillatory motion 134–41

P
parametric functions 77, 79–80
partial fractions 30–3, 91–4, 122, 142, 146
particular integrals 128–32, 150
particular solutions 119–20, 122, 131–2, 134–5, 137–9
points of discontinuity 67, 93–4
points of intersection 43–4, 97–8
polar coordinates 35–47
polar equations 37–40, 42–6
polar graphs 37–40, 43–6, 95–9
pole 35, 45–6